T0213231

Organic Electroluminescence

OPTICAL ENGINEERING

Founding Editor
Brian J. Thompson
University of Rochester
Rochester, New York

Organic
Electroluminescence

Zakya Kafafi
United States Naval Research Laboratory
Washington, D.C., U.S.A.

CRC Press
Taylor & Francis Group
Boca Raton London New York

CRC Press is an imprint of the
Taylor & Francis Group, an **informa** business
A TAYLOR & FRANCIS BOOK

CRC Press
Taylor & Francis Group
6000 Broken Sound Parkway NW, Suite 300
Boca Raton, FL 33487-2742

First issued in paperback 2019

© 2005 by Taylor & Francis Group, LLC
CRC Press is an imprint of Taylor & Francis Group, an Informa business

No claim to original U.S. Government works

ISBN-13: 978-0-8247-5906-3 (hbk)
ISBN-13: 978-0-367-39283-3 (pbk)

Library of Congress Cataloging-in-Publication Data

Catalog record is available from the Library of Congress

Visit the Taylor & Francis Web site at
http://www.taylorandfrancis.com

and the CRC Press Web site at
http://www.crcpress.com

Preface

The 21st century is seeing a big revolution in the way information is displayed electronically. Organic electroluminescent displays on rigid or flexible substrates are envisioned to play a significant if not major role in the area of flat panel displays. They may eventually dominate the market in just a few years. Small, passive, and active matrix organic light-emitting displays that are relatively inexpensive already have penetrated the commercial market in a significant way. Organic electroluminescent displays can be small, such as hand-held or head-mounted devices, or large, such as flat panel screens that can be rolled up or hung flat on a wall. Very bright organic light-emitting diodes (OLEDs) may one day find their way into use as solid state light sources and lasers. The control of spontaneous emission using novel nanostructures and photonic crystals is emerging as a new interesting approach for the realization of organic lasers. This book, *Organic Electroluminescence*, is intended for diversified readers such as graduate and postgraduate students in science and engineering, newcomers to the field, and experts. It can be used as a textbook for a graduate course on organic electroluminescence or as a valuable reference for solid state chemists, physicists, material scientists, and engineers who are interested and/or active in the field.

The book is organized into 11 chapters, which are written by experts in their respective fields of chemistry, physics, materials science, and electrical engineering. Chapters 1 and 2 provide historical background and an overview of OLEDs based on small molecules and polymers. Chapter 1 is a useful introduction to organic electroluminescence that brings the reader up to date with the state-of-the-art performance of organic electroluminescent displays based on small molecules. Chapter 2 defines the basic principles governing electroluminescence, including the critical parameters that control the quantum efficiency and the light output. The theory and mechanisms of carrier injection and transport in organic semiconductors and electron and hole recombination process are discussed in some detail. Chapter 3 addresses the physical properties of organic light-emitting diodes in the space-charge limited regime.

Chapter 4 discusses the development and characterization of amorphous materials with high glass transition temperature and the role they play in carrier injection and transport. Chapter 5 reviews the chemistry of various classes of light-emitting polymers and their uses in OLEDs. Chapter 6 focuses on OLEDs based on phosphorescent rather than fluorescent materials and discusses several mechanisms that give rise to light emission such as energy transfer and direct carrier recombination.

Chapter 7 provides an overview of the different patterning approaches that can be applied to organic electro-active materials for full-color displays and other applications.

Chapter 8 gives a review of the "state-of-the-art" thin film transistors compatible with organic light-emitting devices for large area active matrix. This chapter also addresses the important underlying technologies for the realization of full color, rigid, or flexible organic emissive displays including pixel electronics.

Chapters 9 and 10 summarize future directions for OLEDs in electroluminescent displays and solid state lighting, respectively. Finally, Chapter 11 presents different structures adopted for achieving lasing in organic solids and gives an overview of photo-excited organic lasers using organic thin films, nanostructures, and photonic crystals.

Zakya H. Kafafi
Naval Research Laboratory
Washington, D.C.

Contributors

Marc A. Baldo
Department of Electrical
  Engineering and
  Computer Science
Massachusetts Institute of
  Technology
Cambridge, Massachusetts,
  U.S.A.

Guillermo C. Bazan
Institute for Polymers and
  Organic Solids
University of California
Santa Barbara, California,
  U.S.A.

Zhiliang Cao
Department of Electrical
  Engineering
Center for Solid State
  Electronics Research
Arizona State University
Tempe, Arizona, U.S.A.

Ananth Dodabalapur
University of Texas
Austin, Texas, U.S.A.

Anil R. Duggal
GE Global Research
Niskayuna, New York, U.S.A.

Stephen R. Forrest
Department of Electrical
  Engineering
Princeton University
Princeton, New Jersey, U.S.A.

Miltiadis K. Hatalis
Lehigh University
Bethlehem, Pennsylvania,
  U.S.A.

Zakya H. Kafafi
Organic Optoelectronics Section
Optical Sciences Division
Naval Research Laboratory
Washington, D.C., U.S.A.

Jan Kalinowski
Department of Molecular
  Physics
Gdańsk University of
  Technology
Gdańsk, Poland

Bin Liu
Institute for Polymers and
  Organic Solids
University of California
Santa Barbara, California,
  U.S.A.

Jun Shen
Department of Electrical
  Engineering
Center for Solid State
  Electronics Research
Arizona State University
Tempe, Arizona, U.S.A.

Yasuhiko Shirota
Department of Applied
  Chemistry
Faculty of Engineering
Osaka University
Osaka, Japan

Mark E. Thompson
Department of Chemistry
University of Southern
  California
Los Angeles, California, U.S.A.

Matias Troccoli
Lehigh University
Bethlehem, Pennsylvania,
  U.S.A.

Tetsuo Tsutsui
Department of Applied Science
  for Electronics and Materials
Graduate School of Engineering
  Sciences
Kyushu University
Fukuoka, Japan

Apostolos T. Voutsas
Sharp Labs of America
Camas, Washington, U.S.A.

Takeo Wakimoto
Electronic Technology
  Development Center
Asahi Glass Co., Ltd.
Kanagawa, Japan

Martin B. Wolk
3M Display & Graphics
  Business Laboratory
St. Paul, Minnesota, U.S.A.

Contents

1

Electroluminescence in Small Molecules

CONTENTS

1.1 INTRODUCTION

Small-size color displays (e.g., mobile phones) based on organic light-emitting diodes (OLEDs) have recently penetrated the commercial market. The attraction with this new technology is based on the bright, full-color light emission from OLEDs, which originates from the radiative relaxation of the electronic excited states of π-conjugated molecular systems. The size of the OLED displays is expected to expand in the near future and will include large TV screens, other information displays, and even solid state lighting.

In the last century, much effort has been devoted to developing a variety of artificial lighting sources such as gas lamps, electric light bulbs, fluorescent lamps, neon lamps, cathode-ray tubes, inorganic light-emitting diodes, and semiconductor lasers. Every artificial light source is based on simple mechanisms. For example, incandescence is based on short-wavelength edge emission of blackbody radiation from high temperature substances. Light emission from excited states of atoms or inorganic solids is another mechanism. In contrast, a variety of colors of living species are based on reflection or transmission, which are due to optical transitions between the excited and ground electronic states of molecules. There are animals and plants, such as moonlight mushrooms, luminous bacteria, lantern fish, sea firefly, firefly squid, firefly, and railroad worm that have the ability to continuously emit light and thus are quite visible in the dark. Even the advanced technologies achieved in the 20th century have not been successful in imitating the lighting mechanisms that take place in such living species. Fortunately, however, just at the end of the 20th century, the technology on OLEDs, the artificial version of living light, has been successfully established.

The mechanism for charge injection electroluminescence (EL) in covalently bonded molecular materials is common and is found among many molecular systems, including main-chain π-conjugated polymers. In this chapter, we describe charge injection EL in molecular glass films composed of small molecules. In the case of thin solid films composed of small molecules with unsaturated bonds, there is no need to describe their electronic structure using an extended π-electron system. Their electronic properties can simply be given in terms of their molecular orbitals, in particular the highest occupied molecular orbital (HOMO) and the lowest unoccupied molecular orbital (LUMO). Thus, charge transport across the molecular solid can be described as a hopping process among localized π-electron systems, where the location for opposite charge recombination and light emission is assumed to be on specific molecules. This simple picture for charge-injection EL serves as a useful guide. It is helpful when using different molecular design concepts and provides a reasonable basis for device efficiency considerations.

1.2 RECOMBINATION AND EMISSION IN BULK

It is useful to look at the early work (1960s) on electroluminescence (EL) in organic single crystals to understand the high performance exhibited by OLEDs based on very thin films. Helfrich and Schneider reported bright EL from a single-crystal anthracene, where they used anthracene cation and anion containing electrolyte solutions as anode and cathode, respectively.[1,2] Figure 1.1 shows the data replotted from this study where the emission intensity scales linearly with the injected current density over more than three decades. This result is quite valuable and it provides clear-cut evidence for a charge-injection process, which is the basis for determining the EL quantum efficiency defined as the ratio of photons

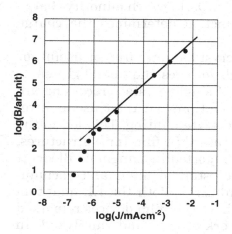

Figure 1.1 The relationship between the brightness and the current density in an OLED made of an anthracene single crystal. (Data from figure 2 in Reference 1 is re-plotted.) The thickness and size of electrodes are 5 mm and 0.2 cm^2, respectively.

emitted per injected charges. Helfrich and Schneider[1,2] reported that the emission arose predominantly from a region near the positive electrode (anode), indicating that electrons are injected from the cathode and transported through the crystal to meet the holes entering from the opposite electrode. This result clearly shows that bulk-controlled charge transport and recombination govern the process of charge injection EL.

Following this early work, numerous reports on electroluminescence from various organic single crystals have appeared.[3–8] It is worthwhile noticing that the reported EL quantum efficiencies decreased as the troublesome and unstable liquid contacts were replaced with their more stable but less efficient solid counterparts. The development of stable solid state contacts with good injection characteristics for both electrons and holes is an important area deserving further attention. Further improvements will not only lead to better EL efficiency but also to overall improved device stability. In molecular systems, which have very low carrier densities, the injected positive and negative charges (referred to as electrons and holes for simplicity) are expected to exist as space charges with no local charge neutrality. This case is clearly different from inorganic semiconductor LEDs, in which minority charge carrier injection at a p–n junction determines the charge recombination process.

OLEDs based on single crystals (μm) are not useful for practical applications. The high voltages, the small light-emitting areas, and the difficulty of single crystal processing are some of the shortfalls that would prevent their use as pixel elements in displays and solid state lighting. This led to the exploration of means to fabricate thin film (nm) structures, and charge injection EL was reported in Langmuir-Blodgett films, vacuum-sublimed polycrystalline films, and vacuum-sublimed amorphous glassy films.[9–14] Both the EL quantum efficiencies and stabilities of those thin-film devices remained low until the breakthrough work of Tang and VanSlyke.[15] In 1987, they reported the fabrication of high-performance OLEDs using a double-layer vacuum-sublimed heterostructure based on organic dyes.[15] Following this demonstration, Adachi and co-workers extended and generalized the concept to multilayer structures,[16-21] and developed high-efficiency devices. It should be stressed, however, that charge recombination and light emission still originate from molecules in the bulk.

1.3 MULTILAYER STRUCTURES

The structure of OLEDs containing two organic layers consists of a transparent indium-tin-oxide (ITO) anode, an organic hole transport layer (HTL), an organic electron transport layer (ETL), and a metal cathode (Mg:Ag alloy film, for example). The HTL fulfills the roles of assisting the injection of holes from ITO and transporting them to the boundary of the two organic layers, while the ETL has the function of assisting the injection of electrons from a metal cathode and their transport throughout the bulk film. Recombination of holes and electrons occurs at the boundary regions between the two organic layers. When the recombination region is located within an ETL, the ETL behaves as an emissive layer (EML). When the recombination occurs within the HTL, on the other hand, the HTL can behave as an EML. Thus these devices are classified into two types; ITO/HTL/ETL(EML)/ Metal and ITO/HTL(EML)/ETL/Metal. A three-layer structure may be also used where an independent thin EML is sandwiched between HTL and ETL (ITO/HTL/EML/ETL/ Metal), in case bipolar materials (which have the ability to transport both electrons and holes) are available. Figure 1.2 depicts these three typical device structures.

Even when HTL materials with low ionization potential are used to match the work function of the ITO anode, an energy barrier to hole injection from the ITO anode to HTL is usually present. In addition, chemical interactions between the ITO surface and the adjacent organic layer may cause degradation of the OLED device. Thus, the insertion of a thin buffer layer between the ITO and the HTL has been found very useful for both lowering the drive voltage and improving the device durability.[22,23] The same is also true for the metal/ organic interface between the metal cathode and the ETL. Many buffer layers have been proposed and used at this

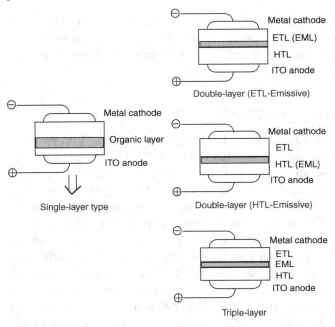

Figure 1.2 Illustrations of single-, double-, and triple-layer device structures. Charge recombination and emission are assumed to occur within the hatched regions.

interface.[24–26] The thickness of the buffer layer used varies depending on the material's resistivity. It is usually less than 10 nm for high-resistivity buffer layers. When doped semiconductor materials are used for the buffer layers, the thickness of the buffer layer may reach 1000 nm.[27,28] Figure 1.3 depicts a variety of multilayer OLED structures with different types of thin or thick buffer layers.

1.4 DESIGN OF HOLE TRANSPORT, ELECTRON TRANSPORT, AND EMISSIVE MOLECULES

One of the aromatic diamines, N,N'-diphenyl-N,N'-bis(3-methyl phenyl)-1,1'-biphenyl-4,4'-diamine (TPD) has been used as a typical HTL material, due to its excellent hole injection and transport properties and its good electron blocking

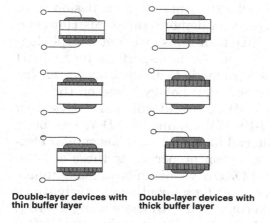

Double-layer devices with thin buffer layer   **Double-layer devices with thick buffer layer**

Figure 1.3 Illustrations of double-layer device structures with thin and thick buffer layers.

capability at the HTL/ETL boundary.[18] Although TPD has superior characteristics for HTL, the stability of vacuum-sublimed thin TPD films is not so high due to its low glass transition temperature (63°C). Other diamine derivatives, including N,N'-diphenyl-N,N'-(2-napthyl)-(1,1'-phenyl)-4,4'-diamine (NPD) and spiro-TPD, have been proposed and used as HTL materials.[22,29–31]

Good HTL materials should satisfy one or more of the general requirements given below.

1. Materials are morphologically stable and form uniform vacuum-sublimed thin films.
2. Materials have small solid state ionization potential.
3. Materials have high hole mobility.
4. Materials have small solid state electron affinity.

These requirements can be used as the general guiding principles for screening new hole transport materials.

Several metal chelates have been proposed and used as ETL materials. Tris(quinolin-8-olato) aluminum (Alq$_3$) and its analogues are known to be one of the most robust ETL materials.[15,32,33] An oxadiazole derivative, 2-(4'-biphenyl)-5-(4''-*tert*-butylphenyl)-1,3,4-oxadiazole (*t*-Bu-PBD) was used as an ETL material in blue-emitting OLEDs, where light emission from the hole transporting layer was demonstrated.[18,19] However, vacuum-sublimed *t*-Bu-PBD thin-films are not morphologically stable and cannot be used for high-performance OLED devices. Several attempts were made to design and synthesize ETL materials with improved morphology based on the electronic properties of *t*-Bu-PBD. An oxadiazole derivative with the dimer structure of *t*-Bu-PBD, named OXD-7, has been found to be a suitable material for ETL,[34] and has similar electron transport and light emission properties to those of *t*-Bu-PBD. Vacuum deposition of OXD-7 yields uniform glassy films. However, OLED devices based on oxadiazole derivatives degrade much faster when compared with those using the Alq$_3$ family. ETL materials, such as 1,2,4-triazole derivative (TAZ),[35] and 2,9-dimethyl-4,7-diphenyl-phenanthroline (BCP) have been examined.[36] One of the best examples for a systematic molecular design lies in the success in the development of distyrylarylene derivatives such as DPVBi for ETL materials.[37,38]

Numerous fluorescent dyes with high photoluminescence quantum yield have been used as emissive centers in OLEDs. However, only a handful of dyes are robust enough to be promising candidates for applications. Laser dyes such as 4-dicyanomethylene-2-methyl-6-*p*-dimethylaminostyryl)-4*H*-pyran (DCM), dimethylquinacridone (QA), and rubrene (Rub) have been successfully used for both increasing the EL quantum efficiency and modifying the emission spectra. Figure 1.4 depicts the molecular structures of several hole and electron transporters, and light emitters, namely, TAPC, TPD, NPD, Alq, *t*-Bu-PBD, OXD-7, TAZ, BCP, DCM, QA, and Rub described above.

The morphological stability of sublimed low-molar-mass materials in their "as deposited" amorphous states is dependent on their tendency to crystallize. Glass-forming materials such as those based on TPD can now be produced by suitable molecular design including the addition of bulky substituents, the generation of asymmetric skeletal structures, and the appendage of dimer-like nonplanar structures and dendritic networks. All these promote the formation of stable glassy films upon vacuum sublimation of the material of interest.

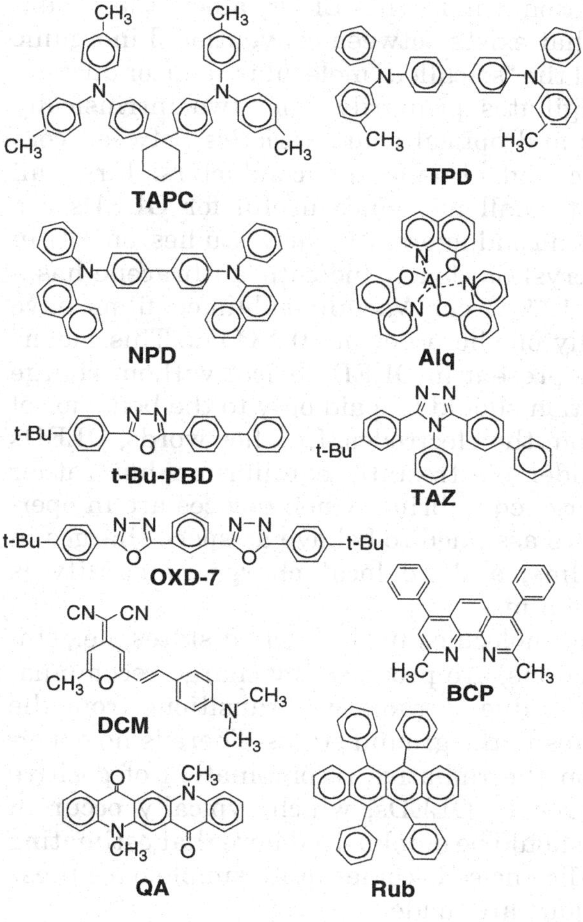

Figure 1.4 Chemical structures of molecules used for hole transport, electron transport, and light emission.

1.5 EL QUANTUM EFFICIENCY AND POWER CONVERSION EFFICIENCY

Based on the basic understanding for the working mechanism for charge-injection-type EL, some important formulae can be derived that quantitatively describe the EL quantum efficiencies. Apparently, OLEDs made of small molecules behave very

similarly to conventional inorganic LEDs, albeit the fundamental difference that exists between conventional inorganic
semiconductors and the "so-called" molecular semiconductors.
This difference originates primarily from two intrinsically
different electronic and optical characteristics between conventional inorganic and organic semiconductors. First, all
solid films made of small molecules useful for OLEDs are
wide-energy-gap semiconductors. As early studies on EL in
anthracene single crystals clearly indicate, anthracene has a
high resistivity of 10^{20} Ω cm. Vacuum-sublimed films have
resistivities typically on the order of 10^{15} Ω cm. This means
that no charges are present in OLED devices without charge
injection and attention should be paid only to the behavior of
injected charges from the electrodes. In other words, OLEDs
are not treated under electrostatic equilibrium but rather
under dynamic charge equilibrium when devices are in operation. All charges are assumed to behave as space charges in
solid molecular films, and no local charge neutrality is
expected within the films.

Second, neutral molecules in the excited states, i.e., singlet and triplet excitons, are produced by charge recombination, and emission is due to radiative transitions from the
neutral excited states to the ground states. There is no experimental evidence on the radiative recombination of positive
and negative charges in OLEDs, which typically occur in
inorganic LEDs. It should be emphasized here that estimating
the EL quantum efficiencies becomes quite simple once these
two basic assumptions are made.

Figure 1.5 shows the elementary processes that take
place upon charge injection into an OLED including recombination of holes and electrons, radiative decay from the electronic excited states, and output coupling of the emitted light.
The external EL quantum efficiency and internal EL quantum
efficiencies are connected by the extraction efficiency of photons. An external EL quantum efficiency, η_ϕ (ext) is defined as
the ratio of numbers of emitted photons outside a device
divided by the number of charges injected into a device. The
internal quantum efficiency, η_ϕ is the ratio of the number of
photons produced within a device divided by the number of

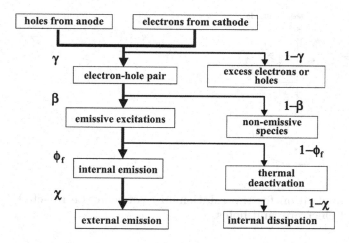

Figure 1.5 A schematic representation of the elementary processes for charge recombination, production of excitons, internal and external light emission in an OLED.

charges injected. Hence the ratio of η_ϕ (ext) to η_ϕ is equal to the extraction efficiency of photons outside an OLED device, which is named the light output coupling factor χ.

$$\eta_\phi(ext) = \chi\, \eta_\phi \qquad (1.1)$$

Using the charge balance factor γ, the production efficiency of emissive excitations β, and the fluorescence quantum yield ϕ_f, the internal quantum EL efficiency η_ϕ is given by Equation 1.2.[39]

$$\eta_\phi = \gamma\beta\phi_f \qquad (1.2)$$

The charge balance factor γ is defined by Equation 1.3 using the number of injected charges experimentally obtained from the measured current density, J, and the electron-hole recombination current J_r.

$$\gamma = \frac{J_r}{J} \qquad (1.3)$$

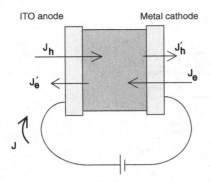

Figure 1.6 Illustration for the relationship between the injected versus escaped electrons and holes.

A quite simple assumption, which represents the mass balance and charge neutrality of holes and electrons depicted in Figure 1.6, explains nicely the meaning of J_r. The amounts of holes and electrons injected from an anode and a cathode are expressed by J_h and J_e, respectively. The amounts of holes crossing through the organic layer/cathode interface and that of electrons crossing through organic layer/anode interface are given by J_h' and J_e', respectively. The following two equations are thus derived:

$$J = J_h + J_e' = J_e + J_h' \tag{1.4}$$

$$J_r = J_h - J_h' = J_e - J_e' \tag{1.5}$$

Although there is no direct means to determine the value of J_r, one can discuss it upon examination of the charge recombination regions. When a narrow charge recombination region is located within an emissive layer, all holes and electrons are assumed to be consumed for recombination and $J = J_r$, i.e., $\gamma = 1.0$. On the other hand, if most electrons go throughout the device without recombination and all the injected holes are used for carrier recombination ($J_h' = 0$, $J_e' > 0$), $J_e \gg J_h$ and $J_r = J_h \ll J_e = J$, and γ is assumed to be far less than unity.[40,41]

The scheme shown in Figure 1.7, introduced by J.C. Scott, is quite helpful in understanding the meaning of the charge

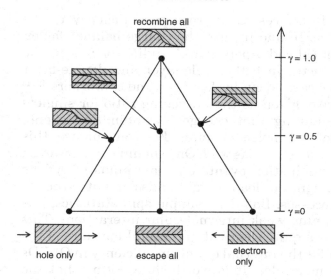

Figure 1.7   A diagram illustrating the charge balance factor γ for different types of hypothetical OLED structures.

balance factor.[43,44] The left and right corners of the triangle represent "hole-only" and "electron-only" devices, respectively. Even when equal numbers of holes and electrons are injected, γ remains zero if all the injected holes and electrons escape to the respective counter electrodes. Thus, any points on the bottom line of the triangle represents γ = 0. The apex of the triangle represents the case when equal numbers of holes and electrons are injected from the respective electrodes, and they all recombine inside the OLED device, which leads to a charge balance factor γ of 1.0.

　　To increase the charge balance factor (efficient charge recombination), attention must be paid not only to attaining balanced injection of holes and electrons but also to confining them within the device emissive region. In the case of single-layer devices, optimization of the charge balance factor is only possible through adjusting the energy barrier to charge injection at the electrode/organic layer interfaces, or balancing the electron and hole mobilities. By using multilayer structures, a charge blocking function can be introduced at the

organic/organic interfaces to improve the efficiency of the
charge recombination and in turn the charge balance factor.
Excellent theoretical and experimental studies concerning the
charge balance factor in both single-layer and double-layer
devices have been carried out by J.C. Scott and co-workers.[42-45]

β is frequently given as the branching ratio for singlets
and triplets, assuming that charge recombination events
occur on well-defined molecular sites. For convenience, this
assumption is adapted here as well. One should note, however,
that charge recombination events on some kinds of defects
and impurities might not lead to light emission but rather to
nonradiative processes. Based on simple spin statistics, β is
taken as π, assuming weak intermolecular interactions. This
assumption gives a quite clear-cut and useful measure of the
25% upper-limit for the internal quantum efficiency in OLEDs
based on small molecules. It is worthwhile noting that the
validity of this assumption has been questioned, primarily in
π-conjugated polymer systems.[46-48]

Before the development of good phosphorescent emitters
at room temperature, the design of highly fluorescent molecules with high fluorescence quantum yield, ϕ_f has been one
of the most important research targets. As discussed in Chapter 6 of this book, excellent phosphorescent molecules have
been proposed[49] and successfully used as the emissive centers
in OLEDs. Hence, the fluorescence ϕ_i and phosphorescence ϕ_p
quantum yields, and the efficiency of singlet–triplet intersystem crossing needs to be characterized to understand fully
the production and relaxation processes governing the emissive excitations. It should be emphasized here that the $(\beta\phi_f)$
product in Equation 1.2 only symbolizes the importance of
the two terms, the efficiency of production of emissive excitations β and the fluorescence quantum yield efficiencies ϕ_f, i.e.,
the emissive transition from the excited to the ground electronic states.

One of the most useful methods to estimate the light
output coupling factor χ is derived from simple ray optics
where all the light with emission angles larger than the total
reflection angle produced within a plate with a refractive

index, n, is assumed never to escape to free space. The light output coupling factor χ is given by Equation 1.6.

$$\chi = 1 - \sqrt{1 - \frac{1}{n^2}} \qquad (1.6)$$

When n is larger than 1.6, Equation 1.6 is further approximated by $(2n^2)^{-1}$. Organic solids composed of aromatic molecules have refractive indices around 1.7, and thus χ is frequently approximated by the number of 0.20. Complex calculations including multilayer device structures, spatial orientation of the light-emitting molecules, polarizations of the emitted light, interference effects due to reflective mirrors, and so on have to be included for the detailed evaluation of the light output coupling factor.[50,51] When randomly oriented molecular emitters and simple multilayer structures are assumed, χ is estimated to be between 20 and 30%. If planar orientation of the emissive centers is possible χ can reach close to 40%.

Assuming $\chi = 0.20$, $\beta = 0.25$, $\gamma = 1.0$, and $\phi_f = 1.0$, a maximum external EL quantum efficiency $\eta_\phi(\text{ext}) = 5.0\%$ can be obtained based on Equations 1.1 and 1.2. It should be noted that this so-called upper limit in $\eta_\phi(\text{ext})$ for OLEDs based on fluorescent molecules provides just a good measure to use as a guide for the basic understanding of the OLED efficiencies and by no means is the exact ceiling value.

The power conversion efficiency in emissive devices is frequently expressed in terms of the luminous power efficiency (lm/W). This expression, which includes the term of the human eye sensitivity (the photopic eye response) with its strong dependence on the emission wavelength, provides no direct physical measure for the device performance. A better measure of the power conversion efficiency, which will be adopted in this chapter, may be given in terms of the electric-to-photon conversion efficiency (W/W) expressed by the ratio of the power of the emitted light to the input electric power.

The relationship between the power efficiency (internal) η_E, which is defined by the ratio of the emitted light power inside a device and the applied electric power (JV), and the quantum efficiency (internal) η_ϕ is given by Equation 1.7,[39] where ε_p and V express the photon energy of the emitted light (eV), obtained by averaging the entire electroluminescence spectrum, and the applied voltage, respectively.

$$\eta_E = \eta_\phi \frac{\varepsilon_P}{e\,V} \qquad (1.7)$$

It should be emphasized that Equation 1.7 is not an approximate but an exact relationship between the power conversion efficiency and the EL quantum efficiency. Equation 1.7 is easily derived from the definitions of the EL quantum efficiency and the power conversion efficiency.

The major factor determining the power conversion efficiency is ε_p/V. In a hypothetical situation when the average photon energy ε_p is equal to the electronic potential provided by the applied voltage, eV, the ratio ε_p/eV is 1.0. Usually this factor is assumed to be far less than unity, because several steps lead to the decrease of the potential eV before the production of the thermally relaxed emissive excitations. Voltage losses due to energy barriers to charge injection, voltage drops due to charge transport resistance, dissipation of electronic energy corresponding to a binding energy of electron-hole pairs at charge recombination, and excess energy due to the thermal relaxations of the hot emissive excitations to the lowest electronic excited state are the main origins for lowering the value of the ε_p/eV factor below unity. Recently, the introduction of high-conductivity doped layers between the cathode/anode and the organic ETL/HTL led to a drastic decrease in the driving voltage of OLEDs, which in some cases brought the value of ε_p/eV close to 1.0.[52–57]

One should note that the applied voltage is only included explicitly in the expression of the power conversion efficiency and the term of the EL quantum efficiency is not directly related to the applied voltage. In another word, the "so-called" energy barriers for charge injection have no direct effect on the EL quantum efficiency. It is the balance of injected holes and electrons that governs the EL quantum efficiency, and the applied voltage needed for injection of electrons and holes

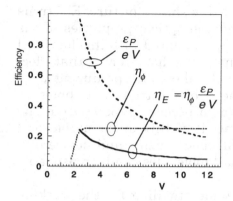

Figure 1.8 The dependence of the power conversion efficiency on the drive voltage (assuming the EL quantum efficiency η_ϕ is independent of the drive voltage and is 0.25, and ε_p is 2.4 eV).

determines the power conversion efficiency. These are the most important key issues for understanding the EL quantum efficiencies in charge-injection type OLEDs. Figure 1.8 shows a typical example for a green OLED device with e_p = 2.4 eV and an internal EL quantum efficiency of 0.25. The power conversion efficiency rapidly decreases with the increase of the drive voltage, whereas the EL quantum efficiency remains unchanged. Hence, the improvement in the charge injection barrier at the anode and the cathode is only effective for increasing the power conversion efficiency through the decrease of the drive voltage. In some cases, modifications of the energy barriers to charge injection may affect the charge balance factor. Minimizing these barriers may result in an increase of the EL quantum efficiency in cases where γ is altered.

1.6 CONCLUDING REMARKS

The optical and electronic properties of isolated small molecules can be simply described according to a molecular orbital picture. The properties of molecular solids that comprise the organic thin films in OLEDs based on small molecules can be understood as the extension of this simple picture. Electrons are localized on molecules and the optical properties of the molecular solids are roughly approximated with this localized electron model. The important point, however, is that electrons and holes move within molecular solids producing large electric current. The movement of electrons and holes in molecular solids is described as a hopping process from one molecular site to another. Based on this quite simple and rough picture, the working EL mechanisms and device efficiency of OLEDs based on small molecules can be reasonably described.

In this chapter, a basic understanding for the working mechanism of charge-injection type electroluminescence of small molecules has been given. Details of the elementary processes of charge injection, transport, and recombination have been briefly described. The focus was on presenting a simple picture of the different processes that occur in an OLED device based on typical organic molecules and simple structures. OLEDs based on small molecules will continue to be very attractive. Their biggest advantages lie in the variety of chemical structures that can be designed and synthesized.

REFERENCES

1. Helfrich, W. and Schneider, W.G., *Phys. Rev. Lett.*, 14, 229, 1965.

2. Helfrich, W. and Schneider, W.G., *J. Chem. Phys.*, 44, 2902, 1966.

3. Kawabe, M., Masuda, K., and Nambu, S., *Jpn. J. Appl. Phys.*, 10, 527, 1971.

4. Lohmann, F. and Mehl, W., *J. Chem. Phys.*, 50, 500, 1969.

5. Williams, D.A. and Schadt, M., *Proc. IEE*, 58, 476, 1970.

6. Bradley, L.L.T., Schowob, H.P., Weitz, D., and Williams, D.F., *Mol. Cryst. Liq. Cryst.*, 23, 271, 1973.

7. Basurto, G. and Burshtein, J.Z., *Mol. Cryst. Liq. Cryst.*, 31, 211, 1975.

8. Glinski, J., Godlewski, J., and Kalinowski, J., *Mol. Cryst. Liq. Cryst.*, 48, 1, 1978.

9. Roberts, G.G., McGinnity, M.M., Barlow, W.A., and Vincett, P.S., *Solid State Commun.*, 32, 683, 1979.

10. Vincett, P.S., Barlow, W.A., Hann, R.A., and Roberts, G.G., *Thin Solid Films*, 94, 171, 1982.

11. Partridge, R.H., *Polymer*, 24, 748, 1983.

12. Era, M., Hayashi, S., Tsutsui, T., and Saito, S., *J. Chem. Soc. Chem. Commun.*, 1985, 557, 1985.

13. Hayashi, S., Wang, T.T., Matsuoka, S., and Saito, S., *Mol. Cryst. Liq. Cryst.*, 135, 355, 1986.

14. Hayashi, S., Etoh, H., and Saito, S., *Jpn. J. Appl. Phys.*, 25, L773, 1986.

15. Tang, C.W. and VanSlyke, S.A., *Appl. Phys. Lett.*, 51, 913, 1987.

16. Adachi, C., Tokito, S., Tsutsui, T., and Saito, S., *Jpn. J. Appl. Phys.*, 27, L269, 1988.

17. Adachi, C., Tokito, S., Tsutsui, T., and Saito, S., *Jpn. J. Appl. Phys.*, 27, L713, 1988.

18. Adachi, C., Tsutsui, T., and Saito, S., *Appl. Phys. Lett.*, 55, 1489, 1989.

19. Adachi, C., Tsutsui, T., and Saito, S., *Appl. Phys. Lett.*, 56, 799, 1990.

20. Adachi, C., Tsutsui, T., and Saito, S., *Appl. Phys. Lett.*, 57, 531, 1990.

21. Adachi, C., Tsutsui, T., and Saito, S., *Optoelectron. Devices Technol.*, 6, 25, 1991.

22. VanSlyke, S.A. et al., *Appl. Phys. Lett.*, 69, 2160, 1996.

23. Shirota, Y. et al., *Appl. Phys. Lett.*, 65, 807, 1994.

24. Wakimoto, T. et al., *IEEE Trans. Electron Devices*, 44, 1245, 1997.

25. Hung, L.S. et al., *Appl. Phys. Lett.*, 70, 152, 1997.

26. Kido, J. and Matsumoto, T., *Appl. Phys. Lett.*, 73, 2866, 1998.

27. Yamamori, A. et al., *Appl. Phys. Lett.*, 72, 2147, 1998.

28. Blochwitz, J. et al., *Appl. Phys. Lett.*, 73, 729, 1998.

29. Adachi, C., Nagai, K., and Tamoto, N., *Appl. Phys. Lett.*, 66, 2679, 1995.

30. Tokito, S., Tanaka, H., Noda, K., Okada, A., and Taga, Y., *Appl. Phys. Lett.*, 70, 1929, 1997.

31. Salbeck, J. et al., *Synth. Met.*, 91, 206, 1997.

32. Hamada, Y. et al., *Jpn. J. Appl. Phys.*, 34, L824, 1995.

33. Kido, J. and Iizumi, Y., *Chem. Lett.*, 963, 1997.

34. Hamada, Y., Adachi, C., Tsutsui, T., and Saito, S., *Jpn. J. Appl. Phys.*, 31, 1812, 1992.

35. Kido, J., Kimura, M., and Nagai, K., *Science*, 267, 1332, 1995.

36. Nakada H. et al., *Polym. Prepr. Jpn.*, 43, 2450, 1994.

37. Tokailin, H. et al.. *SPIE Proc.*, 1910, 38, 1993.

38. Hosokawa, C., Kawasaki, N., Sakamoto, S., and Kusumoto, T., *Appl. Phys. Lett.*, 61, 2503, 1992.

39. Tsutsui, T. and Saito, S., Organic multilayer electroluminescent diodes. In *Intrinsically Conducting Polymer: An Emerging Technology,* Aldissi, M., Ed., Kluwer Academic, Dordrecht, the Netherlands, 1993, 123.

40. Tsutsui, T., *MRS Bull.*, 22, 39, 1997.

41. Tsutsui, T., Aminaka, E., Lin, C.P., and Kim, D.-U., *Philos. Trans. R. Soc. Lond.*, A355, 801, 1997.

42. Scott, J.C. et al., *SPIE Proc.*, 3476, 111, 1998.

43. Scott, J.C., Karg, S., and Carter, S.A., *J. Appl. Phys.*, 82, 1454, 1997.

44. Malliaras, G.G. and Scott, J.C., *J. Appl. Phys.,* 83, 5399, 1998.

45. Ruhstaller, B. et al., *J. Appl. Phys.* 89, 4575, 2001.

46. Cao, Y. et al., *Nature*, 397, 414, 1999.

47. Friend, R.H. et al., *Nature*, 397, 121, 1999.

48. Wohlgennt, M. et al., *Nature*, 409, 494, 2001.

49. Baldo, M.A. et al., *Appl. Phys. Lett.*, 75, 4, 1999.

50. Bulovic, V. et al., *Phys. Rev.*, B58, 3730, 1998.

51. Kim, J.S. et al., *J. Appl. Phys.*, 88, 1073, 2000.

52. Wakimoto, T., Fukuda, Y., Nagayama, K., Yokoi, A., Nakada, H., and Tsuchida, M., *IEEE Trans. Electron Devices*, 44, 1245, 1997.

53. Hung, L.S., Tang, C.W., and Mason, M.G., *Appl. Phys. Lett.*, 70, 152, 1997.

54. Kido, J. and Matsumoto, M., *Appl. Phys. Lett.*, 73, 2866, 1998.

55. Yamamori, A., Adachi, C., Koyama, T., and Taniguchi, Y., *Appl. Phys. Lett.*, 72, 2147, 1998.

56. Blochwitz, J., Pfeiffer, M., Fritz, T., and Leo, K., *Appl. Phys. Lett.*, 73, 729, 1998.

57. Pfeiffer, M., Forrest, S.R., Leo, K., Thompson, M.E., *Adv. Mater.*, 14, 1633, 2002.

2

Emission Mechanisms in Organic Light-Emitting Diodes

JAN KALINOWSKI

CONTENTS

2.1 INTRODUCTION

The various ways light is generated by applying an electric field to organic materials, without involving any intermediate energy forms — the phenomenon known as organic electroluminescence (EL) — render our present understanding of organic EL phenomenological at best. In dealing here with its microscopic mechanisms we wish to strike a note of challenge rather than achievement. The experimental material available reminds us in many respects of the organic crystal field in the pre-thin film era and the theory has been often adequate only to formulate the problems rather than to solve them. The laws of nature seem to be well known, the difficulties of interpretation come from the complexities of the problems arising from the diversity of emitting states, the structure of organic light-emitting diodes (LEDs), and from the great number of variables. We do not try to review all the aspects of the field of organic EL as they have been summarized in recent reviews,[1–9] which contain references to original papers. Instead, a review of the general features of the optical properties of organic LEDs is given without attempting to achieve completeness by quoting all the important papers in the field. The main subject of the present chapter is correlating the optical properties of charge recombination underlying organic LEDs with their electrical characteristics. The ability

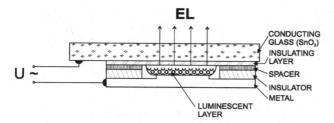

Figure 2.1 The cross section of a powder EL cell containing an organic luminescent powder in a liquid (e.g., castor oil) or solid (e.g., paraffin wax) dielectric medium (luminescent layer). (From Kalinowski, J., *J. Phys. D: Appl. Phys.*, 32, R179, 1999. © Institute of Physics (GB). With permission.)

to convert the supplied electrical power into light in an organic LED is of primary interest, and this fundamental aspect is considered in some detail. Apart from the purely scientific interest, such considerations may be important also for modern LED technology.

2.2 HISTORY AND BASIC CONCEPTS

2.2.1 Early Observations

The EL spectra of anthracene (A) and tetracene-doped anthracene (T : A) of carbon black added EL dielectric cells (Figure 2.1), shown in Figure 2.2, introduce the subject of this section. Electroluminescence is the result of the electric field–imposed formation of emissive states without recourse of any intermediate energy forms, such as heat. For the (T : A)-based cells, the EL spectrum consists of three peaks in the green region, characteristic of the ultraviolet (UV) light-excited emission from tetracene, no anthracene main emission bands can be observed (Figure 2.2A). Even small amounts of tetracene as a common impurity of anthracene are readily detected under EL conditions, although not seen in the photoluminescence (PL) spectra of the same samples (Figure 2.2B). This high sensitivity of EL to small amounts of dopants is typical for all types of organic LEDs.

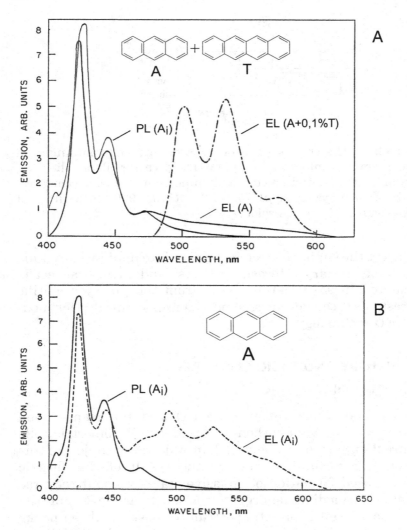

Figure 2.2 The PL and EL emission spectra of anthracene (A) and tetracene-doped anthracene (A + 0.1%*T*) of carbon black added (1%) EL cells. Two different purities of anthracene, A and A_i, were used: A_i is chemically pure anthracene recrystallized many times from benzene (impure), A is anthracene obtained by melting of non-fluorescent dianthracene crystals (high degree of purity). Note the difference in the EL(A), in (part A), and EL(A_i), in (part B), spectra. (Based on Gurnee.[10])

The functioning of early organic EL cells such as that shown in Figure 2.1 is reminiscent of powdered inorganic phosphors excited by an alternating electric field.[11-13] They consist of a transparent front electrode in the form of a glass sheet coated with a thin, degenerately n-type conducting SnO_2 film, separated by a thin layer of a transparent insulating material (e.g., mica or polystyrene) from the luminescent layer followed by a metal electrode. The luminescent layer has been formed in various ways: (1) the cooling of a melt of an organic luminescent material with a certain amount of carbon black (e.g. A + 1% carbon black as employed by Gurnee[10]), (2) the mechanical mixing of luminescent materials with metal powders, the mixture placed in castor oil, paraffin wax, or resin as the embedding dielectric (e.g., anthracene 4:1 mixture with powders of Mn, Fe, Al, or Cu in castor oil as adapted by Lehmann[14]), (3) the evaporation of a suspension of organic materials (e.g., aromatic hydrocarbons deposited from an ethanol suspension as reported by Short and Hercules[15]), (4) organic microcrystals embedded in a dielectric medium (e.g., crystalline carbazole in castor oil, employed by Goldman et al.[16]), and (5) the adsorbing organic luminescent compounds on transparent films from the fluorescent solution, drying them, and embedding them in a dielectric medium (e.g., organic dyes like gonacrin or brilliant acridine orange E adsorbed on cellophane or mica placed with melted paraffin wax, the fabrication procedure used in the pioneer works on organic EL by Bernanose and co-workers[17]). Light emission appears as a result of excitation of the radiative system of the EL cell.

2.2.2 Types of Electroluminescence

Three types of EL can be distinguished based on the mode of excitation of the radiative system:

I. High-field EL. An external electric field (F) excites emitting states directly. In quantum-mechanical terms, this excitation mode is described by the probability of excitation of localized (atomic or molecular) states or by field ionization from the valence band

(VB), when developed (Zener effect) (Figure 2.3A). This mechanism was applied by Bernanose[18] to explain his results on the quadratic field dependence of the EL intensity.

II. Impact EL. At high electric fields, the charge carriers (either generated in the bulk or injected from outside) can gain a substantial amount of energy from the field, E_m, required to impact, excite, or ionize a molecule already over its mean free path, $l_c = E_m/eF$ (Figure 2.3B). To accelerate electrons (e) to 2 to 3 eV and allowing impact generation of excited states, the drift velocity $\cong 10^8$ cm/s must be reached. This is possible at high fields ($\cong 10^6$ V/cm) with field-independent mobilities of 100 cm^2/Vs. Mobilities like that have been observed in some organic single crystals such as naphthalene (for holes) or perylene (for electrons) at temperatures below 77 K.[19] However, the field saturation of the drift velocity at about 10^6 cm/s makes the impact excitation unlikely. Furthermore, the vast majority of organic materials, especially at room temperature, exhibit mobilities much below this value,[20] making this material property an important limiting factor in the operation mechanisms of organic LEDs. This also makes it difficult (but not impossible, taking into account locally enhanced electric fields and an energy distribution of carriers) to understand EL characteristics in pioneering works on organic EL by Bernanose and co-workers[17] and later on by other authors[16,21–28] who considered the phenomenon an analogue of the Destriau effect.[11–13,29,30] The avalanche breakdown in gas microcavities incorporated into the samples has been proposed to alleviate the difficulties.[15]

III. Recombination (or combination) EL. This is the emission of light by a luminescent material, resulting from the fusion of oppositely charged carriers at a semiconductor p–n junction (Figure 2.3C) or in its bulk where they are introduced by injection at electrodes (Figure 2.3D).

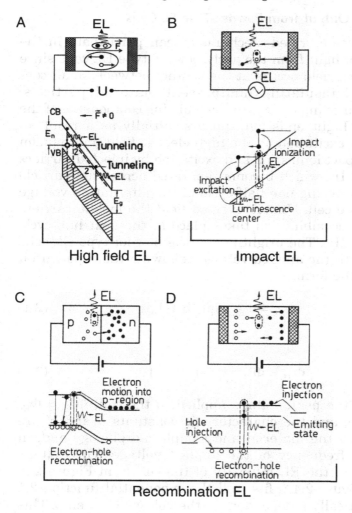

Figure 2.3 The schematic representation of various electronic excitation mechanisms due to AC or DC external electric fields: (A) the tunneling electrons from the valence band (VB) to the conduction band (CB) (Zener effect) and ionization of an acceptor state (–o–) followed by electron–hole recombination, indicated by horizontal and vertical arrows, respectively; (B) excitation or ionization by electron impact; (C) recombination of electrons (•) and (o) holes at a semiconductor *p–n* junction, (D) bulk recombination of electrons and holes injected from electrodes.

2.2.3   Light Output from Powder-Type EL Cells

All the above-described mechanisms can participate in the generation of light from the early powder-type EL cells since a high electric field occurs at the contact between small conducting and insulating luminescent material particles, enabling charge injection into the emitting components of the EL system. Light emission appears initially as a result of either direct excitation by the high electric field (mechanism I) and/or impact ionization and excitation by injected carriers (mechanism II), which become separated across the particle and trapped during one half-cycle of the alternating voltage applied to the cell. Under reversed field the mobile carriers return and recombination takes place in the next half-cycle (mechanism III). The brightness of the powder-type EL cells increases with the applied voltage, following the exponential function of the form:

$$\Phi_{EL} = \Phi_{EL}^{\infty} \exp\left(-b/U\right), \qquad (2.1)$$

or

$$\Phi_{EL} = \Phi_{EL}^{\infty} \exp\left(-b'/U^{1/2}\right), \qquad (2.2)$$

where U is the peak voltage applied to the cell, $\Phi_{EL}^{\infty} = \Phi_{EL}$ with $U \to \infty$, and the characteristic constants b and b' are dependent on the material and sample morphology and, in general, on frequency of the applied voltage. The voltage dependence of the EL intensity of the cell from Figure 2.2, shown in Figure 2.4A, fits well the exponential function 2.1 with b practically independent of the voltage frequency. This result can be explained either by the high-field ballistic ionization process following the field dependent probability of the electron to experience a path, l, greater than l_c, $P = \exp(-E_m/eFl)$, or by the increase in the field increasing probability of impact excitation and ionization by electrons in the voltage-dependent high-field regions of the grains of the luminescent material. It is worthy to note that function 2.2 can be a result of the summation of the EL output of type 2.1 over a large number of emitting particles,[29,30] thus often observed

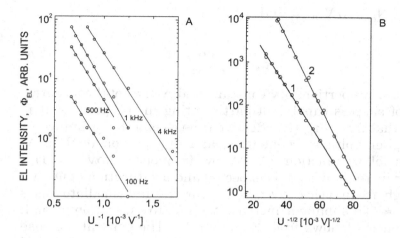

Figure 2.4 The EL intensity (Φ_{EL}) as a function of voltage for the A + T + carbon cell from Figure 2.2, represented by a log Φ_{EL} against $U_\sim^{-1}$ plots for different frequencies (ν) of the alternating voltage, $U_\sim$ (A), and Φ_{EL} plotted vs. $U_\sim^{-1/2}$ (B) for carbazole microcrystals suspended in a castor oil sample at a constant $U_\sim$ frequency 4.5 kHz, following Equation 2.2 with b' different for the orange (1) and blue-green (2) bands of the broad emission spectrum of impure carbazole. (From Goldman, A.G. et al., *Zh. Prikl. Spektr.*, 14, 235, 1971 [in Russian].)

in powder or microcrystalline samples as in the example shown in Figure 2.4B. Since the blue-green band of the EL emission from this system is associated mostly with the emission of diphenyloxide impurity of carbazole, the difference in the slopes of the straight-line plots for the orange (2.1) and blue-green (2.2) components of the EL spectrum, thus in the characteristic parameter b', has been associated with different properties of the emission centers. The frequency (ν_0) response of EL is underlain by the relaxation time (τ_0) of the emitting localities (lighting spots) governed by the release of carriers from traps, although other mechanisms can also be envisaged. From simple kinetics, the number of emitting spots, N_e, increases according to the equation:

$$N_e = N_{e\infty}\,[1 - \exp(-t/\tau_{\text{eff}})], \tag{2.3}$$

where $N_{e\infty} = \alpha N v_0 \tau_{eff}$, and

$$\frac{1}{\tau_{eff}} = \frac{1}{\tau_0} + \alpha v_0 \qquad (2.4)$$

with α a proportionality constant and N the total concentration of sources capable of participating in the process. It is clear that for $t > \tau_{eff}$ the effective time constant τ_{eff} diminishes at higher voltage frequencies, and Φ_{EL} proportional to N_e would follow function 2.3. At low frequencies ($\alpha v_0 \tau_0 < 1$) a linear increase with v_0 is expected and a saturation of $\Phi_{EL}(v_0)$ at high frequencies ($\alpha v_0 \tau_0 > 1$). For the intermediate cases ($\alpha v_0 \tau_0 \approx 1$) a sublinear increase tending to saturation at high frequencies follows from Equation 2.3. The general increase of Φ_{EL} against v_0, apparent in Figure 2.4A, has been reported in the literature to follow some above limiting cases (see, e.g., Goldman et al.[16] and Dubey[26]).

2.2.4   Performance

From the above it is clear that in powder-type EL cells the conversion of electricity to light occurs in their emitter grains, which experience the charge separation during one half-cycle of the voltage (U_-) applied to the cell of thickness d, thus exposed to the nominal electric field strength $F = U/d$. This makes it difficult to calculate the overall conversion efficiency, but it is possible to define the energy conversion efficiency related to the maximum amount of charge that can be generated during one half-cycle of the alternating voltage applied to the cell. It is equal to the capacitance charge of the embedded particles, which obviously depends on geometry and potential drop (thus electrical conductivity) across the particle. If an elementary charge has to traverse during one half-cycle a potential drop across the particle of U_p volts, and an effective particle diameter $2R$, the energy conversion efficiency can be expressed as follows:

$$\eta_p = P \frac{h v}{e U_p} = P \frac{h v}{2 e U} \frac{d}{R}, \qquad (2.5)$$

where $h\nu$ is an averaged one photon energy of the emitted light. Clearly, the conversion efficiency increases at low applied voltages, following a strongly increasing probability, P, of acceleration of an electron to the excitation energy, E_m, and decreases at high voltages, where the potential drop U_p becomes the dominating factor. Indeed, a maximum in the luminous efficiency of about 10 lm/W (corresponding to $\eta \cong$ 2%) has been reported in an inorganic ZnS:Cu,Cl powder cell at $U_{rms} \cong 200$ V.[31,32] At a given U, $h\nu$, and d, η_p increases as the grain radius, R, decreases. For example, a green light-emitting cell ($h\nu = 2.5$ eV) with $F = 2.5 \times 10^4$ V/cm, $R = 5$ μm, and $P = 1$, would reveal η_p at a 10% level. However, the conditions $h\nu > eU_p$, $U_p/2R < F_{breakdown}$, and strongly decreasing P establish the borders for the increase of η_p by reducing U and R.

2.2.5   Single-Crystal EL Structures

A great deal of progress in the understanding of organic EL has been made using EL cells based on single organic crystals (a review of past work has been given by Kalinowski;[33,34] updated discussion by this author can be found in References 35 and 36). The sandwich cell EL structures (Figure 2.5) involve a single crystal and at least one semitransparent electrode. Optical and electrical properties of such structures have been extensively studied.[7,36] The results made clear that the emissive states are mostly molecular singlet excitons formed in the bulk recombination of electrons and holes injected from electrodes; that is, type III of the EL underlies the electric field–imposed emission from organic single crystals. However, in some early works, they have been interpreted in terms of impact ionization excitation by primary electrons injected from a metal contact.[22,23,25,37] The EL spectra of relatively thick (>5 μm) organic crystals can differ from their PL spectra for several reasons: (1) different reabsorption effects caused by a difference in spatial distributions of the emitting states generated by the exciting light in PL and by charge carrier recombination in EL, (2) the presence of bulk dopants or defects acting as more efficient recombination centers than

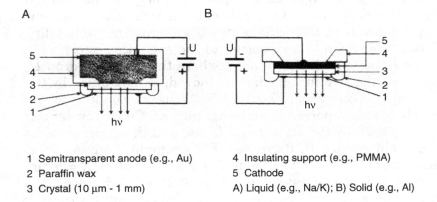

1 Semitransparent anode (e.g., Au)   4 Insulating support (e.g., PMMA)

2 Paraffin wax   5 Cathode

3 Crystal (10 μm - 1 mm)   A) Liquid (e.g., Na/K); B) Solid (e.g., Al)

Figure 2.5 Single-crystal-based EL structures with a liquid (A) and a solid (B) cathode.

acceptors in energy transfer processes, and (3) the splitting of the insulator electronic levels at the interfaces. Interestingly, the difference in the trivial reabsorption effect between PL and EL spectra allowed the spatial distribution of excited states to be inferred, and revealed the role of traps and excitonic interactions. Unexpectedly, the EL emission zone appeared to be stratified, showing from two to three well-resolved regions located at different positions in the crystal dependent on the applied voltage.[35,38,39] A striking example of the stratification of the observed EL emission from an organic crystal is shown in Figure 2.6. This unusual pattern has been rationalized in terms of exciton–exciton and exciton-trapped charge carrier interaction processes. The formation of the near-electrode emission zones has been attributed to the radiative decay of singlet excitons generated directly by free carriers arriving from the counterelectrode and recombining with the oppositely charged carriers injected and trapped in a highly trap-populated region at each of the electrode contacts. The third EL zone appears at a higher voltage and is a result of the radiative decay of singlet excitons produced in the triplet–triplet annihilation process, which becomes strongly

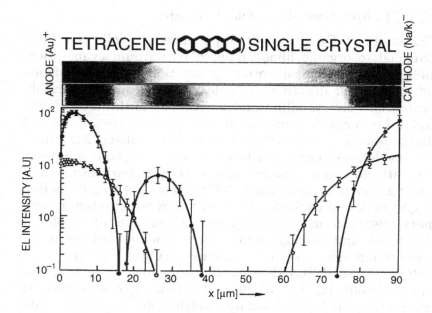

Figure 2.6 The spatial distribution of the EL intensity in a 90-μm-thick tetracene crystal at two different voltages: $U = 750$ V (o) and $U = 950$ V (•). The semitransparent hole-injecting Au anode is located at $x = 0$, and a thick layer of a Na/K alloy forms the electron-injecting contact at the rear crystal side, $x = d = 90$ μm (cf. Figure 2.5). The black field patterns simulate the light intensity distribution for these two voltages. (After Kalinowski.[33])

impeded toward the anode as a result of triplet exciton quenching by increasing concentration of trapped holes. As a result a minimum in the overall emission intensity appears, splitting the anode emission zone into two subzones.[33,35,39] Another emission minimum close to the anode indicates strong quenching of singlets on the gold hole injecting contact (cf. Kalinowski et al.[40]). The spatial pattern of the EL emission in single organic crystals can be, however, completely different, depending on the degree of the "ohmicity" of the injecting electrodes and on the concentration and spatial distribution of traps[35] (cf. Section 2.5).

2.2.6 EL from Single-Layer Film Structures

The study of EL in powder-type dielectric cells has been extended to organic films prepared by vacuum evaporation, melting, and recrystallization or solution cast, sandwiched between carrier injecting electrodes. EL diodes have been fabricated on polycrystalline and amorphous anthracene[24,28,41–45] and other organic compounds.[25,27,46,47] The crucial role of charge injection at electrodes has become realized. An important model, also employed in EL of single crystals and used currently to explain optical and electrical characteristics of vast variety of organic LEDs (cf. Sections 2.4 to 2.6), is due to generation of excited states via the recombination of primary electrons and holes injected at two oppositely placed electrodes, and moving against each other across the luminescent material (see, e.g., Dresner,[41] Vincett et al.,[44] and Kalinowski et al.[45]). Other mechanisms such as impact ionization excitation by primary electrode-injected electrons (mechanism II) in local strong electric fields, however, could not be excluded conclusively. The strong field has been located in the depletion layer of thickness $d_0 = (2\varepsilon_0 \Delta\chi/N_d e^2)^{1/2}$ developed parallel to the metal contact, where $\Delta\chi$ is the difference between Fermi levels in the metal and dielectric (or semiconductor) material, and N_d is the concentration of electron donor centers in the contacting luminescent material. This near-contact field $F_c = (\Delta\chi \pm eU)/ed_0$ is inversely proportional to the thickness of the depletion layer, which for $d_0 << d$ leads to $F_c >> F = U/d$. If $d_c = 0.1d$ with $\Delta\chi << eU$ and typically $U/d = 10^5$ V/cm, F_c is an order of magnitude larger and amounts to 10^6 V/cm. For large values of $\Delta\chi$ and low concentration of electron donors, N_d, the "high-field region" can cover the total thickness of the dielectric layer forming the EL cell, that is, $d_0 \cong d$ and $F_c \cong F$, the situation typical for majority of high-purity organic materials. Consequently, injection-recombination mechanism becomes more appropriate to explain the functioning of thin film EL devices. Field dependence of the charge injection efficiency would underlie the field evolution of their brightness, although an exponential trap distribution can change this picture remarkably. As a matter of fact, the

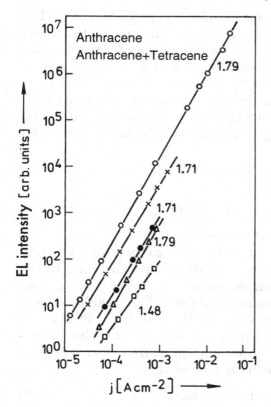

Figure 2.7 Intensity Φ of the EL as a function of the measured current density (j) for five samples of 1-μm-thick single anthracene (A)- and tetracene (T)-doped anthracene films sandwiched between different metallic electrodes: (o) Al$^-$/A/Au$^+$/glass, ($\times$) Al$^-$/A+10^{-3}T/Au$^+$/glass (observation at a tetracene emission band, λ = 500 nm), (•) Al$^-$/A+10^{-4}T/Au$^+$/glass (observation at λ = 500 nm), ($\triangle$) Au$^-$/A/Au$^+$/glass, ($\square$) Al$^-$/A+10^{-4}T/Au$^+$/glass (observation at an anthracene emission band, $\lambda \cong$ 420 nm). The numbers near by the straight lines stand for their slopes. (After Kalinowski et al.[45])

injection-limited-currents have been shown to impose nonlinear brightness-current characteristics (Figure 2.7), assuming the primary injection to be governed by the tunneling current density $j_{pe} = j_{0e} \exp(-\alpha_1 x)$ with $\alpha_1 = (2\sqrt{2}/\hbar)(m_{eff}^e I_A)^{1/2}$ for electrons, and $j_{ph} = j_{0h} \exp(-\alpha_2 x)$ with $\alpha_2 = (2\sqrt{2}/\hbar)(m_{\text{eff}}^h E_g)^{1/2}$ for holes. The preexponential factors in these expressions are the tunneling currents at $x = 0$, and m_{eff}^e and m_{eff}^h are the effective masses of electrons and holes, respectively. I_A and E_g denote injection barrier heights determined by about the electron affinity (I_A) and electrical gap (E_g) of the luminescent material. At high electric fields, when carrier velocities become saturated, the injection currents flowing across the sample can be expressed by[45]

$$j_e \cong j_{0e} \exp[-\alpha_1(\beta/\gamma)]^{1/2} \qquad (2.6)$$

and

$$j_h \cong j_{0h} \exp[-\alpha_2(\beta/\gamma)]^{1/2}, \qquad (2.7)$$

where $\beta = e^2/16\pi\varepsilon_0\varepsilon kT$ and $\gamma = eF/kT$.

In the simplest case of free charge carriers or a discrete set of charge traps, the recombination is proportional to the product of free carrier concentrations (n_f, n_h) or free (say, n_f) and trapped (say, n_{ht}) concentrations of the carriers, $\Phi = k_F k_S^{-1} \alpha_t n_{ef} n_{ht} d$, where k_F is the radiative and k_S the total decay rate constant of emitting singlet excitons, and α_t is the second-order rate constant for recombination of free electrons with trapped holes. At high fields it follows that

$$\Phi = B_1 \exp[-(\alpha_1 + \alpha_2)(\beta/\gamma)^{1/2}] \qquad (2.8)$$

for a discrete set of traps, and

$$\Phi = B_2 \exp\{-[\alpha_1 + (\alpha_2/l_h)](\beta/\gamma)^{1/2}\} \qquad (2.9)$$

for an exponential distribution of traps in energy, $h(E) = (H/kTl_h)\exp(-E/kTl_h)$, where H is the total trap concentration and $l_h > 1$ is the dimensionless quantity characterizing the decrease of $h(E)$ with trapping depth, E. Combining Equations 2.6 and 2.7 with Equations 2.8 and 2.9 and using respective relations between concentration of free and trapped charge carriers[48] yield:

$$\Phi \cong \text{const } j_h^{n_1} \quad \text{and} \quad \Phi \cong \text{const } j_h^{n_2}, \qquad (2.10)$$

where $n_1 = 1 + (\alpha_1/\alpha_2)$ and $n_2 = (1/l_h) + (\alpha_1/\alpha_2)$ for a discrete set and exponential distribution of traps, respectively.

Both Equations 2.8 and 2.9 predict the power-type (Equation 2.10) plots of the experimental data of Figure 2.7. The slopes reflect different recombination mechanisms. Whereas in neat anthracene films, the slope $n = 1.79 \pm 0.05$ suggests the free-trap case or the presence of a discrete set of shallow traps with $\alpha_1/\alpha_2 = 0.79$ corresponding well to $I_A \cong 1.7$ eV, $E_g = 4.1$ eV, $m_{eff}^h = 0.8$ m_e (m_e = free electron mass) and $m_{eff}^e = 1.2$ m_e, $n = 1.48 \pm 0.05$ for anthracene host emission from tetracene-doped samples implies an exponential distribution of hole traps with $l_h = 1.3 \pm 0.1$.[45]

2.2.7 Polymer Single-Layer LEDs

Thin polymer films with large areas can be formed by the precursor polymerization on a substrate or casting and spin coating from solution without any subsequent processing or heat treatment. Intensive studies of neat and doped polyvinylcarbazole (PVK) films have been initiated with a series of works by Partridge.[49] The fluorescence of pure PVK can be ascribed to excimer sites each formed by a pair of carbazole groups in close proximity to one another (cf. Section 2.4). They are formed by the migration of molecular singlet excitons along the polymer chain, and from one polymer chain to another. When doped with low-molecular-weight organic dyes the doped PVK exhibits strong luminescence of the dopants. Comparing the PL and EL spectra of PVK-based EL cells suggests that either emission originates from identical excited states, which in the EL case are initially produced by recombining holes and electrons injected from suitable electrode contacts.[50] However, as in the case of low-molecular-weight-materials-based LEDs, the PL and EL spectra of polymer LEDs can differ due to selective excitation of bimolecular excited states in the recombination process and its electric field alteration (cf. Section 2.4).

Poly(*p*-phenylene vinylenes) (PPVs), poly(3-alkylothiophenes) (*P* 3ATs), poly(*P*-phenylene ethynylenes) (PPEs), derivatives thereof, and other π-conjugated polymers have been attracting widespread attention due to their potential application prospects.[1,51–57] The early conjugated polymer

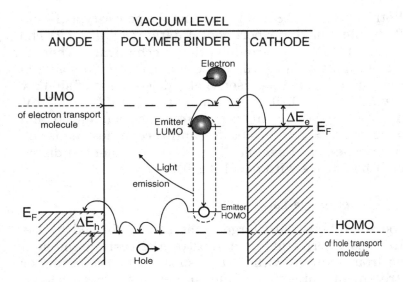

Figure 2.8 Schematic representation of the principal electronic process in MDP-based LEDs. (After Kalinowski et al., *J. Phys. D.* 34, 2274, 2001. © Institute of Physics (GB). With permission.)

(PPV)-based LEDs lasted barely minutes and were observable only in the dark.[51] Present state-of-the-art PPV-based EL devices approach the efficient Alq$_3$-based LEDs (cf. discussion below, and Figure 2.10), showing luminance up to 10^4 Cd/m^2 and lifetime over 10,000 h (see, e.g., Bradley[1]). Both the EL and PL spectrum as well as their EL efficiency depend on degree of the monomer–polymer conversion.[56]

Molecularly doped polymers (MDPs) having a base on electronically inert polymeric binder are of particular interest because they allow us to easily select a variety of dopant molecules with diversified electronic functions (Figure 2.8). The guiding principle for choosing the right composition of an MDP for an EL device is that energetic position of the HOMO/LUMO (Highest Occupied Molecular Orbital/Lowest Unoccupied Molecular Orbital) of a hole/electron-transporting dopant incorporated into such a binder should be as high/low as possible in order to eliminate trapping by accidental impurities of the possible inert binders, some of them frequently used for preparation MDPs to be employed as transport layers in the photoreceptor assemblies of photocopying machines. Poly(methyl methacrylate) (PMMA) and bisphenol-A-polycarbonate (PC) have been used to make MDP-based organic LEDs.[58-64]

2.2.8 Double- and Multilayer LEDs

The worldwide renaissance in research on organic EL has been initiated with the fabrication of highly efficient organic LEDs based on vapor-deposited thin organic films formed in double- and multilayer sandwich structures.[65-69] In these systems light emission results from the generation of excited states via the recombination of electrons and holes within a recombination zone located at one or more layer interfaces, with the electrons and holes injected at electrode contacts transported to the recombination zone through electron (ETL) and hole (HTL), often non-emissive, transporting layers (Figure 2.9). The confinement of the emission in thin emission layer (EML) placed between non-emissive ETL and HTL eliminates quenching of excited states in the ETL contacting with a metal cathode (see, e.g., Burin and Ratner[70]). Either ETL and/or HTL can act as emitters (EML) in the case of DL LEDs. Figure 2.10 shows an example of a DL LED where the EL emission originates from both ETL and HTL, with their relative contribution dependent on the applied voltage. This voltage-induced variation in the EL spectra and following LED color is associated with the location of the recombination zone at the ETL/HTL interface and its field-dependent shift toward the HTL. At low voltages the green emission of Alq$_3$ from the ETL ($\lambda_{max} \cong 530$ nm) dominates. Increasing voltage leads to a varying mixture of the light emitted from the Alq$_3$ ETL and a red-shifted emission ($\lambda_{max} \cong 590$ nm) of the molecules of T5Ohex dispersed in the polymeric HTL. The voltage-induced increase of the red-shifted T5Ohex component is due to increasing electron leakage through the HTL, where electrons hopping between molecules of T5Ohex act as the negatively charged recombination centers to holes transported effectively through the dispersed molecules of TPD. Apparent

Exciton quenching

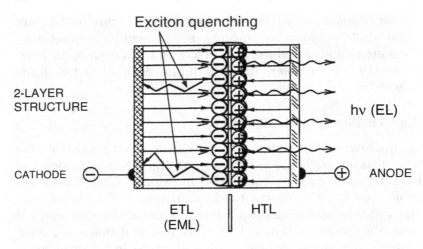

2-LAYER
STRUCTURE

hv (EL)

CATHODE

ANODE

ETL
(EML)

HTL

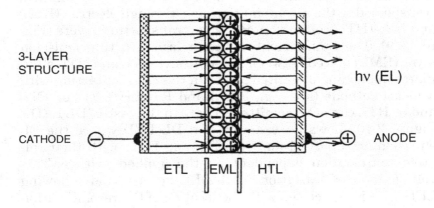

3-LAYER
STRUCTURE

hv (EL)

CATHODE

ANODE

ETL   EML   HTL

Figure 2.9 Double-layer (DL) and triple-layer (TL) thin film EL devices imposing the confinement of electron–hole recombination to narrow zones located nearby organic interfaces far from the electrode contacts.

on the example of Figure 2.10 is the difference between the EL and PL spectra of the LED structure. In particular, no emission from TPD is seen in the EL spectra. DL and multi-layer LEDs show as a rule a much higher EL quantum efficiency as compared with single-layer LEDs. Typically, the

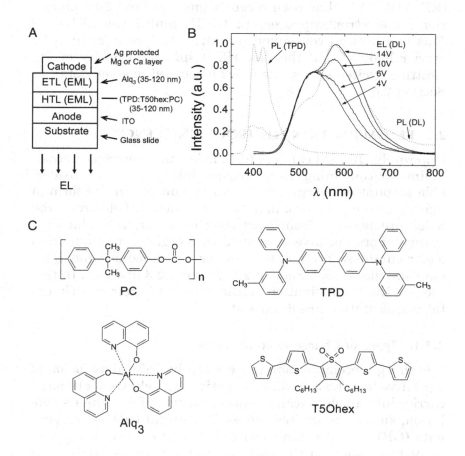

Figure 2.10   Schematic cross section (not to scale) of a DL LED with both ETL and HTL acting as emission layers (A). Normalized PL and EL spectra of the LED. The PL spectrum excited through the ITO anode at λ_{exc} = 350 nm, the EL spectrum evolves with applied voltage (B). Molecular structures of materials used (C): (PC) bisphenol–A-polycarbonate; (TPD) *N,N'*-diphenyl-*N,N'*-bis(3-methyl-phenyl)-[1,1'-biphenyl]-4,4'diamine; (Alq₃) *tris*-(8-hydroxyquinoli-nato) aluminum (III); (T5Ohex) 3,"4"-dihexyl-2,2':5',2":5,"2":5'"', 2''''-quinquethiophene-1"1"-dioxide. (After Kalinowski et al.[63])

external efficiency of 1 to 2% ph/carrier for electrofluorescent (EFL) DL LEDs has been recently improved to $\cong$20% ph/carrier for electrophosphorescent (EPH) multilayer LEDs.[71,72] This is a natural consequence of the interface impeding the carrier leakage and the diversified spin multiplicity of the emitting excited states, the effect discussed in some detail in Section 2.6.

2.3 RECOMBINATION ELECTROLUMINESCENCE

Oppositely charged carriers (e.g., holes and electrons) recombining (or combining) in an organic solid release a considerable amount of energy, which can be emitted in the form of light. This is called recombination radiation.[73] If electrons and holes are injected from electrodes to an organic film or a system of organic layers, we deal with organic recombination electroluminescence and the devices are called organic light-emitting diodes (see Section 2.1, Figures 2.3 and 2.9). In this section, the recombination mechanisms and their experimental manifestation are discussed.

2.3.1 Types of Charge Recombination

The charge recombination process can be defined as fusion of a positive (e.g., hole, h) and a negative (e.g., electron, e) charge carrier into an electrically neutral entity or, following its evolution, successive excited states. The initial (IR) — or geminate (GR) — and volume-controlled (VR) recombination can be distinguished on the basis of charge carriers origin.[74] As illustrated in Figure 2.11, GR is the recombination process following the initial carrier separation from an unstable, locally excited state, forming a nearest-neighbor charge-transfer (CT) state. It typically occurs as a part of intrinsic photoconduction in organic solids due to generation of charge from light-excited molecular states. The probability (P_{GR}) of GR can be expressed by the primary (often assumed to be electric field–independent) quantum yield in carrier pairs for the absorbed photon, η_0, and the ($e–h$) pair dissociation probability, Ω: $P_{GR} = 1 - \eta_0\Omega$. The electric field effect on the effective

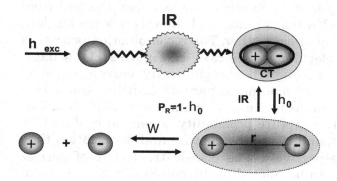

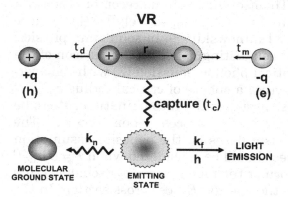

Figure 2.11 (Color figure follows p. 274.) Initial (IR) and volume-controlled (VR) recombination (for explanation, see text).

charge separation can be observed in generation-controlled photoconduction or electric field–induced quenching of luminescence because the external electric field–enhanced dissociation decreases the number of emitting states.[75,76]

If the oppositely charged carriers are generated independently far away from each other (for example, injected from electrodes), the VR takes place, and the carriers are statistically independent of each other. The recombination process is kinetically bimolecular in this case (cf. Figure 2.11). It naturally proceeds through a coulombically correlated electron–hole pair ($e–h$) leading to various emitting states (see Section 2.4) in the ultimate recombination step (mutual carrier capture). The capture probability is defined by $P_c = (1 + \tau_c/\tau_d)^{-1}$, where τ_c and τ_d are the capture and dissociation time of the pair, respectively. The classic treatment of carrier recombination can be related to the notion of the recombination time, τ_{rec}. The recombination time represents a combination of the carrier motion time (τ_m), i.e., the time to get the carriers within the capture radius (which is often mistakenly identified with the Onsager radius $r_{Ons} = e^2/4\pi\varepsilon_0\varepsilon kT$) and the elementary capture time (τ_c), $\tau_{rec}^{-1} = \tau_m^{-1} + \tau_c^{-1}$. Following the traditional description of recombination process in ionized gases, a Langevin-like and Thomson-like volume recombination can be defined if $\tau_c \ll \tau_m$ and $\tau_c \gg \tau_m$, respectively[77,78] (for a more recent description, see Kalinowski[74,79]). In solid state physics, these two cases have been distinguished by comparison of the mean free path for optical phonon emission (λ) with the average distance $4r_{Ons}/3$ across a sphere of critical radius r_{Ons}.[80,81] The Thomson-like case assumes the recombination rate to be limited by the phonon emission process when $\lambda \gg r_{Ons}$. The Langevin-like mechanism bases on the opposite assumption $\lambda \ll r_{Ons}$. Two subcases should be considered when $\lambda = (D\tau_0)^{1/2}$ and $\lambda = v_{th}\tau_0$. In the former the momentum (p) exchange cross section is larger than the energy (E)-loss cross section. In the latter, the reverse is true. Here, D is the diffusion coefficient of a carrier, τ_0 is the lifetime of p and E charge carrier states, respectively, and v_{th} is the thermal velocity of the carriers. Because of low carrier mobility (μ) in organic solids one would expect to deal with the first subcase with $\mu = 1$ cm^2/Vs

$(D = \mu kT/e)$ and mean free path for elastic scattering $\lambda = 10$ Å. This value of λ is clearly much lower than $r_{Ons} \cong 150$ Å ($\varepsilon = 4$), strongly suggesting a Langevin-like model to be appropriate to describe the recombination process in organics. Its signature is a field and practically temperature-independent ratio:[20]

$$\frac{\gamma_{eh}}{\mu_m} = \frac{e}{\varepsilon_0 \varepsilon} = const, \qquad (2.11)$$

where γ_{eh} is the bimolecular (second-order) recombination rate constant, μ_m the sum of carrier mobilities, and ε the dielectric constant. The essence of Equation 2.11, derived from the Smoluchowski expression relating γ_{eh} to the sum particle diffusion coefficient, D, and their interaction radius R via $\gamma_{eh} = 4\pi DR$ if one identifies R with the Onsager radius r_{Ons} and assumes the validity of Einstein's relation, $eD = \mu kT$, is that in the long-time limit charge recombination is a process controlled by diffusion. For molecular solids with typically $\varepsilon = 4$, $\gamma_{eh}/\mu = 4.5 \times 10^{-7}$ Vcm, which span γ_{eh} between 4.5×10^{-5} cm³/s and 4.5×10^{-17} cm³/s, the range corresponding to the limiting values of the carrier mobilities from 10^2 cm²/Vs in the case of some aromatic crystals at low temperature[19,82] down to 10^{-10} cm²/Vs in the case of polymeric films.[83]

The kinetic description of bimolecular reactions in condensed media, based on the solution of Smoluchowski equation, leads to the time (t)-dependent rate constant:[84,85]

$$\gamma(t) = 4\pi DR[1 + R/(\pi Dt)^{1/2}]. \qquad (2.12)$$

Equation 2.12 seems to be consistent with experimental data on time evolution of many chemical reactions in liquids, but is not adequate to describe the reaction kinetics in disordered solids. In disordered solids where the carrier motion is only partially diffusion controlled, carrier hopping across a manifold of statistically distributed in energy and space molecular sites must be defined and taken into account.[86-88] Assuming that the carrier hopping sites are subject to an energy Gaussian distribution $\rho(E) = (2\pi\sigma^2)^{-1/2} \exp(-E^2/2\sigma^2)$, and introducing the average length of hops δ, yields the average hopping frequency:[89,90]

$$v(t) = v_0 \exp\left(-2\delta/\delta_0\right)\left\{\int_0^{C(t)} \exp\left[\left(E/kT\right)-\left(E^2/2\sigma^2\right)\right]dE\right\}^{-1}, \quad (2.13)$$

where v_0 is the effective preexponential factor, δ_0 the charge localization radius at a hopping site,[91] and $C(t) = kT \ln[\sigma^2 v_0 t/(kT)^2]$ is only a weakly varying function of time (t). By substituting Equation 2.13 in the expression for the diffusion coefficient:

$$D = \delta^2 v(t), \quad (2.14)$$

we obtain

$$D(t) = D_0(v_0 t)^{-1+\beta} \quad (2.15)$$

with the dispersion parameter β bearing a weak functional time dependence of the expression for $C(t)$ and possible for an approximation by the time-independent empirical relationship:[90]

$$\beta^{-1} = 1 + \sigma^2/4kT. \quad (2.16)$$

As long as $\sigma^2/4kT > 1$, $\beta < 1$, and $D(t)$ is a decreasing function of time. The Monte Carlo simulations of dispersive transport in organic solids[92] have shown that Equation 2.16 is applicable for σ/kT ranging from 1 to 20. However, $\sigma/kT = 1$–5 may be too small for getting accurate estimations by means of Equations 2.15 and 2.16.

The long-time balance between recombination and drift of carriers as expressed by the γ_{eh}/μ ratio has been analyzed using a Monte Carlo simulation technique and shown to be independent of disorder.[93] Consequently, the Langevin formalism would be expected to obey recombination in disordered molecular systems as well. However, the time evolution of γ_{eh} is of crucial importance if the ultimate recombination event proceeds on the scale comparable with that of carrier pair dissociation. The recombination rate constant becomes then capture rather than diffusion controlled, so that Thomson-like model would be more adequate than Langevin-type formalism for the description of the recombination process.

2.3.2 Experimental Manifestation of Different Recombination Mechanisms

The simplest way to examine GR is to follow the concentration of mobile carriers by measurements of electrical photoconductivity. Electric field effect on the effective charge separation efficiency ($\eta_0\Omega$) determines directly the GR probability, P_{GR}. However, we should keep in mind that the measured photocurrent includes carrier mobilities in addition to the separation efficiency. The carrier mobilities and their possible field dependence must be, therefore, determined independently. Moreover, other mechanisms of charge generation such as injection or photo-injection from electrodes may lead to grave errors in evaluation of $\eta_0\Omega$. Such a situation has been reported for solid films of α-sexithiophene.[94] In Figure 2.12 the field

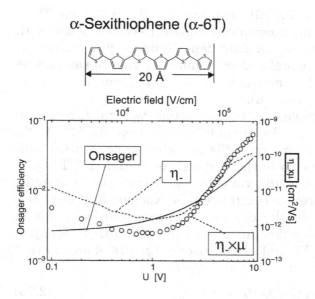

Figure 2.12 Electric field dependence of the photogeneration yield as calculated from Onsager formalism and that (η_-) extracted from the photocurrent measurements taking into account the low-field decrease and high-field increase in the carrier mobility (Equation 2.18). (After Kalinowski et al.[94])

dependence of the photogeneration yield η as calculated from the Onsager formalism is shown (see discussion below; for more details see, e.g., Pope and Swenberg[20]) and that extracted from the photocurrent measurements. The antibatic photocurrent response under negative bias of the illuminated semitransparent electrode has been determined by the bulk generation and recombination of charge according to the following equation:[94]

$$j = [2e\mu(\eta\_I_0/\alpha\gamma_{eh})^{1/2}/d][1 - \exp(-\alpha d/2)]F \qquad (2.17)$$

with the Langevin recombination coefficient, γ_{eh} (Equation 2.11).

A discrepancy between the theoretical (3D-Onsager) curve for $\eta\_$ and experimental data for $\eta\_ \times \mu$ extracted from Equation 2.17 is apparent. The field evolution of mobility:

$$\mu = \mu_0 \exp(\beta_\mu F^{1/2}) \qquad (2.18)$$

with $\beta_\mu < 0$ for $F < 3 \times 10^4$ V/cm and $\beta_\mu = 1.8 \times 10^{-2}$ cm$^{1/2}$V$^{-1/2}$ for $F > 3 \times 10^4$ V/cm, allowed experiment to be reconciled with theory. The nonmonotonic field dependence of μ is characteristic of built-in diagonal and positional disorder of samples as has been well documented on some molecularly doped polymers[83] (cf. Section 2.6).

Another possible way to determine P_{GR}, which is free of these drawbacks, is electric-field modulation (EFM) of PL. Electric-field effect on the effective charge separation efficiency ($\eta_0\Omega$) shows up in the varying population of CT states and, consequently, in the varying concentration of the emitting molecular (Frenkel) excitons. It is expected that the field-induced increase in the charge separation efficiency would translate into PL quenching. The ratio of the PL efficiency in the presence $[\varphi_{PL}(F)]$ and in the absence $[\varphi_{PL}(0)]$ of an external electric field (F) would give directly P_{GR}:

$$\varphi_{PL}(F)/\varphi_{PL}(0) = 1 - \eta_0\Omega = P_{GR}. \qquad (2.19)$$

Assuming Ω to be governed by the Onsager mechanism of dissociation,[95]

$$\Omega_{Ons}(F) = 1 - \xi^{-1}\sum_{j=0}^{\infty} P_j(r_{Ons}/r_0)P_j(eFr_0/kT), \qquad (2.20)$$

where

$$P_j(x) = P_{j-1}(x) - x^j \exp(-x)/j! \qquad (2.21)$$

and

$$P_0(x) = 1 - \exp(-x), \qquad (2.22)$$

the characteristic parameters of the charge separation process, r_0, r_{Ons}, and η_0 can be evaluated by fitting experimental quenching data (Equation 2.19) to the calculation made according to Equation 2.20. The field dependence of Ω_{Ons} appears in a complex manner through the parameter

$$\xi = er_0F/kT \qquad (2.23)$$

determining the expansion terms in the infinite series in Equation 2.20. The expansion coefficients of the series are governed by r_{Ons} and r_0, with the latter a discrete characteristic thermalization length (the primary separation distance; cf. Figure 2.11). In the example of Figure 2.12, the fitting procedure gives $\eta_0 = 0.9$ and $r_0 = 22 \pm 1$ Å; the latter is close to the distance between nearest-neighbor molecular layers in the layered structure of the microcrystalline films made by vacuum evaporation procedure from α-sexithiophene.[96] The field dependence of the EFM effect (quenching) in thin films of α-6T has also been shown to follow the Onsager-type initial recombination process with the same model parameters.[94] A direct comparison of the photocurrent and EFM measurements on films of Alq$_3$ is shown in Figure 2.13. While the EFM at high electric fields fits well η calculated on the basis of the 3D-Onsager model, the fitting of low-field photoconduction data requires the VR of separated carriers to be taken into account according to:[75,76]

$$\eta = \eta_0\Omega_{Ons}(F)[1 - \eta_{VR}(F)], \qquad (2.24)$$

where

$$\eta_{VR}(F) = (1 + 2\varepsilon_0\varepsilon F/en_{tot}d)^{-1}. \qquad (2.25)$$

From the example in Figure 2.13 it follows that the role of the volume recombination (VR) is in reducing η at sufficiently high charge concentration (here above $\cong 10^{15}$ cm^{-3}). The

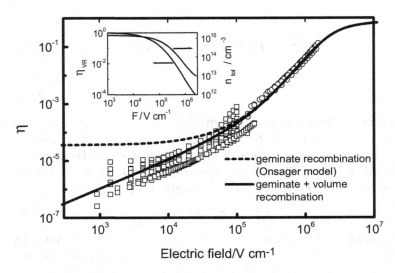

Figure 2.13 Electric field dependence of carrier generation efficiency, η, in a number of solid films of Alq$_3$. Squares represent steady-state photoconduction measurements and circles represent fluorescence-quenching measurements. The dashed line as calculated using the conventional 1938 Onsager theory with r_0 = 1.5 nm and η_0 = 0.8. The solid line is the Onsager model modified by including the field evolution of the bimolecular recombination efficiency (η_{VR}) as shown in the inset along with the field dependence of the total concentration of holes (n_{tot}). (After Kalinowski et al.[75])

role of the VR becomes negligible at high fields ($>10^5$ V/cm), where the field dependence of the photogeneration efficiency, extracted from both electrical and EFM experiments follows well 3D-Onsager's 1938 theory, solely. We note that according to the Onsager's theory the fraction η(0) of the initial quantum yield that escapes geminate recombination at zero field is very small and strongly depends on the initial separation distance, r_0, $\Omega_{Ons}(0)$ = exp($-r_{Ons}/r_0$). Typically, for r_{Ons} = 15 nm, $\Omega_{Ons}(0)$ $\cong 5 \times 10^{-5}$ for $r_0 \cong 1.5$ nm (like in Alq$_3$), and $\Omega_{Ons}(0) \cong 10^{-3}$ for $r_0 \cong 2.2$ nm (like in α-6T). We would expect $\Omega_{Ons}(0)$ to be much higher (thus, P_{GR} much lower) for e–h pairs formed in the VR with distances between r_0 and r_{Ons}. Indeed, the fitting procedure

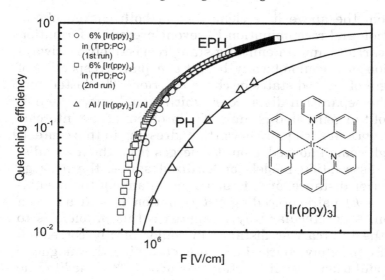

Figure 2.14 Quenching efficiency of the emission as a function of the DC electric field applied to the EPH LED with a 6 %wt content, of fac tris-(2-phenylpyridine)iridium [Ir(ppy$_3$] in (74 %wt TPD:20 %wt PC) emitter (circles, squares) and PH of a neat solid film of Ir(ppy)$_3$ (triangles). The curves are fits to the 3D-Onsager model for charge pair dissociation in external electric fields with $\eta_0 = 1.0$ and $r_0/r_c = 0.18$ for EPH, and $\eta_0 = 0.9$ and $r_0/r_c = 0.095$ for PH. In both cases $\varepsilon = 3$ and $T = 298$ K have been assumed. The molecular structure of Ir(ppy)$_3$ is shown in the inset (for the molecular structures of TPD and PC, see Figure 2.10). (After Kalinowski et al.[97])

applying the Onsager formalism to the field-induced quenching of PH molecular triplet states and EPH with the emitting molecular triplet states preceded by *e–h* pairs, gives $r_0/r_{Ons} = 0.095$ and $r_0/r_{Ons} = 0.18$, respectively (Figure 2.14). More important for organic LEDs is the type of VR mechanism underlying the EL emission. The degree (if any) to which the Thomson-like recombination contributes to the commonly accepted Langevin-like recombination lies at crux of the problem of understanding electricity–light conversion properties of organic LEDs, which is discussed in Section 2.6.

From the above it is clear that in both charge photo-generation and recombination EL event preceding formation of free carriers and an emitting state, respectively, is always formation of a coulombically bound $e–h$ pair irrespective of the origin of excited states or charge carriers. The latter may affect the separation distance, r, which would affect the efficiency of the generated product. The problem of the intercarrier distance in such pairs is worth addressing in this context. Although some knowledge on it emerges from the above-discussed dissociation models of localized states, the average intercarrier distance, and its distribution among the population of $(e–h)$ pairs preceding VR radiation is still an open question. A spectacular way to approach this problem is to follow the intercarrier distance in coulombically correlated pairs of oppositely charged carriers, observing the magnetic field-modulation of their reaction products.[98–102] If the lifetime of $(e–h)$ pairs is long enough for spin evolution, the magnetic field–dependent mixing of singlet $^1(e–h)$ and triplet $^3(e–h)$ pair spin states occurs due to the hyperfine interaction (HFI) of the carriers with their nuclear environments,[103] which modulates the effectiveness of various channels for the pair decay, thus production of emissive states in the VR process. The general scheme of energy levels and formation of emissive singlet (S_1) and triplet (T_1) states is provided in Figure 2.15. Figure 2.15 also implies the existence of overlapping Gaussian energy bands of the $(e–h)$ pairs due to static and dynamic disorder in noncrystalline organic solids.[92,104] External magnetic fields control the conversion rate between singlet $^1(e–h)$ and triplet $^3(e–h)$ pair states when the energy separation between them, $|2J|$, is comparable or smaller than the difference between the Zeeman energies of the pair components or their hyperfine interaction. Several types of magnetic field singlet–triplet mixing mechanisms have been considered dependent on the relation between the Zeeman energy, $\Delta g \mu_B B$ ($\Delta g = g_e - g_h$; μ_B is the Bohr's magneton), due to the different Landé factors, g, for electrons (g_e) and holes (g_h), and the hyperfine coupling of the pair carriers. The Hamiltonian $(\hat{H})$ of the charge pair comprises three terms:

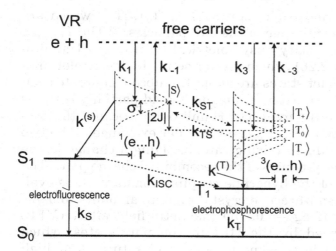

Figure 2.15 The energy-level scheme diagram (not to scale) and electronic transitions leading to fluorescent (S_1) and phosphorescent (T_1) molecular states produced in the bimolecular $e + h$ recombination process (VR) with suitable rate constants (k_1, k_{-1}, k_3, k_{-3}, $k^{(S)}$, $k^{(T)}$). The singlet and triplet states of the pair undergo mixing with the rate constants k_{ST} and k_{TS} ; the $S_1 \rightarrow T_1$ intersystem crossing is characterized by the rate constant k_{ISC}. See text for other details.

$$\hat{H} = \hat{H}_{\text{e-h}} + \hat{H}_{\text{Zeeman}} + \hat{H}_{\text{hyperfine}}, \qquad (2.26)$$

where $\hat{H}_{\text{Zeeman}} = g\mu_B B\left(\hat{S}_e + \hat{S}_h\right)$ with $\hat{S}_e, \hat{S}_h$ representing the spin operator of the electron and hole, respectively, $\hat{H}_{\text{hyperfine}} = \sum_{l,m} a_l \hat{S}_m \hat{I}_l$ represents the hyperfine interaction between l nuclei (often protons) interacting with m relevant electrons, with $\hat{I}_l$ the nuclear spin of the lth nucleus (proton) and $\hat{S}_m$ the electron spin operator of the mth electron. The $\hat{H}_{\text{e-h}}$ term relates to the exchange interaction $\hbar J[(1/2) - 2\hat{S}_e \hat{S}_h]$ provided by a separation (r) exchange parameter $J(r) = J_0 \exp(-\alpha r)$ with a characteristic constant J_0 [$J(r)$ with $r \rightarrow 0$] and $\alpha = 2/r_0$ with $J(r) = 0.135 J_0$ at $r = r_0$. The eigenfunctions of the charge pair Hamiltonian (Equation 2.26) include the singlet states, $|S\rangle$,

which are odd, and triplet states, $|T_+\rangle$, $|T_0\rangle$, $|T_-\rangle$, which are
even under the exchange of the two particles.[105] The $\hat{H}_{hyperfine}$
provides the necessary unsymmetrical term in the Hamilto-
nian (Equation 2.26), which gives rise to singlet–triplet mixing, the odd singlet states are subject to upconversion to even
triplet states.[106] An external magnetic field partially restores
the symmetry of the total pair Hamiltonian because the Zeeman Hamiltonian, $\hat{H}_{Zeeman}$, is even under exchange of the two
particles, and reduces mixing rate, implying a change in the
singlet-to-triplet products of the recombined charge pairs. The
effective value of the hyperfine coupling constant, a, as evaluated for a dye-anthracene crystal system, amounts to $a/g\mu_B$
$\cong 1$ mT. Thus, if $\Delta g \rightarrow 0$, the magnetic field effect (MFE)
becomes governed by the HFI in the pair states, which
appears at very low magnetic fields and saturates at high
fields when the Zeeman term exceeds the HFI energy. The
situation becomes more complex if the e-h interaction energy
is not negligible. This would be important for the short intercarrier distances (r) when $J(r)$ becomes fairly large (larger or
comparable with $\Delta g\mu_B B$ and $\hat{H}_{hyperfine}$). If the degenerate triplet
states fall much below the singlet pair state, the splitting of
$|T_+\rangle$ and $|T_-\rangle$ substates in moderate magnetic field strengths
is not sufficient to level $|T_+\rangle$ and $|S\rangle$ states, the HFI can be
unable to mix these states, no MFE on the recombination
product is expected. This is the reason that MFE on recombination radiation in pure anthracene crystals could not be
observed[107] (see also Groff et al.[103]). However, at a level cross-
ing field $B_c = |2J|/g\mu_B$, the HFI-induced mixing of these states
suddenly sets in, and a sharp change in the products yields
follows above this value of the field (decrease in S_1 and
increase in T_1). We would expect to see decreasing EFL and
increasing EPH efficiency. However, if $k^{(S)} >> k_{ST}$ and k_S comparable with k_{ISC}, the triplet are populated via S_1 and both
EFL and EPH will decrease above this field. This is what has
been recently observed in Alq$_3$- and Ir(ppy)$_3$-based double
layer (DL) LEDs.[108,109] The DL LED with Alq$_3$ emitter, placed
in a steady-state magnetic field (B) (Figure 2.16) shows the

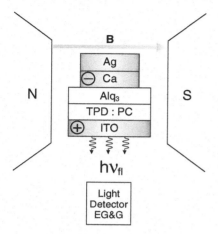

Figure 2.16 Schematic drawing of the experimental setup to measure EL output in a magnetic field (B). This is a topical view of an Alq_3 emitter–based LED placed between the pole pieces (N,S) of an electromagnet such that the magnetic field is parallel to the surface of the sandwich DL EL cell, and the electrofluorescence flux (hv_{fl}) leaves the cell perpendicular to B.

light output following the bell-shaped function of B as it increases from 0 to 0.5 T. A maximum value of the MFE is about 5% and appears at a field $\cong 300$ mT (Figure 2.17A). A similar behavior has also been recently reported for EPH LEDs, where the magnetic field–induced increase in the EPH efficiency up to 6% at about 500 mT and driving current $j \cong$ 3 mA/cm² was followed by a high field decrease of still positive effect continued to the highest accessible field of about 0.6 T (Figure 2.17B). This indicates the above mentioned "singlet route" of the formation of triplet excitons, and a magnetic field shift of the state $|T_+\rangle$ toward the state $|S\rangle$ of the $(e–h)$ pairs, leading to the reversed trend in the mixing at $B \cong 0.5$ T. The broad bell-shaped curves $(\Delta\varphi/\varphi)$ as a function of B are indicative of a distribution of the intercarrier distances dictated by the random nature of the recombination process and by both static and dynamic disorder in the emitters.

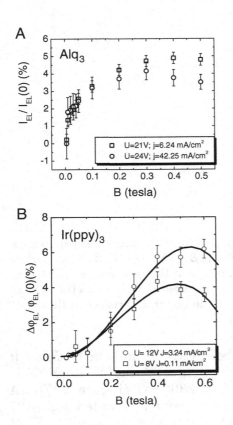

Figure 2.17 Experimental results of the magnetic field dependence of the MFEs on the EL output from a DL EL LED, ITO/75% TPD:25% PC (60 nm)/Alq$_3$ (60 nm)/Ca/Ag (A), and a DL EL LED, ITO/6 wt% Ir(ppy)$_3$:74 wt% TPD:20 wt% PC (60 nm)/100% PBD (50 nm)/Ca/Ag (B). The data have been taken at two different applied voltages and corresponding current densities given in the insets. (After Kalinowski et al.[108,109])

2.4 THE NATURE OF EMISSIVE STATES

A variety of emissive states can be created either by photo-excitation or charge carrier recombination processes. The emissive states illustrated in Figure 2.18 form a set of possible

(a) Photoexcitation

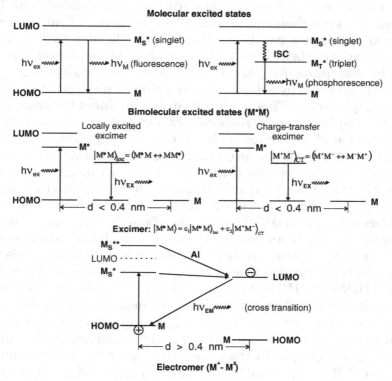

Excimer: $|M^*M\rangle = c_1|M^*M\rangle_{loc} + c_2|M^+M^-\rangle_{CT}$

Electromer (M⁺- M⁻)

(b) Recombination

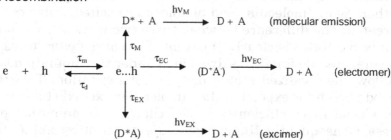

Figure 2.18 Photoexcited (a) and electron–hole recombination-produced (b) excited states in single-component organic solids. D and A in part (b) denote excitation energy donor (D) and energy acceptor (A).

excited states in single-component materials. The term "single-component" refers to a material composed intentionally from one-type chemical species if the presence of non-intentional impurities can be neglected. A molecular (well-localized) excited state is a typical product of photoexcitation. Light absorption ($h\nu_{ex}$) creates excited molecular singlet (M_S^*), which can relax by intersystem crossing to an excited molecular triplet (M_T^*). Their radiative decay produces fluorescence and phosphorescence, respectively. On the other hand, the interaction of an excited molecular state with a ground state molecule (M) leads to formation of *locally excited excimer* or *charge-transfer excimer* if the distance between molecules is shorter than 0.4 nm. If it is larger than 0.4 nm, we can deal with direct transition from the LUMO-located electron of one molecule to the HOMO-located hole of another molecule (*cross transition*); such an "emitting state" has been called *electromer*.[110] Electromer emission occurs when, due to a defect, the electron transfer from LUMO to LUMO and/or the hole from HOMO to HOMO is impeded. The level shift makes the emission red-shifted with respect to molecular emission as well as to the excimer emission. It is important to note that "electromer emission" requires the charge carriers to be separated, for example, by the dissociation of a molecular exciton. In the VR process such separated charge pairs (e–h) are formed when statistically independent carriers approach each other. Again, molecular and bimolecular excited states can be created. The difference between these two ways of creation of emissive states is clearly apparent. The charge pair states, as a rule, precede final emissive states under recombination, and follow the localized states under photoexcitation. This is a good reason for expecting the bimolecular excited states to be produced more efficiently in the charge recombination process. Consequently, the emission spectrum of recombination EL is expected to be, in general, different or at least more complex that under photoexcitation. There are several striking examples confirming this expectation. In Figure 2.19 the PL and EL spectra of an amine derivative (TAPC) in polycarbonate are presented. While in the spectrum of the photoexcited emission (PL) the molecular emission combined with the

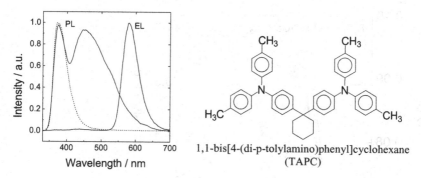

1,1-bis[4-(di-p-tolylamino)phenyl]cyclohexane
(TAPC)

Figure 2.19 Optically excited (PL) and recombination radiation (EL) spectra of TAPC thin films. The dashed curve is the PL spectrum of TAPC in a 10^{-5} M dichloromethane solution. (After Kalinowski et al.[110])

red-shifted band of the excimer emission is apparent, in the recombination radiation (EL) only the single strongly red-shifted emission band could be detected and ascribed to the electromer states.[110] Electromer emission has also been suggested to occur in polymeric LEDs.[111,112] Alternatively, the presence of unintentional admixtures can dominate the EL spectra, while the PL spectra are characteristic of the host material.[113] This could be the reason for a weak but distinctly resolved band at $\cong$540 nm in the EL spectrum of anthracene dispersed in polycarbonate (Figure 2.20). It appears in the tail of excimer-extended (EXs) EL spectra, where the first vibronic band due to the common tetracene impurity is located, with its 0–0 transition at $\cong$500 nm (cf. Figure 2.2) buried under three broad anthracene excimer bands at 457, 465, and 510 nm.[114] However, the identity of the PL and EL spectra of an anthracene-substituted supramolecule of methyl-exopyridine-anthracene (EPAR-Me), peaking at about 540 nm, seems to support the electromer concept for this emission band.[64] Also, the impurity emission has been unable to explain a weak emission from 500 to 540 nm observed in the remote time-resolved experiments on thermoluminescence of γ-irradiated anthracene in squalane.[115]

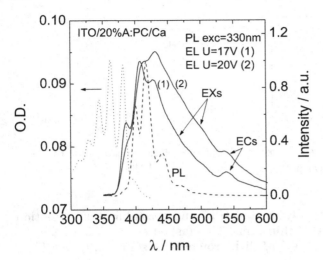

Figure 2.20 Optically excited (PL) and recombination (EL) emission from anthracene (A) dispersed in polycarbonate (PC). EXs denotes emission from excimers. ECs emission from electromers. Absorbance of the 20% A:PC film is shown for comparison. (After Kalinowski et al.[64])

In the case of two- or multicomponent materials with different ionization potentials and electron affinities, the situation becomes more complex (Figure 2.21). Light absorbed by single-type molecules (say, electron donor molecules, D) produces excited donor singlets (D^*). Their radiative relaxation shows up as donor fluorescence ($h\nu_D$). Interacting with the ground state electron acceptors (A), they can form *exciplexes* and *electroplexes*,[63,116,117] the latter being two different molecules, analogue of electromers in single-component solids. The EL spectrum of such molecular systems may reveal several bands corresponding to different emissive species. An example is shown in Figure 2.22. A mixture of the electron-donor (TPD) and electron-acceptor (PBD) molecules can form a single-layer (SL) system (Figure 2.22A) or these molecules can be brought into an intimate contact on the interface of two independent layers forming a DL LED, i.e., TPD in polycarbonate and 100% evaporated PBD (Figure 2.22B). Their

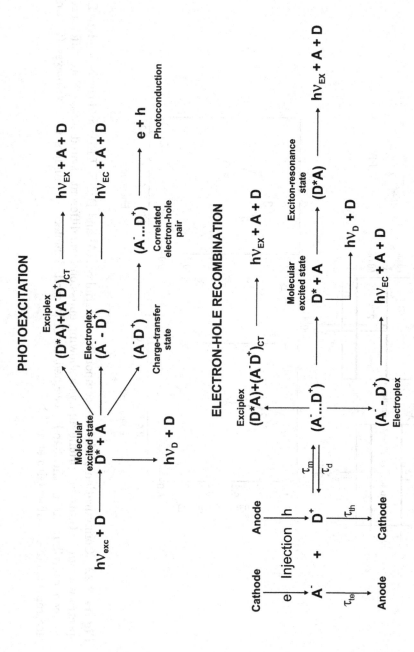

Figure 2.21 Excited states in two-component electron donor — (*D*)–electron acceptor (*A*) — system formed under photoexcitation and in the electron–hole recombination process.

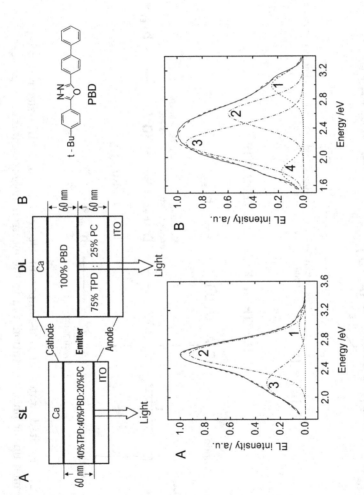

Figure 2.22 Configuration of the two-component material SL (A) and DL (B) LEDs and corresponding EL spectra with their Gaussian profile analysis showing emission from different excited states. Note differences between contributions of the bands ascribed to TPD excited molecular singlets (1), exciplexes (2), and electroplexes (3,4). (After Giro et al.[117])

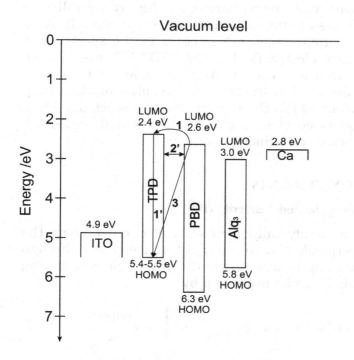

Figure 2.23 The energy level scheme of the material components used for manufacturing the cells described in Figure 2.22. Selected electronic transitions are indicated by lines with arrows. The LUMO and HOMO levels for Alq_3 are added for comparison. (After Kalinowski et al.[63])

emission spectra are different (Figure 2.22A and B, bottom). In the SL system a narrower spectrum is dominated by band 2, which in the broader spectrum of the DL structure decreases. The slightly pronounced shoulder 3 in the previous (SL) spectrum becomes a dominating feature in the DL spectrum. This is an excellent example of the coexistence of exciplex and electroplex emissions. In the energy level scheme of active molecules, we can see how they are formed (Figure 2.23). The locally excited exciplex is created by a LUMO PBD → LUMO TPD electron transfer. The molecular TPD singlet forms an exciplex with the nearest-neighbor molecule of PBD (process 2′). Due to an energy barrier for slightly more distant molecules, a cross transition takes place (process 3). As a result a red-shifted emission 3 appears in the spectrum. A strong attractive electric field at the TPD/PBD interface (cf. Section 2.5) impedes the external electric field–assisted charge-pair dissociation, the electroplex component becoming the major feature of the DL system emission spectrum. For a more detailed discussion of bimolecular excited states the reader is referred to Kalinowski et al.[62–64]

2.5 INJECTION CURRENTS

2.5.1 Injection-Limited Currents (ILC)

The flow of a steady unipolar electrical current along the direction x perpendicular to the injecting electrodes through an insulating sample, thus contributing to the current driving an organic LED, can be expressed by

$$j = j_s(x) + e\mu n(x)\left[F - \frac{e}{16\pi\varepsilon_0\varepsilon x^2}\right] - eD\frac{dn(x)}{dx}, \qquad (2.27)$$

where $n(x)$ is the coordinate dependent concentration of free charge carriers and D the microscopic diffusion coefficient of the carriers, which is directly related to the carrier mobility (μ) through the Einstein relation $D = \mu kT/e$. Equation 2.27 is composed of a hot carrier stream of $j_s(x)$ and two additional terms representing the current flow due to thermalized carriers. The thermalized carriers flow is governed by the macroscopic diffusion proportional to the concentration gradient $[-dn(x)/dx]$, the image force $(-e/16\pi\varepsilon_0\varepsilon x^2)$, and the applied field (F). The latter two form a potential barrier located at $x = x_m$ (see, e.g., Kalinowski[36]). If the diffusion current term is negligible, the drift current in the combined image and external electric fields dominates the collected current, which is determined by those carriers that escape over the image force barrier,

$$j = j_0 \exp(-x_m/l), \qquad (2.28)$$

where j_0 is the current that would flow for the average penetration depth of the carriers $l \to \infty$ (e.g., in the absence of traps and/or scattering process). The field dependence of the collected current in Equation 2.28 enters through the field-dependent position of the barrier:

$$x_m = (e/16\pi\varepsilon_0\varepsilon F)^{1/2}; \qquad (2.29)$$

thus,

$$j = j_0 \exp(-c/F^{1/2}), \qquad (2.30)$$

where $c = l^{-1}(e/16\pi\varepsilon_0\varepsilon)^{1/2}$ for hot carriers penetrating the sample with the mean free path l or $c = (2\pi/h)(em^*\chi_c/2\pi\varepsilon_0\varepsilon)^{1/2}$ assuming the carriers to be one-dimensional Bloch waves damped within the potential threshold (χ_c is the injection barrier referred to the Fermi level of the injecting electrode). The current that can flow through the sample is determined by its value at x_m. The trapping of charge carriers at their way to x_m must then be taken into account, especially at low fields when x_m becomes comparable with the carrier "schubweg," $L = (\mu kT\tau_{trap}/e)^{1/2}$, where τ_{trap} is the carrier trapping time. Then,

$$j = j_0 \exp(-x_m/L) = j_0 \exp(-b_t/F^{1/2}), \qquad (2.31)$$

with

$$b_t = (e^2/16\pi\varepsilon_0\varepsilon\mu kT\tau_{trap})^{1/2} \qquad (2.32)$$

found to fall between 0.1 and 4 (V/cm)$^{1/2}$ for unipolar injection of holes at the illuminated gold interface with anthracene and tetracene single crystals, where L varied in the 20- to 300-Å range.[118,119]

At high electric fields $x_m \to 0$, the image field barrier approaches a triangular shape and the collected current is governed by tunneling, we deal with the classic Fowler-Nordheim treatment,[120]

$$j = BF^2 \exp(-b/F) \qquad (2.33)$$

and

$$b = \left[4(2m_e^*)^{1/2}/3\hbar e\right]\chi_c^{3/2}, \qquad (2.34)$$

where B is a constant and m_e^* is the effective mass of electron inside the triangular barrier (see, e.g., Esaki[121] and Kalinowski[36]).

When $j_s(x)$ is a strongly decreasing function of x ($l \ll x_m$) all hot carriers entering the sample are thermalized at a distance l, and only their fraction due to the thermal activation over the barrier can contribute to the collected current,

$$j = AF^{3/4} \exp(aF^{1/2}), \qquad (2.35)$$

where $A(F) = j_0(el/kT)^{3/4} \exp[-2(\beta/l)^{1/2}] = \text{const}$, and

$$a = 2(\beta e/kT)^{1/2} = (kT)^{-1}(e^3/4\pi\varepsilon_0\varepsilon)^{1/2}. \qquad (2.36)$$

This is the injection current limited by a field-assisted separation of charge from its image in the injecting contact. Since the preexponential factor is a relatively slowly varying function of F, and the constant a is identical with the Schottky parameter a_s, Equation 2.35 can fairly be approximated by the straight-line plot $\ln j - F^{1/2}$, and often interpreted in terms of the Schottky injection into carrier conducting bands. One should, however, remember that the Schottky approach assumes the activated carriers to occupy allowed free electron states within wide conducting band materials. Its application to narrowband insulators (an overwhelming majority of organics) can be misleading (for a comprehensive discussion of this point, see Godlewski and Kalinowski[119]).

In summary, the generally valid Equation 2.27 governing injection-limited current flow in insulators can be reduced to a description of the drift current that evolves with the electric field taking on various functional shapes dependent on the primary carrier injection and motion mechanisms in the insulator. Furthermore, the current-voltage characteristics are affected by the spatial variation of electric field along the current pathway due to either space charge at electrodes or somewhere in the bulk of the sample. The latter is of particular importance for DL LEDs, where due to accumulation of charge at the HTL/ETL interface the field strengths in the anodic and cathodic compartments of the LED structure are different. The positive interface space charge reduces the nominal field (F) within the anode compartment to $F_1 = k_1 F$

by a screening factor $k_1 < 1$. At the same time the cathode field $F_2 = k_2 F$ becomes enhanced by a factor $k_2 = (d/d_2)[1 - k_1(d_1/d)]$ as determined by simple electrostatics arguments. This leads to a stepwise change of the electric field across the thickness (δ) of the interfacial layer in simple proportion to the concentration of charge n_i, $F_1 - F_2 = e\delta n_i/\varepsilon_0\varepsilon$,[7,63,122] and a deviation in the slopes of the Schottky- (or 1D-Onsager)-type $\ln j$ vs. $F^{1/2}$ plots from their theoretical values.[7,62,63] For example, the experimental value of such slopes for common TPD/Alq$_3$ junction-based LEDs, $a_{exp} \cong 2a_s$ caused by a roughly twofold decrease in the anode compartment field with $k_1 = (a_s/a_{exp})^2 \cong 0.4$ (cf. Figure 2.24). Another reason of the difference between a_S and a_{exp} may be due to disorder. A simplified analytical approach to primary carrier injection in disordered systems has been recently reported, assuming hopping theory

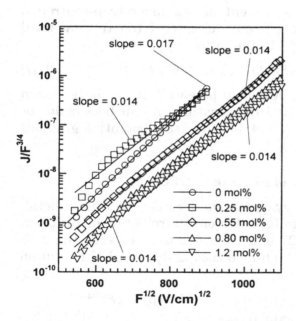

Figure 2.24 1D Onsager plots of the current for the LEDs ITO/TPD/DPP:Alq$_3$/Alq$_3$/MgAg with varying concentration of 6,13-diphenylpentacene (DPP). The mol% concentrations of DPP in EMLs of five different devices are given in the inset. (After Kalinowski et al.[123])

of the carrier transport, which implies the concept of transport energy.[124,125] The injection has been considered a two-step process. On the first step, carriers make initial jumps from the Fermi level of the metal contact into a localized state close to the metal–organic interface. The rate of such a jump is determined by the initial jump distance and by the energy of a target state. On the second step, the carrier random walk within the system of hopping sites is described as a continuum diffusive motion. Then, the 1D-Onsager probability to avoid geminate recombination of a carrier with its image twin has been calculated, assuming the probability to overcome the potential barrier to be independent of the energy of the target localized state for the initial jump. If the transport level is comparable with the height of the electrostatical barrier, the typical 1D-Onsager $\log j - F^{1/2}$ straight-line plots for the current are obtained with the slopes $\partial \ln j / \partial F^{1/2}$ larger than a_S. This approach is reminiscent of the macrotrap-controlled transport[126,127] and even more closely of the trap-modified near-contact barrier.[119] In the latter

$$a = 2\{\beta(eF/kT) - (e/kT)[dV(x)/dx]\}^{1/2}, \qquad (2.37)$$

where $V(x)$ represents the trap potential. It has been shown that at lower fields and temperatures a can be smaller or larger than a_S dependent on the trap potential gradient $dV(x)/dx$.[119]

2.5.2 Diffusion-Controlled Current (DCC)

The current imposed by charge carrier injection from a metallic electrode is said to be diffusion-controlled if the diffusion term in Equation 2.27 is comparable to or exceeds the drift current flow. For the high-field case $[2(\beta\gamma)^{1/2} > 1]$, the solution of Equation 2.27 is[119]

$$j_{DCC} \cong \mu en(x_0)\exp(-\beta/x_0)\pi^{-1/2}(kT/e\beta)^{1/4} (eF)^{3/4}$$
$$\exp[2(\beta e/kT)^{1/2} F^{1/2}], \qquad (2.38)$$

with x_0 the the very near-contact distance where Equation 2.27 is not applicable.

Regarding the constant factor, it is identical to Equation 2.19 in the Emtage and O'Dwyer paper,[128] deriving the diffusion-limited Schottky emission of electrons from a metallic cathode into the conduction band of an insulator. Also, it resembles the general drift current solution (Equation 2.35), though an important difference can be noted due to the mobility appearing in Equation 2.38. This would be of crucial importance if the mobility were an electric field–dependent quantity as often reported function 2.18 (cf. Figure 2.6).

2.5.3 Space-Charge-Limited Current (SCLC)

An indispensable condition of the occurrence of SCLC is that the electrode can supply more carriers per unit time than can be transported through the insulating sample. A contact that behaves in that way is called ohmic contact. At an ideal ohmic contact the electric field vanishes ($F = 0$) because of screening by the injected space charge (the charge concentration $n_0 \to \infty$). In practice, an electrode can only be ohmic if the injection barrier is small enough to ensure that no field-assisted barrier lowering is required to maintain a sufficiently high injection rate. In that case a virtual electrode is established apart from the geometrical contact that serves as a charge reservoir. Its position depends on the applied field, moving into the sample bulk as the applied field decreases.[129] For perfectly ordered or disordered insulating materials, or those containing very shallow traps, the SCL current in a sample of thickness d obeys Child's law (see, e.g., Helfrich[48]):

$$j_{SCL} = (9/8)\varepsilon_0\varepsilon\mu F^2/d. \tag{2.39}$$

In the presence of discrete traps:

$$j_{SCL} = (9/8)\varepsilon_0\varepsilon\Theta\mu F^2/d, \tag{2.40}$$

where μ is the microscopic mobility of the carriers, and Θ is the fraction of free (n_f) to trapped (n_t) space charge $\Theta \cong [N_{eff} \exp(-E_t/kT)]/(H - n_t)$, with N_{eff} the effective density of electronic states, and E_t and H discrete trap depth and trap concentration, respectively. If local traps are distributed in energy according to often-assumed exponential function $h(E)$ = $(H/lkT)\exp(-E/lkT)$ with the total concentration of traps, H, and a characteristic distribution parameter l, the SCL current can be expressed as[130]

$$j_{SCLC} = \frac{N_{eff}\, e\, \mu}{H^l}\left(\frac{\varepsilon_0\, \varepsilon}{e}\right)^l \left[\frac{l^2 \sin[\pi/l]}{(l+1)\pi}\right]^l \left(\frac{2l+1}{l+1}\right)^{l+1} \frac{F^{l+1}}{d^l}. \quad (2.41)$$

It is a power function of the applied field with the power $l + 1 > 2$ dependent on the steepness of the trap energy distribution, l. All the power functions of the applied field, Equations 2.39, 2.40, and 2.41, are subject to modification by a field-dependent mobility, $\mu(F)$, which can be modified by field-dependent trapping effects through $\Theta = N_{eff}\, n_{tot}^{\ell}/H^{\ell}$, where n_{tot} = $n_f + n_t \cong \varepsilon_0 \varepsilon F/ed$ (see, e.g., Helfrich[48]).

2.5.4 Double Injection

To manufacture an organic LED functioning on the basis of electron–hole recombination processes a system of organic films has to be provided with two injecting contacts, one injecting electrons, another injecting holes (cf. Figure 2.10). Description of electrical properties of such systems is much more complex than that for systems with one injecting contact because the recombination current adds to the drift and diffusion currents flowing between electrodes through the sample.[131] The total current density may be orders of magnitude larger than with single injection, although the positive and negative net space charges situated at the respective electrodes are roughly equal to the one-carrier space charge connected with current flow. The generally valid equation governing double-injection current in the presence of space charge has been derived by Parmenter and Ruppel.[132] For two ohmic electrodes and trap-free (or shallow trap) case:

$$j = (9/8)\varepsilon_0 \varepsilon \mu_{eff} F^2/d. \quad (2.42)$$

This is identical to Equation 2.39 except for the mobility, which in the double-injection case is a complex combination of individual electron (μ_e) and hole (μ_h) microscopic mobilities including so-called recombination mobility ($\mu_0 = \varepsilon_0 \varepsilon \gamma_{eh}/2e$),[48]

$$\mu_{eff} = \mu_0 \, \nu_e \, \nu_h \left(\frac{2}{3}\right)^2 \left[\frac{\left(\frac{3}{2}[\nu_e + \nu_h] - 1\right)!}{\left(\frac{3}{2}\nu_e - 1\right)! \left(\frac{3}{2}\nu_h - 1\right)!} \right]^2$$

$$\times \left[\frac{(\nu_e - 1)! \, (\nu_h - 1)!}{(\nu_e + \nu_h - 1)!} \right]^3 \qquad (2.43)$$

where ν_e and ν_h are the dimensionless parameters defined as the ratio μ_e/μ_0 and μ_h/μ_0, respectively. Let us note that the introduction of μ_0 accounts for the recombination effect on the current. Its relation to the carrier mobilities allows us to distinguish three limiting cases for the SCL double injection current flow:

1. Injected plasma (or weak recombination) case for $\nu_e \gg 1$ and $\nu_h \gg 1$, with a large space overlap, interpenetrating electrons and holes mostly reach opposite electrodes. Equation 2.43 reduces to $\mu_{eff} \cong (2/3)[2\pi(\mu_e\mu_h/\mu_0)(\mu_e+\mu_h)]^{1/2}$.
2. Volume-controlled current (or strong recombination) case for $\nu_e \ll 1$ and $\nu_h \ll 1$. This is the case of negligible space-charge overlap with $\mu_{eff} \cong \mu_e + \mu_h$.
3. One-carrier SCLC flow for $\nu_e \gg 1$, $\nu_h \ll 1$ (or $\nu_e \ll 1$, $\nu_h \gg 1$). Equation 2.43 may then be reduced to $\mu_{eff} \cong \mu_e$ (or μ_h).

The current is practically one carrier SCL current, the less mobile carriers recombining very near the electrode from which they are injected.

It is interesting to note that in the Langevin recombination mechanism with the γ_{eh} expressed by the sum of the individual carrier mobilities (Equation 2.11), the interrelation between $\nu_{e,h}$ and μ_0 switches to the relation between electron and hole mobilities themselves. For example, the strong recombination case can be defined by either $v_e = \mu_e/\mu_0 = 2(1 + \mu_h/\mu_e)^{-1} \ll 1$, that is, $\mu_h/\mu_e \gg 1$ or $v_h = \mu_h/\mu_0 = 2(1 + \mu_e/\mu_h)^{-1} \ll 1$, that is $\mu_e/\mu_h \gg 1$, if Equation 2.11 for γ_{eh} is taken into account. An excellent system to study the strong recombination double injection case is DL LED with an interface blocking passage of charge carriers across the LED.

The presence of active charge traps can change the above picture entirely. The relation between the average trapping time (τ_{trap}) of the carriers, the average thermal release time (τ_{rel}) of the carriers from traps, and their average recombination time (τ_{rec}) becomes important. For active (sufficiently deep) traps the condition $\tau_{rel} \gg \tau_{trap}$ is usually fulfilled; otherwise trapping would be ineffective. The situation becomes even more complex if traps are concentrated in front of the electrodes, the case very probable due to the exposition of sample surface to the ambient atmosphere during sample handling and deposition of the electrodes. A striking example of a complex recombination zone due to inhomogeneous trapping effects has been presented in Figure 2.6.

2.6 EL QUANTUM EFFICIENCY

2.6.1 Basic Definitions

Quantum EL efficiency is one of the most important critical figures of merit for OLEDs and by definition relates the photon [$h\nu(J)$] flux per unit area [$\Phi_{EL} = I_{EL}/h\nu$ (photon/cm^{2}s), where I_{EL} is the light energy (radiant) flux per unit area (J/cm^{2}s)], and the carrier stream per cm^2 (j/e) of the device as the ratio:

$$\varphi_{EL} = e\,\Phi_{EL}/j \quad \text{(photon/carrier).} \quad (2.44)$$

This quantity is a measure of the degree of the conversion of the current into light. The quantum efficiency (QE) defined by Equation 2.44 can be associated with other performance parameters such as the dimensionless energy conversion efficiency:

$$\eta = \frac{E_{EL}}{U\,i}, \quad (2.45)$$

where E_{EL} is the light-energy (radiant) flux (Watt) and Ui is the electrical power (Watt) supplied to the device,

$$\varphi_{EL} = \frac{eU}{h\nu}\eta\,. \tag{2.46}$$

It follows from Equation 2.46 that for the applied voltages $U > 3$ V, $\varphi_{EL} > \eta$ whenever emission occurs within the visible light range. We note that the definition Equation 2.44 assumes monochromatic emission at a constant photon energy, $h\nu$. Commonly, the radiant flux I_{EL} is measured over the total emission band $f(h\nu)$, and the averaged photon energy $< h\nu >$ must be used to obtain Φ_{EL},

$$\langle h\nu \rangle = \frac{\int\limits_0^\infty h\nu f(h\nu)\,d(h\nu)}{\int\limits_0^\infty f(h\nu)\,d(h\nu)}\,. \tag{2.47}$$

Furthermore, for the face-detected emission (as usually is the case) the light output coupling factor ξ reduces the measured Φ_{EL} to $\Phi_{EL}^{(ext)}$, so that we deal with the external quantum EL efficiency

$$\varphi_{EL}^{(ext)} = \xi\,\varphi_{EL} = e\,\Phi_{EL}^{(ext)}\big/j\,, \tag{2.48}$$

which can be a small fraction of the internal EL quantum efficiency, φ_{EL}. Also, the angular intensity pattern of the emitted light must be taken into account. Thus, the precise estimation of the EL QE requires experimentally measured angular distribution of the EL intensities at all wavelengths within the emission spectrum (note that emission spectra differ in general when measured at different observation angles). Commonly, the Lambertian emission pattern is assumed, and

$$\varphi_{EL}^{(ext)} \cong \frac{\pi e\xi}{j\langle h\nu \rangle}\int\limits_0^\infty I_{EL}(h\nu)\,d(h\nu)\,. \tag{2.49}$$

For the integral in Equation 2.49 the total energy flux per unit area as measured by a radiometer at the normal direction to the emissive surface is substituted. This approximation can lead to substantial (up to 30%) deviations from the external value of the EL efficiency.[133]

In optoelectronic applications, photometric quantities are often used to express the degree of the current conversion into light. The luminous efficiency with the Lambertian emission pattern is

$$\varphi_{EL}^{(ext)}(L) = \frac{\pi L_0}{j}\left(\frac{C\,d}{A}\right),\qquad(2.50)$$

where L_0 is the luminance L (Cd/m^2) at normal incidence. The luminous efficiency of 1 Cd/A corresponds to 4π lm/A. The photopic vision function $V(\lambda)$ must be invoked to translate luminous to physical quantities (see, e.g., Helbig[134]),

$$\varphi_{EL}^{(ext)} = \frac{\pi e \xi L_0}{h K_m j \langle h\nu\rangle}\int_0^\infty f(h\nu)\,V(h\nu)\,d(h\nu).\qquad(2.51)$$

The constant $K_m = 673$ lm/W is here the luminous efficiency at $\lambda = 555$ nm, where the function $V(\lambda)$ reaches its maximum. For example, a very high luminous efficiency 28 Cd/A $\equiv$ 350 lm/A for an electrophosphorescent LED based on the Ir(ppy)$_3$:CBP emitter translates into the power conversion efficiency $\eta \cong 30$ lm/W ($\equiv 0.056$) and the quantum efficiency $\Phi_{EL}^{(ext)} \cong 10\%$ photon/carrier at $L \cong 100$ Cd/m^2.[135]

2.6.2   Steady-State EL

In a simplified, often-used picture of homogeneously distributed emissive singlets producing electrofluorescence, their concentration, S, under steady-state electron–hole recombination conditions, can be expressed by a simple equation:

$$\frac{dS}{dt} = P_S\,\gamma\,n_h\,n_e - k_S\,S = 0\qquad(2.52)$$

and

$$S = \left(P_S \, \gamma / k_S \right) n_h \, n_e \,, \tag{2.53}$$

where γ is the second-order recombination rate constant to be identified with γ_{eh} defined in Section 2.3, $k_S = k_F + k_n$ is the total decay rate constant including all nonradiative decays with the overall rate constant k_n, and P_S is the probability that as result of the e–h recombination event a singlet excited state will be created.

The evaluation of $\Phi_{EL}^{(ext)}$ is now straightforward and leads to

$$\Phi_{EL}^{(ext)} = \xi \, P_S \, \varphi_{FL} \, \gamma \, n_h \, n_e \, d \,, \tag{2.54}$$

where the $\varphi_{FL} = k_F / k_S$ determines the relative contribution of radiative decay events of excited singlet states, identified often with the FL efficiency.

Equation 2.54 relates the EL output to the uniform throughout the emitter concentrations of holes (n_h) and electrons (n_e). The latter are naturally associated with hole (j_i^h) and electron (j_i^e) currents injected at the electrodes through the following kinetic equations:

$$\frac{j_i^h}{e\,d} - \frac{n_h}{\tau_t^h} - \gamma \, n_e \, n_h = 0 \tag{2.55a}$$

$$\frac{j_i^e}{e\,d} - \frac{n_e}{\tau_t^e} - \gamma \, n_e \, n_h = 0 \,. \tag{2.55b}$$

Here, $\tau_t = d/\mu F$ is the carrier transit time dependent on the carrier mobility, μ, and electric field, F, operating in the sample. The bimolecular decay of holes and electrons can be expressed by the recombination time

$$\tau_{rec}^{e,h} = \left(\gamma \, n_{h,e} \right)^{-1} \tag{2.56}$$

to be compared with the monomolecular decay time, τ_t, of carrier discharge at opposite electrodes. Two limiting cases leading to simplified interrelations between $\Phi_{EL}^{(ext)}$ and injection currents have been distinguished based on such a comparison.[136] These are as follows:

1. Injection–Controlled EL (ICEL) for $\tau_t \ll \tau_{rec}$. Then, according to Equations 2.55, $n_h \cong j_i^h/e\mu_h F$, $n_e = j_i^e/e\mu_e F$, and

$$\Phi_{EL}^{(ext)} = \xi \, \varphi_{FL} \, P_S \, \gamma \frac{j_i^e \, j_i^h}{e^2 \, \mu_e \, \mu_h \, F^2} \, d \, . \qquad (2.57)$$

The important message that follows from Equation 2.57 is that the light output for organic LEDs operating in the ICEL mode cannot be related directly to the driving current (j) which must not be identified with the recombination current

$$j_R = \gamma \frac{j_i^e \, j_i^h}{e^2 \, \mu_e \, \mu_h \, F^2} \, d \qquad (2.58)$$

and which is not known from electrical measurements. For the ICEL mode, the EL is a "side" effect of the current flow and its intensity is proportional to the product of the electron and hole injection currents. Several subcases of ICEL can be considered dependent on the carrier injection mechanisms from the electrodes[36] (see also Section 2.5).

2. Volume-Controlled EL (VCEL) $\tau_t \gg \tau_{rec}$. In this case the first-order decay terms in Equations 2.55 can be neglected; the carriers decay totally in the bimolecular recombination process (a weak leakage of carriers to electrodes), $j_i^e = j_i^h = j$, and

$$\Phi_{EL}^{(ext)} = \xi \, P_S \, \varphi_{FL} \, \frac{j}{e} \, . \qquad (2.59)$$

Here, the measured current j is simply the recombination current, j_R, and as long as ξ, P_S, and φ_{FL} do not depend on j, $\Phi_{EL}^{(ext)}$ remains directly proportional to the driving current. The slope of the linear plot of $\Phi_{EL}^{(ext)}$ with j determines the product $\xi P_S \varphi_{FL}$.

The variety of results on EL intensity vs. current flowing through device has been observed experimentally (see Kalinowski[36] and references therein). The $\Phi_{EL}^{(ext)}$ has been shown to increase both linearly and nonlinearly with increasing current. The latter can be either sublinear and supralinear dependent on the applied voltage range. On the analytic side, the interplay between the recombination and leakage current seems to account for the variety of observations. It is useful to distinguish between SL and DL LEDs because interfacial energy and mobility barriers at two component organic layers in the DL devices increase the recombination current greatly, leading to the VCEL operation mode or at least to its approximation. One expects the linear increase of $\Phi_{EL}^{(ext)}$ with increasing drive current. In SL devices it is more difficult to avoid the leakage of charge carriers to opposite electrodes, unless perfectly ohmic electrodes are available. They usually operate in the ICEL mode, and $\Phi_{EL}^{(ext)}$ becomes a complex function of the drive current. A careful analysis of the experimental data allows us then to distinguish three segments of such a dependence. The nonlinear increase of the EL intensity vs. j in the low-current regime passes to a linear segment for the intermediate currents regime and tends to saturation at high current densities. This behavior reflects the gradual voltage evolution of the recombination to the leakage current ratio $(j_R/j_L \; ; j = j_R + j_L)$. Increasing at low voltages and remaining constant at moderate electric fields, it shows remarkable decrease at high voltages (large current densities). Deviations from the linear relationship $\Phi_{EL}(j)$ also occur in DL organic LEDs. This is illustrated in Figure 2.25 showing the EL output vs. current density in a double-logarithmic scale for one of the most-studied DL organic LEDs based on the TPD/Alq$_3$ junction. The low-current density supralinear increase followed by a slightly current-increasing EL output at moderate currents, rolls off smoothly as the cell current exceeds 100 mAcm^{-2}. All three segments of the $\Phi_{EL}(j)$ curve have been approximated by the power-type functions with the powers given by the log Φ_{EL} − log j straight-line plots. This behavior can be explained by the ICEL mode operation conditions predicting $\Phi_{EL}(j)$ to follow expression 2.57. By definition, the measured cell current $j \cong j_i^{(e)} + j_i^{(h)}$. Thus, Equation 2.57 provides the EL output variation with the measured current density (j) in a form

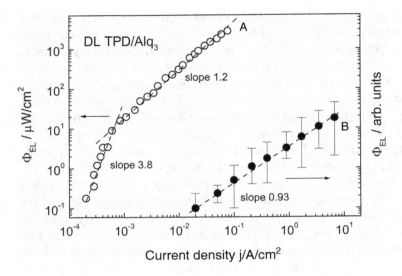

Figure 2.25   EL intensity vs. current density for DL LEDs based on the TPD/Alq$_3$ junction: (A) ITO/TPD(60 nm)/Alq$_3$(60 nm)/Mg/Ag structure (left-hand scale in absolute units);[36] (B) ITO/TPD(20 nm)/Alq$_3$(40 nm)/Mg:Ag.[137]

$$\Phi_{EL}^{(ext)} \sim \frac{j^2}{\mu^* \, j_{SCLC}}, \tag{2.60}$$

where $\mu^* = \left(\mu_h^{-1} + \mu_e^{-1}\right)^{-1}$. Since the current increase is imposed by the increasing voltage applied to the cell, both j and j_{SCLC} are the voltage-increasing quantities. Moreover, since by definition $j < j_{SCLC}$, the Φ_{EL} must be a function of the injection efficiency, φ_{inj}, defined by the ratio

$$\varphi_{inj} = \frac{j}{j_{SCLC}} \leq 1 . \tag{2.61}$$

The injection efficiency appears to be a crucial factor for the functional shape of $\Phi_{EL}(j)$. The irreproducible behavior of injection contacts enables understanding of the variety of $\Phi_{EL}(j)$ characteristics. The data in Figure 2.25 would indicate a strongly increasing φ_{inj} at low fields (low current densities) and its much slower increase at high field (high current densities above 1 mA/cm²). In the upper limit of the attainable currents (>100 mA/cm²) the $\Phi_{EL}(j)$ approaches linearity and even a sublinear behavior may be seen from its log–log plot (the data with arbitrary units of Φ_{EL}). The field-induced enhancement of the injection efficiency is difficult to rigorously analyze because different preparation conditions modify the contact in a microscopically uncontrollable manner. Therefore, the interpretation of the strong initial gradient $\partial\Phi_{EL}(j)/\partial j$ remains an open question, although a more detailed discussion of the processes that underlie the EL efficiency alleviates its understanding. A relatively simple explanation for the gradient $\partial\Phi_{EL}(j)/\partial j = n < 2$ (but $n > 1$) can be proposed if a weakly varying $\varphi_{inj}(j)$ and $j_h \cong j_e$ obeying the thermionic injection mechanism (Equation 2.35) is assumed in addition to the exponential field increase of the mobility according to Equation 2.18. Under these premises, with the accuracy to a weakly varying function of F, and with $\beta_\mu^{(e)} \cong \beta_\mu^h = \beta_\mu$,

$$\Phi_{EL}(j) \sim j^{2-\beta_\mu/a} . \tag{2.62}$$

The experimental value $(2 - \beta_\mu/a) = 1.2$ requires $\beta_\mu = 0.76 \times 10^{-2}(cm/V)^{1/2}$ if $a = 0.95 \times 10^{-2}(cm/V)^{1/2}$ is assumed as calculated from Equation 2.36 with $\varepsilon = 2.4$ for TPD (note that a slightly lower value for a is obtained in Alq₃ due to its higher ε; but still the equality $j_i^{(e)} = j_i^{(h)}$ holds due to a little difference in the injection barriers at ITO/TPD and Mg/Alq₃ interfaces; cf., e.g., Kalinowski[36]). This value of β_μ is in good agreement with the data obtained from the field dependence of the TOF-measured mobility for holes[138,139] and electrons[138,140] in Alq₃. A field-decreasing φ_{inj} could explain a slightly sublinear plot of $\Phi_{EL}(j)$ at high current densities. Some additional effects such as the current decreasing of the light output coupling factor (ξ) or the emission efficiency (φ_{FL}) can contribute effectively to the saturation tendency at high voltages. Variations in the factor ξ can occur as a result of the field evolution of the recombination zone. Its high-field (large current densities) confinement in the near-cathode region may change the ratio of EL intensities in the direction normal to the substrate face to that emitted from the edge of the substrate ($\Phi_{EL}^{(surface)}/\Phi_{EL}^{(edge)}$).[141,142] Stronger effects on Φ_{EL} can be expected due to a reduction of φ_{FL}, caused by singlet exciton quenching on charge carriers, structural defects, and metallic contact itself. The quenching of singlet excitons in organic LEDs based on optimal 60-nm-thick Alq_3 emitters has been evaluated to create emission losses between 8 and 30%.[143] These are comparable with the losses caused by high-field-assisted dissociation of singlet excitons or their electron–hole pair precursors.[76,97,144] Triplet excitonic interactions have been revealed in EL-current relationships for aromatic crystals.[7,36] The total EL flux from these crystals comprises a fast and delayed component attributed to the radiative decay of the prompt (PEL)- and delayed (DEL)-formed singlet excitons. If the DEL component originated from singlet excitons created in the process of triplet–triplet annihilation a superlinear increase of Φ_{EL} with increasing current would appear even for devices operation in the VCEL mode:

$$\Phi_{EL} = \Phi_{PEL} + \Phi_{DEL} = \varphi_{FL} \, P_S \, \tau_S \, j + \frac{\varphi_{FL} \, \gamma_{TT}^{(S)} \, \tau_T^2 \left(2 P_S + P_T\right)}{e^2 \, d} \, j^2. \quad (2.63)$$

The effective power of the $\Phi_{EL}(j) \sim j^n$ dependence can vary between one and two, depending on the prompt and delayed component contributions, which are determined by the effective lifetimes of singlet (τ_S) and triplet (τ_T) excitons along with the probabilities of their formation in the $e–h$ recombination process (P_S, P_T). The DEL component depends in addition on the $T–T$ interaction constant [$\gamma_{TT}^{(S)}$] leading to a singlet exciton, and the sample thickness, d. The EL intensity as a function of the measured current for a number of aromatic crystals has been shown to follow a supralinear relationship and could be approximated by a power dependence within limited ranges of the current.[36] The explanation is that the total EL output reflects the averaged signals from the PEL and DEL components according to Equation 2.63; the increasing DEL increases the power "n." A decrease in the effective power for high currents seen for neat tetracene crystals[36] would suggest the triplet–triplet ($T + T$) and triplet-charge carrier ($T + q$)

interactions to reduce the effective triplet exciton lifetime $(\tau_{eff}^{(T)})$ according to

$$\frac{1}{\tau_{eff}^{(T)}} = \frac{1}{\tau_T} + \gamma_{TT} \cdot T^2 + \gamma_{Tq} \, Tn_q \,, \tag{2.64}$$

where τ_T is the intrinsic triplet exciton lifetime, γ_{TT} and γ_{Tq} the overall triplet–triplet (T^2) and triplet-charge carrier (Tn_q) second-order interaction rate constant, respectively, and n_q the total concentration of charge carriers. To explain variations of "n" from crystal to crystal the dependence of P_S on the trap depth of one of the recombining carriers has been invoked.[35]

2.6.3 Field Dependence of the EL QE

The field evolution of the EL QE ($\Phi_{EL}^{(ext)}$) reveals typically a nonmonotonous function showing a maximum at a certain (usually high) value of the applied field. A number of experimental examples are shown in Figure 2.26 and Figure 2.27. Independent of the LED structure the initial field (or current) increase in EL QE is followed by the rolloff preceded by more or less broad maximum dependent on the carrier injection efficiency from the electrodes and transport properties of the materials forming the EL device. Caution is appropriate concerning the generalization of some literature results showing the high-field saturated (see, e.g., Malliaras and Scott[151]) or increasing (see, e.g., Campbell et al.[152]) EL efficiencies, since they can be simply due to the limited range of the applied field or to modified external interactions of excitons with metal electrodes. To understand this behavior, the field dependence of the quantities defining $\Phi_{EL}^{(ext)}$ must be examined. By comparing Equation 2.48 with Equation 2.59 the external EL quantum efficiency

$$\varphi_{EL}^{(ext)} = \xi \, P \, \varphi_r \tag{2.65}$$

can be simply expressed by the probability of creation of a singlet ($P = P_S$) or triplet ($P = P_T$) exciton and the efficiency of their radiative decay (φ_r). It is important to note that

Figure 2.26 The EL external quantum efficiency as a function of applied field (A) and luminance efficiency as a function of applied voltage (B) for various organic LEDs. 1: ITO/[6 wt% Ir(ppy)$_3$:74 wt% TPD:20 wt% PC](50 nm)/100% PBD(50 nm)/Ca/Ag; 2: ITO/TPD(60 nm)/[0.5 wt% quinacridone (QAC):Alq$_3$](50 nm)/Mg; 3: ITO/TPD(60 nm)/Alq$_3$(60 nm)/Al-CsF; 4: ITO/TPD(60 nm)/Al-LiF; 5: ITO/TPD(60 nm)/Alq$_3$(60 nm)/Mg; 6: ITO/(75 wt% TPD:PC)(60 nm)/Alq$_3$(60 nm)/Mg; 7: ITO/(75 wt% TPD:PC)(60 nm)/Alq$_3$(55 nm)/Mg; 8: ITO/TPD(60 nm)/(0.5% QAC:Alq$_3$)(50 nm)/Al; 9: ITO/polyethylenedioxythiophene (PEDOT)(20 nm)/terphenyl-PPV(80 nm)/low-work function cathode; 10: ITO/PEDOT(20 nm)/polyspiro[2,2′,7,7′-tetrakis-(2,2-diphenylamino)spiro-9,9′-bifluorene (Spiro-TAD); 2,2′,7,7′-tetrakis-(2,2-diphenylvinyl)-spiro-9,9′-bifluorene (Spiro-DPVBi)](80 nm)/low-work function cathode; 11: ITO/PEDOT(20 nm)/Spiro-TAD (20 nm)/[Alq$_3$:DCM](10 nm)/Alq$_3$(30 nm)/cathode (not specified); 12: ITO/polystyrene sulfonate (PSS):PEDOT/polyfluorenes or PPV/low-work function metal cathode (ink-inject printed organic layers); 13: as in item 12 with the spin coating–prepared organic layers; 14: ITO/PEDOT(20 nm)/[4 wt% PtOEP:(PMMA:Alq$_3$(1:1))(100 nm)]/Ca/Al; 15: ITO/PEDOT (20 nm)/[4 wt% PtOEP:(PMMA:PBD)(1:1)(100 nm)]/Ca/Al.

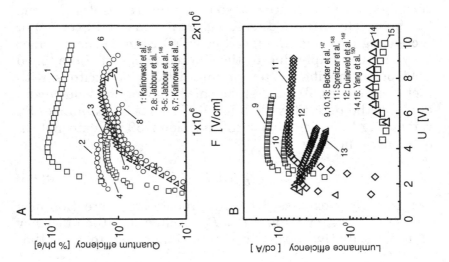

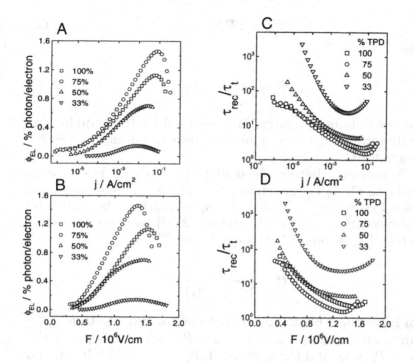

Figure 2.27 External EL quantum efficiency plotted against current density (A) and against applied field (B) driving ITO/(%TPD: PC)(70 nm)/Alq$_3$(60 nm)/Mg/Ag DL LEDs with different concentrations of TPD in the HTL (given in the figure). Corresponding recombination-to-transit-time ratio calculated from Equations 2.66 and 2.68) using the data for φ_{EL} and $\xi = 0.6$, $\varphi_r = 25\%$ and $P = P_S = 0.25$, and plotted against the current (C) and applied field (D). (After Kalinowski et al., *J. Phys. D.* 34, 2282, 2001. © Institute of Physics (GB). With permission.)

Equation 2.48, thus Equation 2.65, assumes the recombination probability $P_R = 1$. This is indeed the case when the driving current coincides with the recombination current, or in other words, for the upper limit of the VCEL operating LEDs. Whenever, the LED function obeys the ICEL operation mode, Equation 2.48 is no longer valid, and

$$\varphi_{EL}^{(ext)} = \xi P P_R \cdot \varphi_r \,, \qquad (2.66)$$

where

$$P_R = \frac{k_{rec}}{k_{rec} + k_t} \langle 1 \qquad (2.67)$$

is the recombination probability defined by the bimolecular recombination (k_{rec}) and monomolecular (k_t) decay first-order rate constants.

To obtain a better physical picture of the phenomena underlying P_R, it is convenient to replace the rate constants by their inverses $\tau_{rec} = k_{rec}^{-1}$ and $\tau_t = k_t^{-1}$ which have been defined as the recombination time (Equation 2.56) and carrier transit time (see Equations 2.55), respectively. Then,

$$P_R = \frac{1}{1 + \left(\tau_{rec}/\tau_t\right)} \,. \qquad (2.68)$$

From Equation 2.68 it is clear that maximizing P_R (thus $\varphi_{EL}^{(ext)}$) requires minimizing the ratio τ_{rec}/τ_t, $P_R \to 1$ as $\tau_{rec}/\tau_t \to 0$. For $\tau_{rec} = \tau_t$, $P_R = \frac{1}{2}$, and for $\tau_{rec}/\tau_t \gg 1$, $P_R \cong \tau_t/\tau_{rec}$. The ICEL and VCEL modes can be redefined on the basis of P_R. The ICEL mode, when the major carrier decay is due to the electrode capture, assumes $P_R < \frac{1}{2}$, i.e., $\tau_{rec}/\tau_t > 1$; the VCEL mode under the carrier decay dominated by the recombination applies for $P_R > \frac{1}{2}$, i.e., $\tau_{rec}/\tau_t < 1$. The P_R and thus τ_{rec}/τ_t ratio can be determined from the absolute value of $\varphi_{EL}^{(ext)}$ if ξ, P, and φ_r are provided independently (see Equation 2.66). Figure 2.27C and D show the variations of τ_{rec}/τ_t with drive current and applied field. As expected, the minima occur at the field strengths corresponding to those for the maxima of the EL QE (Figure 2.27A and B). The ratio τ_{rec}/τ_t exceeds unity within the entire range of electric fields attained, although for the highest concentrations of TPD in the HTL it approaches unity. This indicates that all these LEDs operate in the ICEL regime approaching the demarcation value of $\tau_{rec}/\tau_t = 1$ ($P_R = \frac{1}{2}$), below which the VCEL mode sets in. The absolute values of the τ_{rec}/τ_t ratio as well as their electric field gradient may be a subject to some uncertainties associated with the assumption of the field independence of ξ, P, and φ_r. However, the apparent trend of the ratio to decrease with concentration of TPD in HTL suggests its values (thus, $\varphi_{EL}^{(ext)}$) to be associated with the injection efficiency of holes. In fact, using the definition equations for τ_{rec} (Equation 2.56) and τ_t in Equations 2.55a and b, we find

$$\frac{\tau_{rec}}{\tau_t} = \frac{\mu_{e,h} F}{\gamma n_{h,e} d},$$
(2.69)

which for comparable concentrations of holes and electrons $(n_h \cong n_e)$ translates into

$$\frac{\tau_{rec}}{\tau_t} = \frac{e\mu_{e,h}\left(\mu_e + \mu_h\right)F^2}{\gamma\,jd} = \frac{8}{9}\frac{e}{\varepsilon_0 \varepsilon}\frac{\mu_{e,h}\left(\mu_e + \mu_h\right)}{\mu_{eff}\,\gamma}\frac{j_{SCL}}{j},$$
(2.70)

where j_{SCL}/j is the inverse of the injection efficiency defined by Equation 2.61 and μ_{eff} is the effective mobility of the carriers (Equation 2.43) under double injection current (j). Assuming $\mu_{e,h}$ and γ to be field-independent parameters the ratio τ_{rec}/τ_t is inversely proportional to the injection efficiency $\varphi_{inj} = j/j_{SCL}$. The decreasing tendency in the τ_{rec}/τ_t ratio with increasing concentration of TPD in HTL is compatible with this prediction since the injection efficiency increases as the concentration of the electron donor centers (TPD) at the contact with ITO increases.[61]

Equation 2.70 expresses that the field dependence of the ratio τ_{rec}/τ_t is governed by the field dependence of the mobilities, μ_e, μ_h, recombination coefficient, γ, and injection efficiency, j/j_{SCL}. For the Langevin recombination mechanism the γ is governed by the carrier motion (see Equation 2.11) so that Equation 2.70 can be simplified to

$$\frac{\tau_{rec}}{\tau_t} = \frac{8}{9}\frac{\mu_{e,h}}{\mu_{eff}}\frac{j_{SCL}}{j}.$$
(2.71)

Two different expressions for μ_{eff} for the weak recombination case and for the strong recombination case follow from

Equation 2.43. They would correspond to the ICEL operating LEDs ratio:

$$\left(\frac{\tau_{rec}}{\tau_t}\right)_{ICEL} \cong \frac{2}{3\sqrt{\pi}} \frac{\mu_{e,h}}{\sqrt{\mu_e \mu_h}} \frac{j_{SCL}}{j} \tag{2.72}$$

and to the VCEL operating LED ratio:

$$\left(\frac{\tau_{rec}}{\tau_t}\right)_{VCEL} \cong \frac{8}{9} \frac{\mu_{e,h}}{\mu_e + \mu_h} \frac{j_{SCL}}{j} . \tag{2.73}$$

The experimental values of τ_{rec}/τ_t in Figure 2.27, all exceeding unity, and of $\varphi_{inj} = j/j_{SCL} << 1$ as shown in Figure 2.28 indicate that the ICEL operation mode occurs typically for TPD/Alq$_3$ junction-based LEDs, and τ_{rec}/τ_t should obey Equation 2.72. A higher value of $\varphi_{EL}^{(ext)}$ for the (75% TPD:PC) HTL, as compared with the 100% TPD HTL LED, can be explained by different morphology of recrystallizing pure TPD and TPD mixed with PC binder layers. Due to the mass transport for crystallization of initially amorphous pure TPD films, bare ITO glass islands are formed[153] reducing the effective area for hole injection. The polymer (PC) suppresses the crystallization process in the (TPD:PC) layers, making them less rough[154] and covering the accessible area of the substrate uniformly. Thus, the injection current from ITO is a result of trade-off between the effective injection area and concentration of electron donor centers (TPD) in the HTL.[63] Taking $(\tau_{rec}/\tau_t) \cong 5$ at 10^6 V/cm (100% TPD/Alq$_3$ in Figure 2.27D), $\mu_{e,h}/\sqrt{\mu_e \mu_h} \cong 10^{-2}$ follows from Equation 2.72 using $j_{SCL}/j = 1.4 \times 10^3$ from Figure 2.28A (curve 4). This implies the $\mu_e << \mu_h$ choice for the $\mu_{e,h}$ in the numerator of the ratio $\mu_{e,h}/\sqrt{\mu_e \cdot \mu_h} = \sqrt{\mu_e/\mu_h}$, and $\mu_e/\mu_h \cong 10^{-4}$. This is not the case either for μ_e and μ_h in Alq$_3$ ($\mu_e/\mu_h \cong 10^2$; see Kepler et al.[138]) or μ_e in Alq$_3$ and μ_h in TPD ($\mu_e/\mu_h \cong 10^{-2}$; Kepler et al.;[138] Stolka et al.[155]). Thus, the only possibility to rationalize this ratio is attributing $\mu_h \cong 10^{-7}$ cm^2/Vs to holes in Alq$_3$ and $\mu_e \cong 10^{-11}$ cm^2/Vs to electrons in TPD (at $F = 10^6$ V/cm). The latter is a

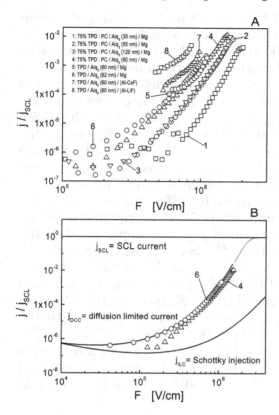

Figure 2.28 Comparison of the electric field dependence of injection efficiency for the TPD/Alq$_3$ junction-based DL LEDs using the experimental data for j published in the literature and j_{SCLC} calculated from Equation 2.42 with Equation 2.18, assuming $\mu_{eff} = \mu_e + \mu_h$ with $\mu_h = \mu_h(TPD) + \mu_e(Alq_3) \cong \mu_h(TPD) \cong 5.2 \times 10^{-4} (cm^2/Vs) \exp(0.0187F^{1/2})$, and $\varepsilon = 3$. (A) The data obtained for the following structures: 1. ITO/75% TPD:PC(60 nm)/Alq$_3$(35 nm)/Mg/Ag. 2. ITO/75% TPD:PC(60 nm)/Alq$_3$(55 nm)/Mg/Ag. 3. ITO/75% TPD:PC (60 nm)/Alq$_3$(120 nm). 4. ITO/75% TPD:PC(60 nm)/Alq$_3$(60 nm)/Mg/Ag.[63] 5. ITO/100% TPD(60 nm)/Alq$_3$(60 nm)/Mg.[146] 6. ITO/100% TPD(47 nm)/Alq$_3$(62 nm)/Mg:Ag.[63] 7. ITO/100% TPD (60 nm)/Alq$_3$(60 nm)/Al-CsF/Al. 8. ITO/100% TPD(60 nm)/Alq$_3$(60 nm)/Al-LiF AC.[146] (B) Theoretical fit (solid lines) according to the diffusion-controlled injection current DCC (Equation 2.38), $j = 5 \times 10^{-6} \exp(0.017F^{1/2})$ and Schottky-type injection current (Equation 2.35), $j = 5 \times 10^{-6} \exp(0.0095F^{1/2})$. Circles and up triangles are the data 4 and 6 from part A.

very low value, difficult to measure under usual TOF conditions as reported in the literature (see, e.g., Stolka et al.[155]).
We note that applying Equation 2.73 to the same data gives
$\mu_e \cong 10^{-2}\mu_h$, which could relate the electron mobility in Alq$_3$ ($\cong$
10^{-5} cm^2/Vs; see Barth et al.[156]) to the hole mobility in TPD
($\cong 10^{-3}$ cm^2/Vs; Stolka et al.[155]). The latter can be easier acceptable for the lowest $\tau_{rec}/\tau_t \to 1$ obtained with the (75%
TPD:PC)/Alq$_3$ structure at $F \cong 1.4 \times 10^6$ V/cm (Figure 2.27D)
and with $\varphi_{inj} \cong 10^{-2}$ (Figure 2.28A). An important message
follows from Figure 2.28B. The figure reveals a distinct difference between the field dependence of the injection efficiency for the currents controlled by diffusion (see Section 2.5)
and those limited by the field-assisted thermionic injection
(see Section 2.5). Both j_{DCC} and j_{ILC} are proportional to the
product $F^{3/4} \exp(\beta_{eff} F^{1/2})$, but $\beta_{eff}^{(DCC)} = a + \beta_\mu$ as compared with
$\beta_{eff}^{(ILC)} = a$. Clearly, $\beta_{eff}^{(DCC)} > \beta_{eff}^{(ILC)}$ because the former contains a
term β_μ characteristic of the Poole–Frenkel-type electric field
increase of the carrier mobility (cf. Equation 2.18). The field
dependence of the j_{DCC}/j_{SCL} ratio agrees very well with the
experimental data following the DCC behavior of j with $\beta_{eff}^{(DCC)}$
= 0.017 (cm/V)$^{1/2}$ (Kalinowski et al.[123]). This does not exclude
the occurrence of the ILCs in other EL structures. It is worthy
to note here that the (τ_{rec}/τ_t) ratio differs from zero ($P_R < 1$)
even for $j = j_{SCL}$. Yet

$$\left(\frac{\tau_{rec}}{\tau_t}\right)_{SCLC} \cong \frac{1}{1+\mu_{he}/\mu_{eh}} \qquad (2.74)$$

according to Equation 2.73, and $\tau_{rec}/\tau_t \to 0$ ($P_R \cong 1$) only if $\mu_{h,e}$
>> $\mu_{e,h}$. For $\mu_{h,e} = \mu_{e,h}$, $\tau_{rec}/\tau_t \cong \frac{1}{2}$ ($P_R \cong \frac{2}{3}$), and for $\mu_{h,e} << \mu_{e,h}$,
$\tau_{rec}/\tau_t \cong 1$ ($P_R = \frac{1}{2}$). Let us consider the case of an SL LED
based on TPD, where $\mu_h >> \mu_e$. Surprisingly, the P_R for holes
and electrons is different. Equations 2.74 and 2.68 yield τ_{rec}/τ_t
$\cong 1$ and $P_R = \frac{1}{2}$ for holes, and $\tau_{rec}/\tau_t \to 0$ and $P_R \cong 1$ for electrons.
However, it can be understood if the spatial distribution of
the injected charge will be taken into account. The formal
condition for the SCL current is the concentration of the
injected charge at the injecting contact to attain infinity (cf.
Section 2.5), so that $\tau_{rec} = (\gamma\, n)^{-1} \to 0$ only at the contacts.

Since fast holes reach the opposite contact (cathode) after a short time ($\tau_t \to 0$ for $\mu_h \to \infty$), and slower electrons reach the anode after a much longer time ($\tau_t \to \infty$ for $\mu_e \to 0$), the ratio τ_{rec}/τ_t for holes equals 1, and tends to 0 for electrons. This reflects in the position and width of the recombination zone as mentioned already in Section 2.2. It follows from experiment that the injection efficiency and carrier mobilities increase or at least not diminish at high electric fields ($F > 10^5$ V/cm) in amorphous or polycrystalline organic layers forming thin-film LEDs (see e.g., Schein and Brown;[82] Borsenberger et al.;[83] Kepler et al.;[138] Tsutsui et al.[140]). This should give a monotonic decrease in τ_{rec}/τ_t (Equation 2.71), unless $\mu_{e,h}$ is a much stronger field-increasing function than μ_{eff}. As a consequence P_R and $\varphi_{EL}^{(ext)}$ are expected to be monotonically increasing functions of applied field. Indeed, such behavior can be observed in the lower-field segment of the $\varphi_{EL}^{(ext)}$ (F) curves. But even for a maximum injection efficiency ($j = j_{SCL}$) the field-increasing QE can occur. Such a case has been assumed for Alq$_3$ and poly[2-methoxy-5-(2-ethylhexyloxy)-1,4-phenylenevinylene] (MEH-PPV) emitter-based LEDs with Mg:Ag cathodes and poly(3,4-ethylenedioxythiophene):poly(4-styrenesulphonate) (PEDOT:PSS)-covered ITO anodes.[157] The third limiting case of double injection, $\mu_{eff} \cong \mu_e$ (see Section 2.5) for Alq$_3$ LEDs has been considered to be applicable with the electron mobility modified by trapping, i.e., μ_e replaced by $\Theta\mu_e$. Since Θ increases with applied field (see Section 2.5), the ratio (Equation 2.74), $(\tau_{rec}/\tau_t) \cong \mu_h / \Theta\mu_e$, appears to be a decreasing function of F and would lead to increasing $\varphi_{EL}(F)$. In this way the field-increasing φ_{EL} does not require the assumption of diversification of the emission efficiency of the excited states formed at trapping sites and that of the excited states created on unperturbed molecules of Alq$_3$, made in that work.[157] The question arises what is the reason for the high field decrease of $\varphi_{EL}^{(ext)}$ (F). One of the reasons could be the transition from the Langevin to Thomson description of the volume recombination process (see Section 2.3). The recombination coefficient γ in Equation 2.70 cannot any longer be expressed by the mobility of charge carriers (see Equation 2.11) and τ_{rec}/τ_t follows a field-increasing function of the mobility in the numerator of

Equation 2.70 or a field-decreasing γ. The Thomson-like recombination occurs whenever the capture time (τ_c) in the ultimate step of the recombination process becomes comparable with the dissociation time (τ_d) of an initial (coulombically correlated) charge pair (CP). Such a recombination scheme, depicted in Figure 2.11, allows P_R to be expressed using these time constants by the product of the probabilities to form a coulombically correlated $(e–h)$ pair $P_R^{(1)}$ and its decay by the mutual carrier capture $P_R^{(2)}$. However, to complete this picture the overall recombination probability should also include exciton–charge carrier interaction $[P_R^{(3)}]$.

Then

$$P_R = P_R^{(1)} P_R^{(2)} P_R^{(3)} = \left(1+\tau_m/\tau_t\right)^{-1}\left(1+\tau_c/\tau_d\right)^{-1}\left(1+\tau_S/\tau_{Sq}\right)^{-1}, \quad (2.75)$$

where $P_R^{(3)} = (1 + \tau_S/\tau_{Sq})^{-1}$ for excited molecular singlets is determined by the ratio of the singlet exciton lifetime τ_S to their quenching time (τ_{Sq}) due to the interaction with charge carriers (q). The direct relaxation of the CP states has been assumed to be very slow as compared with τ_d and τ_c. Singlet exciton–charge carrier interaction has been considered the process contributing to the electrofluorescence roll-off at high electric fields,[158] and has been shown to modify the electric field-induced PL quenching rate.[159] A comparison of the electric field dependence of the PL quenching rate in Alq$_3$ when using non-injecting Al electrodes[75,76,144,159] and electron-injecting Mg:Ag cathode[159] allows us to evaluate the singlet exciton–charge carrier interaction rate constant, $\gamma_{Sq} \cong 10^{-9}$ cm^3 s^{-1} for electrons. This yields $\tau_{Sq} \cong 10^{-7}$ s. The fluorescence quenching by injected holes has been suggested to occur in thin ($\sim$15 nm) layers of donor-type materials of tetra(N,N-diphenyl-4-aminophenyl)ethylene (TTPAE).[160] However, a more exact analysis of the data leads to a conclusion that the effect is due to the field-induced dissociation of singlet excited states rather than to their hole quenching. The field-induced dissociation of singlet excitons in Alq$_3$ also seems to be responsible for the quenching effects in the devices with the Al cathode and negatively biased ITO electrode. This point has been discussed and the effect supported by the charge photogeneration measurements.[75,76] The field-assisted dissociation seems to commonly appear whenever the excited states are produced in the presence of high electric fields regardless of their spin multiplicity (singlets or triplets), type of excitation (optical or electrical) (cf. Section 2.4) and electronic properties of the material (evidence for electric field-assisted dissociation of excited singlets in conjugated polymers has been reported in several works[161–163]). Various dissociation models have been discussed to explain the experimental results, but excellent agreement with experiment is provided by the 3D-Onsager theory of geminate recombination as demonstrated for high electric fields.[76] Employing this model to $P_R^{(2)}$ and substituting $P_R \cong P_R^{(1)} P_R^{(2)}$ (that is, assuming $P_R^{(3)} = 1$) in Equation 2.66, the field dependence of the EL quantum efficiency (QE) can be calculated for different current conditions. The results are presented in Figure 2.29. The role of the field-assisted disso-ciation on ohmic injection (SCLC in the figure) is to reduce the low-field constant value of the QE starting from approx-imately 10% at 10^5 V/cm up to an order of magnitude at 5×10^6 V/cm for $r_0/r_c = 0.15$, that is, for the initial intercarrier separation of $e–h$ pairs $r_0 \cong 2.3$ nm ($r_c \cong 15$ nm with $\varepsilon = 3.8$). In the case of either diffusion-limited current (DCC) or Schottky-type injection current, a low-field decrease in QE is observed; then QE passes through a series of minima and maxima whose positions are sensitive to the average initial intercar-rier separation, r_0. For the DCC case where one well-pro-nounced maximum occurs around 0.8 MV/cm, the field evolution above 10^5 V/cm resembles the typical experimental evolution.[152] In contrast, only weak features on the QE(F) curves for the Schottky-type injection-underlain device cur-rents can be distinguished with a general decreasing trend in QE. This prediction is in reasonable agreement with vari-ation of the relative EL efficiency as a function of applied bias at room temperature for a 90-nm-thick film of TPD provided with a weakly injecting Al cathode.[152] Quantitative compari-son between theory and experiment for DL TPD/Alq$_3$ junction-based LEDs indicates that the dissociation quenching alone is unable to reproduce the functional dependence of the QE on the applied field (Figure 2.30A). However, dissociation

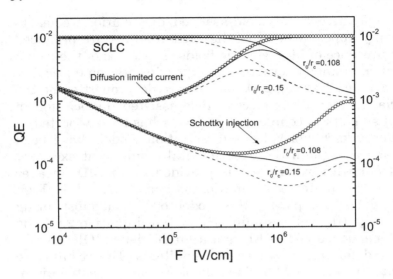

Figure 2.29 The field dependence of the EL QE in TPD/Alq$_3$ junction-based LEDs working in the VCEL mode as calculated from Equations 2.66, 2.71, and 2.75 with $\xi = 0.2$, $P = P_S = 0.25$, $\varphi_r = 0.25$ and $\varepsilon = 3.8$ at $T = 298$ K. The following assumptions have been made in the calculation: $P_R^{(1)} = (1 + \tau_m/\tau_t)^{-1}$ with $\tau_m/\tau_t \cong 2$ $\mu_h(\text{Alq}_3)/\mu_h(\text{TPD})(j_{\text{SCL}}/j)$ and field-dependent mobilities $\mu_h(\text{Alq}_3)$ and $\mu_h(\text{TPD})$ in Alq$_3$ and TPD, respectively, and (j_{SCL}/j) taken from Figure 2.28B; $P_R^{(2)} = (1 + \tau_c/\tau_d)^{-1}$ with $\tau_c/\tau_d = \tau_c v_0 \Omega_{\text{Ons}}(F)$, $\Omega_{\text{Ons}}(F)$ given by Equation 2.20 at different ratios of r_0/r_c (cf. Section 2.3) and $\tau_c v_0 = 10$. The small circles marked curves are due to the QE determined solely by the $P_R^{(1)}$. Note the difference in the curve shapes for three different injection mechanisms (SCLC, DCC, and Schottky-type injection). (After Kalinowski and Stampor.[164])

quenching combined with quenching due to singlet exciton–charge carrier interaction yields good agreement (Figure 2.30B). The latter can be reached at $\tau_c v_0 = 15$, where v_0 is the usual frequency factor equal to 10^{12} to 10^{13} s^{-1}. This allows evaluation of the capture time τ_c on $1.5 \times (10^{-12}$ to $10^{-11})$ s scale. The rate constant assumed in this way, $\gamma_{Sq} \cong 10^{-9}$ s, agrees in the order of magnitude with the estimated value for singlet exciton–electron interaction in Alq$_3$.[159]

EL quantum efficiency can be correlated with the width of the recombination zone, w, defined as a distance traversed by a carrier during the recombination time,* τ_{rec},[165]

$$w = \mu_{e,h} F \tau_{rec.} \tag{2.76}$$

Combining Equation 2.76 and the carrier transit time between electrodes (d), $\tau_t = d/\mu_{h,e} F$, yields $\tau_{rec}/\tau_t = w/d$, and the recombination probability (Equation 2.68) becomes directly correlated with the recombination zone as follows:

$$P_R = (1 + w/d)^{-1}. \tag{2.77}$$

Inserting Equation 2.77 into Equation 2.66 yields

$$\varphi_{EL}^{(ext)} = \frac{\xi P \varphi_r}{1 + w/d}. \tag{2.78}$$

According to the definition equation (Equation 2.76), w may greatly exceed the device thickness, d, leading to a very low value of $\varphi_{EL}^{(ext)} = (d/w)\xi P\varphi_r$ whenever $w \gg d$. On the other hand, the upper limit of $\varphi_{EL}^{(ext)} = (1 + 2/d)^{-1}\xi P\varphi_r$ (nm) is smaller than $\xi P\varphi_r$ because, due to the discrete structure of materials, w must be limited to ~2 nm corresponding to an average dimension of the molecules forming low-molecular-weight organic layers of EL device. The often-employed expression for $\varphi_{EL}^{(ext)} = \xi P\varphi_r$, assuming $P_R = 1$, is unjustified as it would require $w \to 0$, thus, an unphysical assumption of a continuous homogeneous medium with the recombination time for carriers $\tau_{rec} \to 0$. It is necessary to point out that the $w/d = \tau_{rec}/\tau_t$ ratio, affecting $\varphi_{EL}^{(ext)}$ through Equations 2.66, 2.68, and 2.78, may differ if different assumptions are made regarding the relative contributions of the electron and hole currents flowing in the device. Table 2.1 demonstrates how the EL quantum efficiency changes with varying ratio (w/d) of the

---

*We note that such defined recombination zone width must not be identified with the geometrical limits imposed on the charge recombination by the device structure.

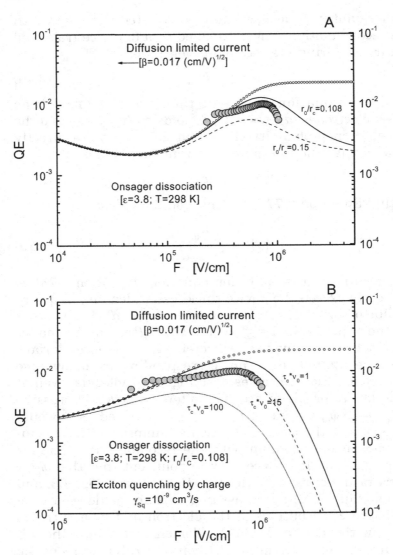

Figure 2.30 The field dependence of the EL QE for a diffusion-limited current driven ITO/TPD/Alq$_3$/Mg:Ag LED. The lines represent theoretical predictions of QE(F) according to Equations 2.66, 2.71, and 2.75 with different model parameters (r_0/r_c; $\tau_c v_0$). The small circle curves show the case with $P_R = P_R^{(1)}$ only (no quenching). The shaded circles stand for the experimental data of Kalinowski et al.[123]

Table 2.1 EL QEs Calculated According to Equation 2.78 as a Function of the Ratio of Recombination Zone Width to the Sample Thickness (w/d) at Different Values of ξ and φ_r $(\equiv \varphi_{PL})$

	$\varphi_{EL}^{(int)}$			
	$\xi = 1$ $\varphi_{PL} = 1$ 25%	$\xi = 1$ $\varphi_{PL} = 0.25$ 6.25%	$\xi = 1$ $\varphi_{PL} = 1$ 25%	$\xi = 1$ $\varphi_{PL} = 0.25$ 6.25%
w/d	$\varphi_{EL}^{(ext)}$ (%)	$\varphi_{EL}^{(ext)}$ (%)	$\varphi_{EL}^{(ext)}$ (%)	$\varphi_{EL}^{(ext)}$ (%)
$P_S = 0.25$	$\xi = 0.2$ $\varphi_{PL} = 1$ I	$\xi = 0.2$ $\varphi_{PL} = 0.25$ II	$\xi = 0.35$ $\varphi_{PL} = 1$ III	$\xi = 0.35$ $\varphi_{PL} = 0.25$ IV
0	5.00	1.25	8.75	2.20
0.10	4.55	1.14	8.00	2.00
0.15	4.30	1.09	7.60	1.91
0.20	4.17	1.04	7.29	1.83
0.25	4.00	1.00	7.00	1.76
0.30	3.85	0.96	6.73	1.69
0.35	3.70	0.93	6.48	1.63
0.40	3.57	0.89	6.25	1.57
0.50	3.33	0.83	5.83	1.47
0.60	3.13	0.78	5.47	1.38
0.70	2.94	0.74	5.15	1.29
0.80	2.78	0.69	4.86	1.22
0.90	2.63	0.66	4.60	1.16
1.00	**2.50**	**0.63**	**4.38**	**1.10**
1.10	2.38	0.60	4.17	1.05
1.20	2.27	0.57	3.97	1.00
1.50	2.00	0.50	3.50	0.88
1.80	1.79	0.45	3.13	0.79
2.00	1.67	0.42	2.92	0.73
3.00	1.25	0.31	2.19	0.55
5.00	0.83	0.21	1.46	0.37
6.00	0.71	0.18	1.25	0.31
7.00	0.63	0.16	1.09	0.28
8.00	0.55	0.14	0.97	0.24
9.00	0.50	0.13	0.88	0.22
10.00	0.45	0.11	0.80	0.20
15.00	0.31	0.08	0.54	0.14
20.00	0.24	0.06	0.42	0.10
30.00	0.16	0.04	0.28	0.07
50.00	0.10	0.03	0.17	0.04
100.00	0.05	0.01	0.09	0.02
1000.00	0.005	0.001	0.008	0.002

Note that $\varphi_{EL}^{(ext)}$ $[w = d; w/d = 1] = (1/2)\, \varphi_{EL}^{(ext)}$ (max) for $w = 0$.

width of the recombination zone w to an electrofluorescent (P = P_S = 0.25) device thickness d. Limiting the ratio w/d to the range $0(w = 0) - 1000(w >> d)$ implies variation of $\varphi_{EL}^{(ext)}$ between 1.25 and 10^{-3}% photon/carrier for typical ξ = 0.2 and φ_r = 0.25. The above-described lower limit for w = 2 nm reduces the former value to 1.22% photon/carrier for common organic layer thickness d = 100 nm, but to about 1% photon/carrier for d = 10 nm. The extension of the recombination zone over all the device thickness d ($w = d$) reduces this value by a factor of two ($\varphi_{EL}^{(ext)} \cong 0.63$% photon/carrier). These data indicate that to optimize the EL quantum efficiency from an LED we must minimize the recombination zone width as related to the device thickness. Furthermore, if we assume the recombination zone width to be independent of thickness, d, Equation 2.78 provides a simple method to determine w by means of experimentally measured $\varphi_{EL}^{(ext)}$ as a function of d. A plot of $1/\varphi_{EL}^{(ext)}$ vs. d^{-1},

$$\frac{1}{\varphi_{EL}^{(ext)}} = A + \frac{B}{d} , \tag{2.79}$$

where $A = (\xi P \varphi_r)^{-1}$ and $B = Aw$, is then expected to be a straight line with the slope-to-intercept ratio yielding directly the width of the recombination zone, w. In addition, the light output coupling factor ξ, can be determined from the intercept at $d^{-1} \rightarrow 0$, if P and φ_r are known independently. This approach has been successfully applied to the ITO/TPD(d_h)/Alq$_3$(d_e)/Mg: Ag structures.[166] Figure 2.31 shows external quantum efficiency as a function of electric field measured at various Alq$_3$ thickness. For thin emitting layers (d_e < 20 nm) $\varphi_{EL}^{(ext)}$ appears to be field independent, whereas for thick Alq$_3$ films (d_e > 25 nm) a nonmonotonic evolution of $\varphi_{EL}^{(ext)}$ with electric field is observed and is consistent with the previous results of Figures 2.26 and 2.27. The inverse of $\varphi_{EL}^{(ext)}$ can be fitted to a straight-line plot vs. d_e^{-1} (Figure 2.31B) giving the field-dependent w, 70 nm at 0.75 MV/cm and 40 nm at 1 MV/cm. The value $\xi \cong$ 0.43 of the light-output coupling factor follows from the intercept $A = (\xi P \varphi_r)^{-1}$ with $P = P_S$ = 0.25 and φ_r = φ_{PL} = 25% at F = 0.75 MV/cm and φ_r = 17% at 1.0 MV/cm. The reduced value

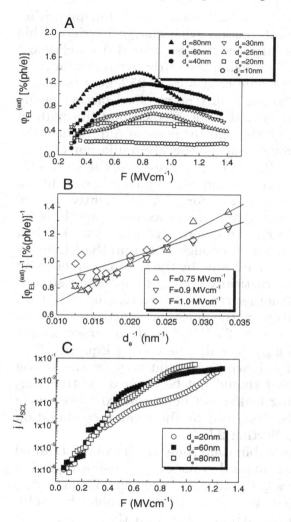

Figure 2.31 The external EL quantum efficiency as a function of electric field (A) of the bilayer devices ITO/TPD(d_h)/Alq$_3$(d_e)/Mg:Ag of the total thickness $d = d_h + d_e = 120$ nm with varying Alq$_3$ thickness (d_e). (B) The inverse $\varphi_{EL}^{(ext)}$ as a function of the inverse of the emitter thickness (d_e) at three different electric fields (uptriangles, downtriangles, diamonds). The data fit to Equation 2.79 is given by straight solid lines. (C) The carrier injection efficiency as a function of applied field for the devices with three different emitter thickness. (After Kalinowski et al.[166])

of φ_r has been used at high fields because of exciton quenching by electric field and interaction with charge carriers. This value of ξ is twice as large as that estimated from classical ray optics to be $1/2n^2$,[167,168] using the refractive index for Alq_3, $n = 1.7$. An alternative would be $\varphi_r = \varphi_{PL} = 25\%$ kept constant, and ξ varying with electric field. The physical picture of the carrier recombination is that it proceeds along the path of holes migrating through the emitter layer of Alq_3 toward the Mg:Ag cathode. However, due to the accumulation of both holes and electrons at the TPD/Alq_3 interface, the most efficient recombination occurs in the Alq_3 region adjacent to the interface and, whenever $w < d_e$ forms a sufficiently narrow zone far from the cathode, the singlet exciton quenching by this metallic electrode can be ignored (cf. Figure 2.9). When Alq_3 thickness becomes small enough (<25 nm) the EL quantum efficiency becomes dominated by the excitonic quenching at the cathode and it practically does not depend on electric field (Figure 2.31A). Another reason for decreasing $\varphi_{EL}^{(ext)}$ with decreasing d_e is a drop in the injection efficiency for thin emitter devices (see Figure 2.31C). The drop comes simply from the assumption $\mu_{eff} = \mu_e + \mu_n$ when using Equation 2.42 for the calculation of j_{SCL}, which is valid only for the strong recombination case and should not be applied for thin Alq_3 layers when increasing leakage of holes toward the cathode renders the flow to be described by the weak-recombination limit, as discussed in Section 2.5.

Obviously, the recombination zone width can be obtained directly from the experimental value of $\varphi_{EL}^{(ext)}$ using Equation 2.78, if the parameters ξ, P, and φ_r are provided. We note that Equation 2.78 applies to the conditions when the field-assisted dissociation of excited states and their interaction with charge carriers can be neglected ($P_R^{(2)} = P_R^{(3)} = 1$ in Equation 2.75). Under such conditions the field dependence of $\varphi_{EL}^{(ext)}$ (F) is an increasing function of F following the field-decreasing width of the recombination zone. This approach has been employed to study the effect of electric field on the recombination zone width with doped and undoped emitter layers.[123,158]

2.7 FINAL REMARKS

In this chapter we have attempted to present the most important electronic processes underlying organic EL and their application to characterizing optical properties of organic LEDs, in a way that reveals the limitation as well as strength of various models of bimolecular recombination, formation, and nature of emissive states, charge injection, material imperfections, and film morphology. The general characteristics of organic EL can be derived from theoretical models. Quantitative properties of specific organic LEDs can rarely be obtained from theory alone. In the application of the theoretical analyses, the choice of approximations is most important. Although the importance of the molecular and bimolecular excited states has been demonstrated, questions concerning the exact identity of the primary entities responsible for the emission have yet to be answered. To understand the role of the local intermolecular structures on interband or crossing transitions contributing to the EL spectra, more-detailed studies are needed. Some of the intricacies have been unraveled by means of the new investigative methods such as magnetic field effects on light output and current driving organic LEDs, and the *electromer* and *electroplex* concepts. The magnetic field effects found to be useful probes in the study of detailed electron behavior in local intermolecular structures have certainly not been utilized to the full extent.

The principal contribution of this chapter has been in presenting a unified framework for correlating experimental data on the EL quantum efficiency of organic LEDs (Section 2.6), preceded by a historical sketch of organic EL (Section 2.2) and a comprehensive description of the recombination EL (Sections 2.3 to 2.5). It is perhaps appropriate here to mention some of the common pitfalls, which have in past led to incorrect interpretations of EL quantum efficiency (EL QE) data.

1. Not all, and rather rarely, organic LEDs operate under space-charge-limited conditions. Thus, EL QE increases initially with applied voltage because of increasing injection efficiency, which usually has not been recognized.

2. The injection current mechanisms differ at low and high electric fields, and this has not been taken into account in theoretical predictions of the EL QE throughout the entire accessible field range.

3. No electric field–dependent mobility has been invoked to complete theoretical predictions of the field-dependent EL QE.

4. The effect of high electric fields on the decay of the excited states must not be neglected, since various quenching mechanisms like exciton–exciton and exciton–charge carrier interactions or exciton dissociation set in, leading to a reduction in the EL QE.

5. The assignment of the recombination rate of the oppositely charged carriers to their diffusivity only (Langevin recombination), although common at low fields, may not be appropriate at high electric fields, where the capture time becomes comparable with the dissociation time of the coulombically correlated charge pair and the Thomson recombination mechanism becomes important.

6. The improper definition of the recombination zone limited to geometrical dimensions of the LED on the one extreme and reduced to zero (equivalent to recombination probability equal to unity) on the other. In reality, it must be limited by at least one intermolecular distance on the lower limit and tend to infinity on the upper limit, if it follows the natural definition of the recombination zone width as a distance traversed by the charge carrier within the recombination time.

Predictive contributions in the study of EL and its utilization in organic LEDs has not as yet been as great as in inorganic semiconductor-based devices because of the complexity of the organic substances. However, the predictive content of theoretical studies on organic EL is remarkably increasing. We may yet attain the long-sought goal of design of new EL phenomena and of improved organic LED performance from theoretical analyses of electronic processes in organic solids.

ACKNOWLEDGMENTS

This work was supported in part by the U.S. Office of Naval Research. The author acknowledges the invaluable contribution of his past and present co-workers of the Gdańsk (Poland) and Bologna (Italy) EL groups. The author is pleased to acknowledge the assistance of his wife Krystyna and son Sebastian with the technical preparation of the manuscript.

REFERENCES

1. Bradley, D.D.C., *Synth. Met.*, 54, 401, 1993.

2. Salbeck, J., *Ber. Bunsenges. Phys. Chem.*, 100, 1667, 1996.

3. Miyata, S. and Nalwa, H.S. (Eds.), *Organic Electroluminescent Materials and Devices*, Gordon & Breach, Amsterdam, 1997.

4. Heeger, A.J., *Solid State Commun.*, 107, 673, 1998.

5. Tsutsui, T., In *The Phosphor Handbook*, S. Shionoya and W.M. Yen, Eds., CRC Press, Boca Raton, FL, 1998, chap. 9, sec. 3.

6. Cacialli, F. In *The Polymer Systems*, D.L. Wise, G.E. Wnek, D.J. Trantolo, T.M. Cooper, and J.D. Gresser, Eds., Marcel Dekker, New York, 1998, chap. 4.

7. Kalinowski, J., *J. Phys. D: Appl. Phys.*, 32, R179, 1999.

8. Mitschke, U. and Bäuerle, P., *J. Mater. Chem.* 10, 1471, 2000.

9. Nguen, T.-P., Molinie, P., and Destruel, P., In *Advanced Electronic and Photonic Materials and Devices*, Vol. 10, H.S. Nalwa, Ed., Academic Press, New York, 2001, chap. 1.

10. Gurnee, R.F., In *Proc. Organic Crystal Symp.*, NRC Canada, Ottawa, 1962, 109–116.

11. Destriau, G., *J. Chim. Phys.*, 34, 117, 327, 462, 1937.

12. Destriau, G., *Philos. Mag.*, 38, 774, 880, 1947.

13. Destriau, G. and Ivey, H.F., *Proc. IRE*, 43, 1911, 1955.

14. Lehmann, W., *J. Electrochem. Soc.*, 109, 540, 1962.

15. Short, G. and Hercules, D.M., *J. Am. Chem. Soc.*, 87, 1439, 1965.

16. Goldman, A.G., Kurik, M.V., Viertzimakha, Ja.I., and Korolko, B.N., *Zh. Prikl. Spektr.*, 14, 235, 1971 [in Russian].

17. Bernanose, A.B., Comte, M., and Vouaux, P., *J. Chim. Phys.*, 50, 64, 1953 [in French].

18. Bernanose, A., *J. Chim. Phys.*, 52, 369, 1955.

19. Karl, N., In *Defect Control in Semiconductors*, Vol. 2, K. Sumino, Ed., North Holland, Amsterdam, 1990, 1725.

20. Pope, M. and Swenberg, C.E., *Electronic Processes in Organic Crystals*, Clarendon Press, Oxford, 1982.

21. Namba, S., Yoshizawa, M., and Tamura, H., *Jpn. J. Appl. Phys.*, 28, 439, 1959 [in Japanese].

22. Pope, M., Kallmann, H.P., and Magnante, P., *J. Chem. Phys.*, 38, 2042, 1963.

23. Zvyagintsev, A.M., Steblin, V.I., and Tchilaya, G.S., *Zh. Prikl. Spektr.*, 13, 165, 1970.

24. Vityuk, N.W. and Mikho, V.V., *Fiz. Tehn. Poluprov. (Phys. Tech. Semicond.)*, 6, 1735, 1972 [in Russian].

25. Steblin, V.I., *Zh. Prikl. Spektr.*, 19, 724, 1973 [in Russian].

26. Dubey, P.D.R., *Indian J. Phys.*, 52A, 74, 1978.

27. Sinha, N.P., Misra, Y., Tripathi, L.N., and Misra, M., *Phys. Stat. Sol.* (a), 59, 101, 1980.

28. Sinha, N.P., Misra, Y., Tripathi, L.N., and Misra, M., *Solid State Commun.*, 39, 89, 1981.

29. Fischer, A.G., In *Luminescence of Inorganic Solids*, P. Goldberg, Ed., Academic Press, New York, 1966, chap. 10.

30. Vereshchagin, I.K., *Electroluminescence of Crystals*, Izd. Nauka, Moscow, 1974 [in Russian].

31. Lehmann, W., *Illum. Eng.*, 51, 684, 1956.

32. Piper, W.W. and Williams, F.E., *Solid State Phys.*, 6, 95, 1958.

33. Kalinowski, J., *Excitonic Interactions in Organic Molecular Crystals*, Technical Univ. of Gdańsk Publ., Gdańsk, 1977 [in Polish with English abstract].

34. Kalinowski, J., *Mater. Sci. (PL)*, 7, 44, 1981.

35. Kalinowski, J., *Synth. Met.*, 64, 123, 1994.

36. Kalinowski, J. In *Organic Electroluminescent Materials and Devices*, S. Miyata and H.S. Nalwa, Eds., Gordon & Breach, Amsterdam, 1997, chap. 1.

37. Zvyagintsev, A.M. and Steblin, V.I., *Zh. Prikl. Spektr.*, 17, 1009, 1972 [in Russian].

38. Kalinowski, J., Godlewski, J., and Gliński, J., *J. Lumin.*, 17, 175, 1978.

39. Kalinowski, J., *Proc. SPIE*, 1910, 135, 1993.

40. Kalinowski, J., Godlewski, J., and Gliński, J., *Acta Phys. Pol.* A65, 413, 1984.

41. Dresner, J., *RCA Rev.*, 30, 302, 1969.

42. Dresner, J. and Goodman, A.M., *Proc. IEEE*, 58, 1868, 1970.

43. Elsharkawi, A.R. and Kao, K.C., *J. Phys. Chem. Solids*, 38, 95, 1977.

44. Vincett, P.S., Barlow, W.A., Hann, R.A., and Roberts, G.G., *Thin Solid Films*, 94, 171, 1982.

45. Kalinowski, J., Godlewski, J., and Dreger, Z., *Appl. Phys.*, A37, 179, 1985.

46. Kampas, F.J. and Goutermann, M., *Chem. Phys. Lett.*, 48, 233, 1977.

47. Hayashi, S., Etoh, H., and Saito, S., *Jpn. J. Appl. Phys.*, 25, L773, 1986.

48. Helfrich, W., In *Physical and Chemistry of the Organic Solid State*, Vol. 3, D. Fox, M.M. Labes, and A. Weissberger, Interscience, New York, 1967, chap. 1.

49. Partridge, R.H., *Polymer*, 24, 733, 739, 748, 755, 1983.

50. Kido, J., In *Organic Electroluminescent Materials and Devices*, S. Miyata and H.S. Nalwa, Eds., Gordon & Breach, Amsterdam, 1997, chap. 10.

51. Burroughes, J.H., Bradley, D.D.C., Brown, A.R., Marks, R.N., Mackay, K., Friend, R.H., Burn, P.L., and Holmes, A.B., *Nature*, 347, 539, 1990.

52. Braun, D. and Heeger, A.J., *Appl. Phys. Lett.*, 58, 1982, 1991.

53. Friend, R.H., *Synth. Met.*, 51, 357, 1992.

54. Grem, G., Leditzky, G., Urlich, B., and Leising, G., *Adv. Mater.*, 4, 36, 1992.

55. Karg, S., Riess, W., Dyakonov, V., and Schwoerer, M., *Synth. Met.*, 54, 427, 1993.

56. Zhang, C., Braun, D., and Heeger, A.J., *J. Appl. Phys.*, 73, 5177, 1993.

57. Riess, W., In *Organic Electroluminescent Materials and Devices*, S. Miyata and H.S. Nalwa, Eds., Gordon & Breach, Amsterdam, 1997, chap. 2.

58. Kido, J., Kohda, M., Okuyama, K., and Nagai, K., *Appl. Phys. Lett.*, 61, 761, 1992.

59. Kido, J., Kohda, M., Hongawa, K., Okuyama, K., and Nagai, K., *Mol. Cryst. Liq. Cryst.*, 227, 277, 1993.

60. Johnson, G.E., Mc Grane, K.M., and Stolka, M., *Pure Appl. Chem.*, 67, 175, 1995.

61. Kalinowski, J., Giro, G., Di Marco, P., Fattori, V., and Di-Nicoló, E., *Synth. Met.*, 98, 1, 1998.

62. Kalinowski, J., Cocchi, M., Giro, G., Fattori, V., and Di Marco, P., *J. Phys. D*, 34, 2274, 2001.

63. Kalinowski, J., Cocchi, M., Giro, G., Fattori, V., and Di Marco, P., *J. Phys. D*, 34, 2282, 2001.

64. Kalinowski, J., Giro, G., Cocchi, M., Fattori, V., and Zamboni, R., *Chem. Phys.*, 277, 387, 2002.

65. Tang, C.W. and VanSlyke, S.A., *Appl. Phys. Lett.*, 51, 913, 1987.

66. Tang, C.W., VanSlyke, S.A., and Chen, C.H., *J. Appl. Phys.*, 65, 3610, 1989.

67. Adachi, C., Tokito, S., Tsutsui, T., and Saito, S., *Jpn. J. Appl. Phys.*, 27, L269, 1988.

68. Adachi, C., Tokito, S., Tsutsui, T., and Saito, S., *Jpn. J. Appl. Phys.*, 27, L713, 1988.

69. Adachi, C., Tsutsui, T., and Saito, S., *Appl. Phys. Lett.*, 55, 1489, 1989.

70. Burin, A.L. and Ratner, M.A., *J. Phys. Chem.*, 104, 4704, 2000.

71. D'Andrade, B.W., Baldo, M.A., Adachi, C., Brooks, J., Thompson, M.E., and Forrest, S.R., *Appl. Phys. Lett.*, 79, 1045, 2001.

72. Ikai, M., Tokito, S., Sakamoto, Y., Suzuki, T., and Taga, Y., *Appl. Phys. Lett.*, 79, 156, 2001.

73. Kalinowski, J., In *Electronic Phenomena in Organic Solids*, J. Kahovec, Ed., Wiley-VCH, Weinheim, 2004, p. 25; *Macromolecular Symposia*, 212.

74. Kalinowski, J., *Mol. Cryst. Liq. Cryst.*, 355, 231, 2001.

75. Kalinowski, J., Stampor, W., and Szmytkowski, J., *Polish J. Chem.*, 76, 249, 2002.

76. Szmytkowski, J., Stampor, W., Kalinowski, J., and Kafafi, Z.H., *Appl. Phys. Lett.*, 80, 1465, 2002.

77. Langevin, P., *Ann. Chem. Phys.*, 28, 289, 433, 1903.

78. Thomson, J.J., *Philos. Mag.*, 47, 337, 1924.

79. Kalinowski, J., Cocchi, M., Fattori, V., Di Marco, P., and Giro, G., *Jpn. J. Appl. Phys.*, 40, L282, 2001.

80. Lax, M., *Phys. Rev.*, 119, 1502, 1960.

81. Morris, R. and Silver, M., *J. Chem. Phys.*, 50, 2969, 1969.

82. Schein, L.B. and Brown, D.W., *Mol. Cryst. Liq. Cryst.*, 87, 1, 1982.

83. Borsenberger, P.M., Magin, E.H., Van der Auweraer, M., and De Schryver, F.C., *Phys. Status Solidi a*, 140, 9, 1993.

84. Montroll, E.W., *J. Chem. Phys.*, 14, 202, 1946.

85. Hong, K.M. and Noolandi, J., *J. Chem. Phys.*, 68, 5172, 1978.

86. Scher, H. and Montroll, E.M., *Phys. Rev.*, B12, 2455, 1975.

87. Hamill, W.H., *J. Phys. Chem.*, 82, 2073, 1978.

88. Chekunaev, N.I., Berlin, Yu.A., and Flerov, V.N., *J. Phys. C*, 15, 1219, 1982.

89. Scher, H., Shlesinger, M.I., and Brendler, J.T., *Phys. Today*, 44, 26, 1991.

90. Berlin, Yu.A., Chekunaev, N.I., and Goldanskii, V.I., *Chem. Phys. Lett.*, 197, 81, 1992.

91. Gill, W.D., *J. Appl. Phys.*, 43, 5033, 1972.

92. Bässler, H., *Phys. Status Solidi b*, 107, 9, 1981.

93. Albrecht, U. and Bässler, H., *Phys. Stat. Solidi (b)* 191, 455, 1995.

94. Kalinowski, J., Stampor, W., Di Marco, P., and Garnier, F., *Chem. Phys.*, 237, 233, 1998.

95. Onsager, L., *Phys. Rev.*, 54, 554, 1938.

96. Servet, B., Horowitz, G., Ries, S., Lagorsse, O., Alnot, P., Yasser, A., Deloffre, F., Srivastava, P., Hajlaoui, R., Lang, P., and Garnier, F., *Chem. Mater.*, 6, 1809, 1994.

97. Kalinowski, J., Stampor, W., Myk, J., Cocchi, M., Virgili, D., Fattori, V., and Di Marco, P., *Phys. Rev.*, B66, 235321, 2002.

98. Staerk, H., Kühnle, W., Treichel, R., and Weller, A., *Chem. Phys. Lett.*, 118, 19, 1985.

99. Staerk, H., Busmann, H.-G., Kühnle, W., and Treichel, R., *J. Phys. Chem.*, 95, 1906, 1991.

100. Tanimoto, Y., Okada, N., Itoh, M., Iwai, K., Sugioka, K., Takemura, F., Nakagaki, R., and Nagakura, S., *Chem. Phys. Lett.*, 136, 42, 1987.

101. Ito, F., Ikoma, T., Akiyama, K., Kobori, Y., and Tero-Kubota, S., *J. Am. Chem. Soc.*, 125, 4711, 2003.

102. Kalinowski, J., Szmytkowski, J., and Stampor, W., *Chem. Phys. Lett.*, 378, 380, 2003.

103. Groff, R.P., Merrifield, R.E., Suna, E., and Avakian, P., *Phys. Rev. Lett.*, 29, 429, 1972.

104. Bässler, H., *Phys. Status Solidi b*, 175, 15, 1993.

105. Carrington, A. and McLachlan, A.D., *Introduction to Magnetic Resonance with Applications to Chemistry and Chemical Physics*, Harper & Row, New York, 1967.

106. Geacintov, N.E. and Swenberg, C.E., In *Organic Molecular Photophysics*, Vol. 2, J.B. Birks, Ed., Wiley, London, 1975, chap. 8.

107. Schwob, H.P. and Williams, D.F., *Chem. Phys. Lett.*, 13, 581, 1972.

108. Kalinowski, J., Cocchi, M., Virgili, D., Di Marco, P., and Fattori, V., *Chem. Phys. Lett.*, 380, 710, 2003.

109. Kalinowski, J., Cocchi, M., Virgili, D., Fattori, V., and Di Marco, P., *Phys. Rev. B,* 70, 205303, 2004.

110. Kalinowski, J., Giro, G., Cocchi, M., Fattori, V., and Di Marco, P., *Appl. Phys. Lett.*, 76, 2352, 2000.

111. Jiang, X., Jen, A.K.Y., Carlson, B., and Dalton, L.R., *Appl. Phys. Lett.*, 81, 3125, 2002.

112. Jiang, X., Lin, S., Lin, M.S., Herguth, P., Jen, A.K.Y., Fong, H., and Sarikaya, M., *Adv. Funct. Mater.*, 12, 745, 2002.

113. Uckert, F., Tak, Y.-H., Müllen, K., and Bässler, H., *Adv. Mater.*, 12, 905, 2000.

114. Hofmann, J., Seefeld, K.P., Hofberger, W., and Bässler, H., *Mol. Phys.*, 37, 973, 1979.

115. Al-Jarrah, M., Brocklehurst, B., and Evans, M., *J. Chem. Soc. Faraday Soc.*, 2, 72, 1921, 1976.

116. Grandlund, T., Petterson, L.A.A., Anderson, M.R., and Inganäs, O., *J. Appl. Phys.*, 81, 8097, 1997.

117. Giro, G., Cocchi, M., Kalinowski, J., Di Marco, P., and Fattori, V., *Chem. Phys. Lett.*, 318, 137, 2000.

118. Godlewski, J. and Kalinowski, J., *Phys. Status Solidi a*, 56, 293, 1979.

119. Godlewski, J. and Kalinowski, J., *Jpn. J. Appl. Phys.* Part 1, 28, 24, 1989.

120. Fowler, R.H. and Nordheim, L., *Proc. R. Soc.* (London), A119, 173, 1928.

121. Esaki, L., In *Tunelling Phenomena in Solids*, E. Burstein, and C. Lundqvist, Eds., Plenum, New York, 1969, chap. 5.

122. Tak, Y.-H., and Bässler, H., *J. Appl. Phys.*, 81, 6963, 1997.

123. Kalinowski, J., Picciolo, L.C., Murata, H., and Kafafi, Z.H., *J. Appl. Phys.*, 89, 1866, 2001.

124. Arkhipov, V.I., Wolf, U., and Bässler, H., *Phys. Rev. B*, 59, 7514, 1999.

125. Baldo, M.A., Lamansky, S., Burrows, P.E., Thompson, M.E., and Forrest, S.R., *Appl. Phys. Lett.*, 75, 4, 1999.

126. Kalinowski, J., Godlewski, J., and Mondalski, P., *Mol. Cryst. Liq. Cryst.*, 175, 67, 1989.

127. Kalinowski, J., Godlewski, J., Di Marco, P., and Fattori, V., *Jpn. Appl. Phys.* Part 1, 31, 818, 1992.

128. Emtage, P.R. and O'Dwyer, J.J., *Phys. Rev. Lett.*, 16, 356, 1966.

129. Pope, M. and Solowiejczyk, Y., *Mol. Cryst. Liq. Cryst.*, 30, 175, 1975.

130. Godlewski, J. and Kalinowski, J., *Solid State Commun.*, 25, 473, 1978.

131. Lampert, M.A. and Mark, P., *Current Injection in Solids*, Academic Press, New York, 1970, chaps. 10 to 14.

132. Parmenter, R.H. and Ruppel, W., *J. Appl. Phys.*, 30, 1548, 1959.

133. Tsutsui, T., and Yamamoto, K., *Jpn. J. Appl. Phys.* Part 1, 38, 2799, 1999.

134. Helbig, E., *Principles of Photometric Measurements (Grundlagen der Lichtmesstechnik)*, Akademische Verlag, Leipzig, 1972 [in German].

135. Baldo, M.A. and Forrest, S.R., *Phys. Rev. B*, 64, 85201, 2001.

136. Kalinowski, J., Di Marco, P., Camaioni, N., Fattori, V., Stampor, W., and Duff, J., *Synth. Met.*, 76, 77, 1996.

137. Burrows, P.E., Shen, Z., Bulović, V., McCarty, D.M., Forrest, S.R., Cronin, J.A., and Thompson, M.E., *J. Appl. Phys.*, 79, 7991, 1996.

138. Kepler, K., Beeson, P.M., Jacobs, S.J., Anderson, R.A., Sinclair, M.B., Valencia, V.S., and Cahill, P.A., *Appl. Phys. Lett.*, 66, 3618, 1995.

139. Naka, S., Okada, H., Onnagawa, H., Kido, J., and Tsutsui, T., *Jpn. J. Appl. Phys.*, 38, L1252, 1999.

140. Tsutsui, T., Tokuhisa, H., and Era, M., *Proc. SPIE*, 3281, 230, 1998.

141. Bulović, V., Khalfin, V.B., Gu, G., and Burrows, P.E., *Phys. Rev. B*, 58, 3730, 1998.

142. Lu, M.-H. and Sturm, J.C., *J. Appl. Phys.*, 91, 595, 2002.

143. Kalinowski, J., Fattori, V., and Di Marco, P., *Chem. Phys.*, 266, 85, 2001.

144. Stampor, W., Kalinowski, J., Di Marco, P., and Fattori, V., *Appl. Phys. Lett.*, 70, 1935, 1997.

145. Jabbour, G.E., Kawabe, Y., Shaheen, S.E., Wang, J.F., Morell, M.M., Kippelen, B., and Peyghambarian, N., *Appl. Phys. Lett.*, 71, 1762, 1997.

146. Jabbour, G.E., Kippelen, B., Armstrong, N.R., and Peyghambarian, N., *Appl. Phys. Lett.*, 73, 1185, 1998.

147. Becker, H., Büsing, A., Falcon, A., Heun, S., Kluge, E., Parham, A., Stössel, P., Spreitzer, H., Treacher, K., and Vestweber, H., *Proc. SPIE*, 4464, 49, 2002.

148. Spreitzer, H., Vestweber, H., Stössel, P., and Becker, H., *Proc. SPIE*, 4105, 125, 2001.

149. Duineveld, P.C., De Kok, M.M., Buechel, M., Sempel, A.H., Mustaers, K.A.H., Van de Weijer, P., Camps, I.G.J., Van den Bigge Iaar, T.J.M., Rubingh, J.-E.J.M., and Haskal, E.I., *Proc. SPIE*, 4464, 59, 2002.

150. Yang, X.H., Neher, D., Scherf, U., Bagnich, S.A., and Bässler, H., *J. Appl. Phys.*, 93, 4413, 2003.

151. Malliaras, G.G. and Scott, J.C., *J. Appl. Phys.*, 85, 7426, 1999.

152. Campbell, A.J., Bradley, D.C.C., Laubender, and Sokolowski, M., *J. Appl. Phys.*, 86, 5004, 1999.

153. Han, E., Do, L., Niidome, Y., and Fujihara, M., *Chem. Lett.* (Japan), 969, 1994.

154. Uemura, T., Okuda, N., Kimura, H., Okuda, Y., Ueba, Y., and Shirakawa, T., *Polym. Adv. Technol.*, 8, 437, 1997.

155. Stolka, M., Janus, J.F., and Pai, D.M., *J. Phys. Chem.*, 88, 4707, 1984.

156. Barth, S., Müller, P., Riel, H., Seidler, P.F., Riess, W., Vestweber, H., and Bässler, H., *J. Appl. Phys.*, 89, 8791, 2001.

157. Segal, M., Baldo, M.A., Holmes, R.J., Forrest, S.R., and Soos, Z.G., *Phys. Rev. B*, 68, 75211, 2003.

158. Kalinowski, J., Murata, H., Picciolo, L.C., and Kafafi, Z.H., *J. Phys. D Appl. Phys.*, 34, 3130, 2001.

159. Ichikawa, M., Naiton, R., Koyama, T., and Taniguchi, Y., *Jpn. J. Appl. Phys.*, 40, L1068, 2001.

160. Hieda, H., Tanaka, K., Naito, K., and Gemma, N., *Thin Solid Films*, 331, 152, 1998.

161. Tasch, S., Kranzelbinder, G., Leising, G., and Scherf, U., *Phys. Rev.*, B55, 5079, 1997.

162. Khan, M.I., Bazan, G.C., and Popovic, Z.D., *Chem. Phys. Lett.*, 298, 309, 1998.

163. Hertel, D., Soh, E.V., Bässler, H., and Rothberg, L.J., *Chem. Phys. Lett.*, 361, 99, 2002.

164. Kalinowski, J. and Stampor, W., unpublished.

165. Kalinowski, J., Di Marco, P., Fattori. V., Giulietti, L., and Cocchi, M., *J. Appl. Phys.*, 83, 4242, 1998.

166. Kalinowski, J. Palilis, L.C., Kim, W.H., and Kafafi, Z.H., *J. Appl. Phys.*, 94, 7764, 2003.

167. Saleh, B.E.A. and Teich, M.C., *Fundamentals of Photonic*, Wiley, New York, 1991.

168. Greenham, N.C., Friend, R.H., and Bradley, D.D.C., *Adv. Mater.*, 6, 491, 1994.

3

Physical Properties of Organic Light-Emitting Diodes in Space Charge–Limited Conduction Regime

JUN SHEN AND ZHILIANG CAO

CONTENTS

3.1 INTRODUCTION

This chapter deals with a subset of carrier transport problems in organic light-emitting diodes (OLEDs): those problems within the space charge–limited (SCL) conduction regime. In particular, we emphasize the case in which trapped charges exist in high concentrations. Typical organic materials (e.g., Alq_3) used in modern OLEDs have few intrinsic carriers. During normal operation, high concentrations of electrons and holes are injected from opposite electrodes and then recombine within a suitable light-emitting layer. These electrons and holes do not usually have a uniform distribution within the organic layers. Carrier concentrations toward the respective injecting electrodes are usually much higher than those within the body of the organic layer, and space charges are formed. A large body of theoretical and experimental work has been devoted to the study of carrier transport and recombination within the SCL conduction regime.[1–7] The space charge can involve traps, namely, energy states that are within the energy gap and are away from the conducting bands by more than a few kT, where k is the Boltzmann constant and T is the absolute temperature. Carriers in these states (traps) have low escaping probability and tend to be stationary even when an electric field is applied.

In addition to the carrier transport process in an OLED, carrier injection and recombination processes are equally important, although they are not the focus of this chapter. An earlier review concerning these processes by Helfrich provided a good foundation to the subject.[8] Several groups have recently made important contributions to the understanding of carrier injection into the organic layers. For example, contacts with LiF or its composites have been shown to improve the carrier injection.[9,10] Campbell and Smith experimentally showed that the electron Schottky barrier cannot be smaller than about 0.6 eV for elemental metal contacts.[11] This finding is counter to the common belief that the Schottky model, which derives the barrier height solely from the work-function difference, works well for organic/metal contacts. Interestingly, the same group found that the Fermi-level is not pinned at the MEH-PPV/metal contact.[12] Kalinowski theoretically and experimentally studied the space-resolved electroluminescence (EL) in single crystals of anthracene and tetracene as a function of applied electric field.[13] He emphasized the importance for the EL spatial pattern of the charge injection ability of electrodes and distributions of electron and hole traps, as the spatial distribution of emitting states in electroluminescent cell is dependent on the trap-dependent electron–hole recombination process involving exciton–charge carrier interaction. In general, carrier injection into organic materials has been described using thermionic emission and Fowler–Nordheim tunneling,[3,12,14–19] and modifications of the above models by taking into account of the interface recombination current,[20,21] polaron effects,[12,22] and disorder.[23–25] While accurate determination of the recombination coefficient including all processes is still not readily available, it has been proposed that the Langevin recombination mechanism generally applies to the organic systems.[26,27] Based on the Langevin formalism, the recombination coefficient is proportional to the carrier mobility in the following way:

$$K_b = \frac{q}{\varepsilon}\left(\mu_n + \mu_p\right), \qquad (3.1)$$

where μ_n and μ_p are the electron and hole mobility, respectively, and ε is the dielectric constant. Under high electric fields, the ratio $K_b/(\mu_n + \mu_p)$ increases with the field strength.[28] Equation 3.1 needs to be modified in a similar way as done by Onsager, and the consequence to OLEDs needs to be studied further.[29,30] Kalinowski has provided evidence and arguments that the recombination process in OLEDs can be limited by the Thomson mechanism.[31] The electron–hole recombination coefficient was derived by Kalinowski et al.[32] from the transient behavior of the EL intensity. It is in good agreement with the value from the Langevin theory with the recombination process controlled by the motion of holes in the hole-transporting layer.

The remainder of the chapter is organized as follows. In Section 3.2, some basic characteristics of the trap charge–limited conduction (TLC) process are reviewed and discussed. In Sections 3.3 and 3.4, features involving doping and discrete traps are illustrated. Results of heterojunction devices are presented in Section 3.5. Finally, a summary is provided in Section 3.6. Theoretical models are given in the appendix.

3.2 CHARACTERISTICS OF TRAP CHARGE–LIMITED CONDUCTION

According to their conduction properties, OLED materials can probably be classified as either semiconductors or insulators. Their typical band gap is about 3 eV, and the device is usually undoped in terms of conduction type. Thus, the current carriers have to be injected from the electrodes. Most of the TCL conduction features can be understood by considering the case where only one kind of carrier (say, holes) is injected. When traps are not present, the current–voltage relationship obeys the classic Child's square law: $J \sim V^2$.[1–3] This relationship deviates from the linear form because the carrier distribution is not a constant in the solid; rather, it decays rapidly into the solid from the injecting electrode. [The exact carrier concentration $n(x)$ as a function of the distance x from the injecting electrode is $n(x) = (3/4)(\varepsilon\varepsilon_0/e)(V/L^2)\sqrt{L/x}$).][8] There is a so-called space charge buildup near the electrode, giving the process the name "space charge limited conduction." In this case, the space charge consists of conducting carriers. When significant traps are present, the trapped charge concentration can be many orders of magnitude larger than that of the conducting carriers. There will also be space charge buildup near the electrode; however, in that case, the dominant space charges are the trapped charges. The conduction process is thus called TCL conduction, in order for it to be distinguished from the previously mentioned trap-free SCL case. When the carrier (e.g., hole) injection level is not very large, the Poisson and current continuity equations can be simplified as: $d\varepsilon/dx \approx qp_t/\varepsilon$ and $J_p \approx qp\mu_p\varepsilon$ (here ε is the electric field, p and p_t are the free and trapped hole concentrations, respectively, and μ_p

is the hole mobility). From the simplified equations, we can see that the trapped charge (p_t) directly determines the electric field distribution and thus the potential and energy band profiles while the free holes (p) alone contribute to the current density. Of course, the two equations are still coupled because of the thermal balance between the free and trapped holes. This is in contrast to the SCL case where only free holes are present in both Poisson and current continuity equations.

The trapped hole distribution and trends of trapped and free hole concentrations for a typical total trap density $(H_p = 10^{19} \text{ cm}^{-3})$ are plotted in Figure 3.1 and Figure 3.2. From

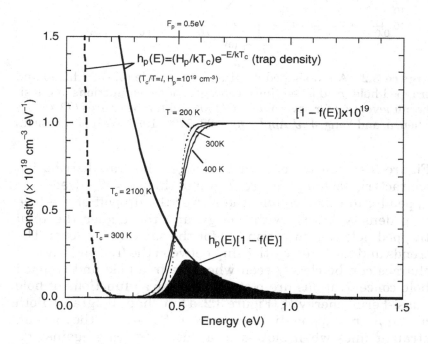

Figure 3.1 Hole trap density $[h_p(E)]$, hole Fermi–Dirac probability function $[1 - f(E)]$, and trapped hole distribution $[h_p(E)(1 - f(E))]$ as functions of energy. (From Shen, J. and Yang, J., *J. Appl. Phys.*, 83, 7706, 1998. With permission.)

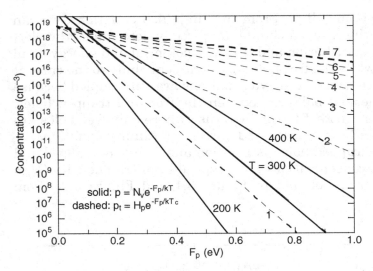

Figure 3.2 A fan-shaped diagram of free hole (p, solid lines) and trapped hole (p_t, dashed lines) concentrations as functions of quasi-Fermi level (F_p) for several l (= $T_c/300$) and temperatures (T). (From Shen, J. and Yang, J., *J. Appl. Phys.*, 83, 7706, 1998. With permission.)

Figure 3.1, we can see that the larger the trap-distribution characteristic temperature T_c (and thus $l = T_c/T$, see the appendix to this chapter), the slower the dropoff of the hole trap density [$h_p(E)$] with energy, and the higher the total trapped hole concentration (p_t) for the same Fermi level. The trends and the interrelationship between the free and trapped charges can be clearly seen when the free hole and trapped hole concentrations are plotted together as functions of hole quasi-Fermi energy F_p (Figure 3.2, a fan-shape diagram). Both p_t and p are exponential functions of F_p, so that they are all straight lines when plotted in a linear-log scale against F_p. The slopes of the lines are quite different. For the free holes (p), the slope is $1/kT$; and for the trapped holes, the slope is smaller by a factor of $1/l (l = T_c/T)$. When l is large (say, $l = 7$), p_t changes much more slowly with F_p than does p. Or equivalently, we can say that the free hole concentration rises much more rapidly than the filling of traps when an external bias is applied and increased, even though the trapped hole concentration is many orders of magnitude larger. Now we use the simple "leaky capacitor" analogy to obtain the qualitative J–V relationship. Given an applied voltage V, the stored charge in the organic layer is $Q = CV$, where $Q \approx qp_tAL$ (with A the area and L the length) and $C \approx \varepsilon A/L$. Since $p_t = H_p$ $\exp(-F_p/kT_c)$, we can obtain $F_p \approx -kT_c \ln[\varepsilon V/(qH_pL^2)]$, or F_p changes linearly with $\ln(V)$. One can thus easily obtain the dependence $p \sim (V/L^2)^l$ (with $l = T_c/T$) using $p \sim \exp(-F_p/kT)$ and the above $F_p \sim V$ dependence. If we ignore diffusion, the current density is thus

$$J_p \sim qp\mu_p(V/L) \sim V^{l+1}/L^{2l+1}, \tag{3.2}$$

the previously derived TCL result.[1-3] To put the above derivation process into more meaningful words, we say again that the traps determine how the internal profiles (F_p, etc.) change with external bias by the charging process, while the free carriers participate in conduction. The equilibrium condition between the trapped and free carriers in turn determines the current–voltage relationship. From Figure 3.2, we can also see clearly that there are two limiting cases where traps are no longer effective in the above roles: (1) when $l \leq 1$ (the two slopes are parallel), or (2) when $p > p_t$. In either case, we call the conduction process "trap-free" because the free-carrier space charge becomes dominant and J–V dependence reduces back to the SCL square-law form with l decreasing or p increasing. For $l \leq 1$, Godlewski et al.[33] derived an analytical expression for the current density two decades ago. Here we have put quotation marks on the term trap-free because in reality, the traps are still present and they are not even necessarily full.

To understand the mechanisms governing the OLED performance under TCL conditions, numerical models have been developed to solve the coupled Poisson and current continuity equations, assuming traps are present[34] (also see the appendix). The model has been used to study the effects of several important parameters (carrier injection level, device thickness, temperature, etc.).

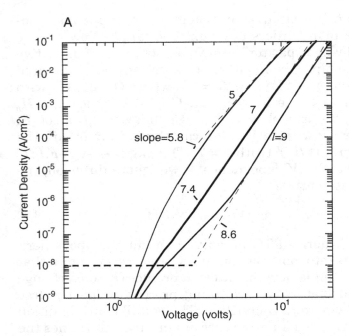

Figure 3.3 (A) Calculated TCL current-voltage (J–V) characteristics for three l values (l = 5, 7, and 9). The parameters we used for this simulation are H_p = H_n = 10^{19} cm^{-3}, T = 300 K, ε_r = 3.4, n_i = 1 cm^{-3}, E_g = 2.4 eV, K_b = 10^{-4} cm^3/s, μ_n = 5 ×10^{-5} cm^2/V.s, μ_p=$0.01\mu_n$, n_0 = p_0 = ρ_0 = 5 ×10^{20} cm^2, L = 60 nm. (B) Profiles of the energies, electric field, and charge concentrations at 10 V of the device structure in A with l = 7. (From Shen, J. and Yang, J., *J. Appl. Phys.*, 83, 7706, 1998. With permission.)

Figure 3.3A shows the simulated current–voltage (J–V) characteristics of a single-layer OLED with typical Alq$_3$ material parameters. Power-law J–V relationships are obtained, in agreement with previous analytical results and consistent with experimental observations.[6] The analytical derivation determines the power factor to be l + 1 (Equation 3.2), slightly larger than the corresponding extracted slopes from the simulation. At high voltages, the slopes tend to decrease slightly because the device approaches the "trap-filled" limit when the injection level is high. At low voltage (V < 2 V), our simulated

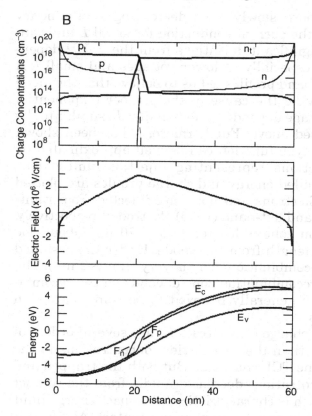

Figure 3.3 (continued)

TCL currents continue to decrease more or less following the same trend. Experimentally, a much gradual current–voltage (*J–V*) characteristic has been observed to intercept the TCL (*J~V^{l+1}*) curves at about 2 V (illustrated by the dashed line at $J = 10^{-8}$ A/cm² in Figure 3.3A); this voltage is often called the turn-on voltage.

The film thickness dependence of the *J–V* characteristics was also simulated. The current density follows the analytical power-law dependence $J \sim 1/L^{2l+1}$ for $L < 600$ Å.[34] For smaller *L*, the current is significantly smaller than the analytical trend. Experiments by Burrows et al.[6] show that the current increases even more slowly with decreasing L. Boundary effects may limit the current conduction for small L and cannot be ignored. Significant deviations from the J–V^{l+1} dependence also occur, especially at lower voltages and small l. A closer look at the band profiles led us to believe that the diode p–n junction may be the cause of the steeper (exponential type) current–voltage dependence for smaller L, similar to the SCL case discussed above. Furthermore, it has been shown that a power-law-type function can be well approximated by exponential functions representing injection-limited currents.[35,36] The detailed energy and charge profiles are plotted in Figure 3.3B. Electrons and holes are injected from right-hand ($x = 60$ nm) and left-hand ($x = 0$) electrodes, respectively. The recombination zone is located at $x \sim 20$ nm, about one third of the total length from the anode. Under the assumed conditions, the recombination zone is very thin (~ 2 nm). The thickness of the recombination zone depends on the recombination rate and is generally believed to be thin (≤ 5 nm) in organic materials.

The trapped charge concentrations are several orders of magnitude larger than the free carrier concentrations in the bulk, signifying the TCL character. But both the trapped and free charge concentrations decrease rapidly from the contact into the bulk, which is characteristic of the space charge limit with significant charge buildup near the contact. As a result of the charge buildup near the contact, a potential energy barrier exists at about 5 nm away from each contact. The electric field is approximately linear in the bulk and reaches maximum at the recombination zone.

The two orders of magnitude difference between the free electron and hole concentrations immediately before they meet at the recombination zone results from their mobility difference ($\mu_n = 100\mu_p$) and the current continuity requirement. But the trapped electron and hole concentrations differ much less in magnitude (about a factor of two). The mobility discrepancy between electrons and holes has much less effect on the trapped charges because they do not contribute directly to the current.

The fact that the location of the recombination zone occurs at $x = 20$ nm instead of being at the anode interface is a direct consequence of the small difference (a factor of 2 instead of 100) between the trapped electron (n_t) and trapped hole (p_t) concentrations. As mentioned previously, the transport properties of OLEDs within the TCL regime can be qualitatively understood using a leaky capacitor model. The free carriers, which support the current, play only a minor role because of their much smaller concentrations as compared to the trapped charges. The charge neutrality condition requires the equality between the total positive and negative charges. Approximately, we should have

$$L_p \times P_t(\text{ave}) = L_n \times N_t(\text{ave}), \tag{3.3}$$

where L_p and L_n are the lengths of the p- and n-type space charge regions, respectively, and $P_t(\text{ave})$ and $N_t(\text{ave})$ are the average trapped charge densities in the p- and n-type regions, respectively. Apparently, the location of the recombination zone is largely determined by the trap charges in the TCL case. In a symmetrical device (with the same parameters for electrons and holes) and under balanced injection, the recombination zone is located at the center, with equal amount of positive or negative charges on either side. In an asymmetrical device $(\mu_n = 100\mu_p)$, $P_t(\text{ave}) \sim 2N_t(\text{ave})$ and apparently the length of the n-type region must be twice that of the p-type region, which pins the location of the recombination zone to a distance one third of the total length from the anode. Determining the location of the recombination zone has been a hot topic since it directly affects device performance.[4,37–39] The general belief is that the initial electron and hole binding happens near the hole-transporter/Alq$_3$ interface because the mobility of electrons is two orders of magnitude larger than that of holes. After the electron–hole pair forms an exciton, it can further diffuse a couple of hundred angstroms into the Alq$_3$ layer and then recombine radiatively.[4,37–39] Various studies of the recombination zone have been discussed in some detail by Kalinowski.[31] Our result, based on a single-layer device, indicates that the initial electron–hole binding can actually occur significantly inside (20 nm in this case) the Alq$_3$ layer, in spite of the mobility discrepancy.

The thickness of the recombination zone depends on the recombination rate and is generally believed to be thin (≤ 5 nm) in organic materials. The effects of the recombination coefficient (K_b) on the TCL carrier transport process have been studied.[34] The simulation results indicate that the current increases when K_b decreases. The recombination zone becomes very wide when K_b is very small, approaching the injected plasma limit as discussed above.[2,7,17] In that limit, the electrical field strength is smaller because of its effective partial cancellation by the mutual spreading of positive and negative charges. When K_b is large enough and the recombination zone is small compared to the total device thickness, the current is no longer sensitive to K_b.

3.3 EFFECTS OF DOPING

Doping is an important and useful tool for device improvement as well as property characterization. It has been used to improve the emission efficiency and intensity of OLED materials.[4,37,39–43] From the theoretical point of view, we are interested in understanding the physical mechanisms associated with the doped layer in affecting OLEDs in the TCL regime. We would like to know why, in some cases, the doped layer can completely quench the host emission spectrum and give rise to its own, and in others, the host spectrum seems to be unperturbed by doping. Also, we will try to see the difference in the role of doping in the TCL regime vs. that in inorganic semiconductors such as Si or GaAs. We describe the doped layer in one of the following four different ways: (1) the doped layer has a higher concentration of traps than the host, (2) the doped layer changes the trap distribution characteristic, (3) the doped layer introduces a discrete trap level in the energy gap of the host, and (4) the doped layer has a higher recombination rate than the host. Other possibilities, such as the combination of several of these effects, can be similarly understood once we elucidate the physics involved. Some additional features that fall outside of the above descriptions are explained by other mechanisms, such as Gaussian disorder (diagonal and off-diagonal) and charge–dipole interaction.[28,44] For instance, the concentration dependence of recombination zone can be explained in terms of the disordered mechanism since the concentration determines which disorder (diagonal or off-diagonal) dominates and whether the recombination process is dependent on the disorder-controlled carrier mobility.[36] Apparently, the trapped charges determine the location of the recombination zone in the TCL case. Now, if a layer of additional n-type trapped charge is introduced into the OLED, the charge balance can be disturbed and Equation 3.3 is modified to the following form:

$$L_p \times P_t(\text{ave}) = (L_n - d) \times N_t(\text{ave}) + d \times N_{dt}(\text{ave}), \quad (3.4)$$

where d is the thickness of the doped layer, $N_{dt}(\text{ave})$ is the average trapped electron density in the doped layer. We can see that a significant change in the amount of trapped charges in the doped layer can shift the location of the recombination zone, as illustrated in Figure 3.4. The typical thickness of the OLED emitter is about 100 nm. If we assume $P_t(\text{ave}) = N_t(\text{ave})$, $d = 5$ nm, then the recombination zone shifts are as large as 10 and 2.5 nm for $N_{dt}(\text{ave}) = 5N_t(\text{ave})$ and $2N_{dt}(\text{ave})$, respectively.

Figure 3.5 shows the calculated energy and charge profiles of an OLED under TCL conditions. Electrons and holes are injected from the right-hand ($x = 60$ nm) and left-hand ($x = 0$) electrodes, respectively. The n-doped layer is placed at $x_d = 42.5$ nm and its thickness is 5 nm. Here we have assumed that the trap density in the doped layer (H_{nd}) is five times higher than that in the host (H_{n0}). The energy profiles of a similar structure without the doped layer are also shown in dot-dashed lines. The most significant change in the device is the shift of the recombination zone toward the doped layer. It is also interesting to see that the recombination zone is not *at* the doped layer in this particular case. As we discussed in Section 3.2, the trapped charges directly determine the location of the recombination zone under TCL conditions (Equation 3.3).

Without doping, the averaged trap charges are $P_t(\text{ave}) \approx 2 \times 10^{18}$ cm^{-3} $\approx 2N_t(\text{ave})$ (we can "eyeball it" from the p_t and n_t curves). According to Equation 3.3, we have $L_p = L_n/2$, in

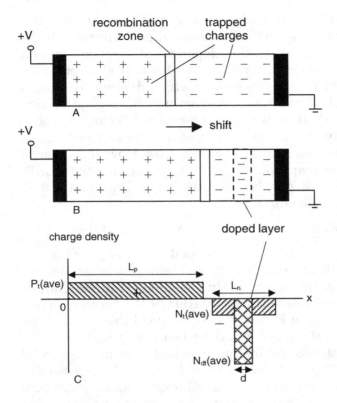

Figure 3.4 Illustration of the trap charging effects with (A) and without (B) a doped layer. Trap filling in the organic layer forms a *p*-type (+) and an *n*-type (−) regions. When additional trap charges (e.g., *n*-type) are introduced in the doped layer, the *p*- and *n*-type regions will expand or shrink accordingly, to satisfy the overall charge neutrality condition. This charge redistribution results in a recombination zone shift as indicated by the arrow. The charge densities in the respective regions are plotted schematically in C. (From Yang, J. and Shen, J., *J. Appl. Phys.*, 84, 2105, 1998. With permission.)

good agreement with the numerical result (the location of the recombination zone is at x ∼ 20 nm), which confirms the intuitive trap-charging picture discussed earlier. The free carrier concentrations, on the other hand, differ by a factor of 100 ($p = 100n$) near the recombination zone (Figure 3.5A,

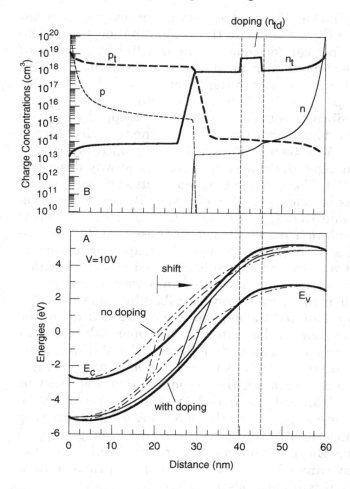

Figure 3.5 Calculated energy (A) and charge (B) profiles at 10 V of a single-layer device with the following host material parameters: $T = 300$ K, $T_{cn0} = T_{cp0} = 2100$ K ($l_n = l_p = 7$), $H_{p0} = H_{n0} = 10^{19}$ cm^{-3}, $\varepsilon_r = 3.4$, $n_i = 1$ cm^{-3}, $E_g = 2.4$ eV, $K_{b0} = 10^{-4}$ cm^3/s, $\mu_n = 5 \times 10^{-5}$ cm^2/V.s, $\mu_p = 0.01\mu_n$, $n_0 = p_0 = \rho_0 = 5 \times 10^{20}$ cm^{-3}, $L = 60$ nm. The doping density is $H_{nd} = 5 \times 10^{19}$ cm$^{-3} = 5H_{n0}$. (From Yang, J. and Shen, J., *J. Appl. Phys.*, 84, 2105, 1998. With permission.)

the same result with or without doping). This carrier concentration difference arises from the mobility difference ($\mu_n = 100\mu_p$) and the current continuity requirement ($j_n \sim n\mu_n$, $j_p \sim$

$p\mu_p$, and $j_p = j_n$) when the electric field is approximately equal on either side (it *is* true near the recombination zone). The fact that the free carrier concentrations differ by a factor of 100 while the respective trapped charge concentrations differ by only a factor of 2 is due to the l factor ($l = T_c/T = 7$ in this case): $p_t/n_t = (p/n)^{1/l} = (100)^{1/7} \approx 2$. This result also implies the thermal equilibrium between the free and trapped charges.

When the doped layer is introduced, the trap charges are redistributed. As shown in Figure 3.5, the doped layer at $x_d = 42.5$ nm changes the charge balance in the way described by Equation 3.4. The average trapped electron density in the *doped* layer is increased to $\sim 7 \times 10^{18}$ cm^{-3}. By using Equation 3.4, we can find that $L_p \sim 30$ nm $\sim L_n$, again in agreement with the simulation result. Based on these results, we conclude that if a doped layer changes the trap charge concentration, the trap charge balance is modified and the location of the recombination zone is shifted in a way to satisfy the new charge neutrality condition described by Equation 3.4.

We also calculated the dependence of the location of the recombination zone on the trap density. Apparently, the shift distance depends on the amount of the additional trap charge. For example, when the trap density in the n-trap doped layer on the n-side is increased, the recombination zone will be further shifted toward the doped layer. The recombination zone is more or less pinned at the doped layer for higher trap density because further shift will cause the doping traps to be unoccupied. But if the n-trap doping layer is on the p-side, it will not have any effect on the recombination zone because the doped traps will stay mostly empty.

From the work of Tang and co-workers,[37] we can see that the uniform doping somehow changed the ratio between the radiative and nonradiative recombination rates and had no effect on the I–V characteristics. If the process is indeed a TCL process, then our model indicates the possibility that no additional traps are introduced by doping (because otherwise I–V would change). After comparison with their work, we can see that the location of the recombination zone of a single layer is predicted to be at one third of the total length inside the Alq$_3$ layer, which, apparently, cannot be directly applied to the bilayer case. On the other hand, following the basic principles discussed above, and if we believe the hole transporter also has some hole traps, then the recombination zone would have to be at the Alq$_3$/hole transporter interface because of the blockage to the electrons and the hole charging in the hole transporter. If we believe that the thin doping layer can be described by one or a combination of the aforementioned possibilities, then the doping layer could actually pull the recombination zone away from the Alq$_3$/hole transporter interface, very similar to what was described in the experimental results. In the other words, the thin doping layer can also affect the location of the recombination zone. Of course, we cannot exclude the exciton diffusion interpretation as well as the recombination zone broadening possibility at this time. Another similar experiment was done by Schöbel et al.[4] Their result can be consistent with our doping picture if doping in their device does introduce a significant number of additional hole/electron traps. Several other doping experiments can be similarly analyzed.

3.4 EFFECTS OF DISCRETE TRAPS

The trap energy level distribution is generally described in one of the following ways: (1) an exponential distribution, (2) discrete levels, and (3) a Gaussian distribution.[2,3,45] Characteristic of the exponential trap distribution is a high power-law current–voltage relationship, which has been discussed in Section 3.2. In this section, we discuss the effects of discrete traps on the double-carrier injection OLED conduction processes. Discrete trap levels are more often associated with traps that are farther apart from each other, preventing them from strong coupling. Because of the discrete nature, the current-voltage characteristic usually has a sharp transition region, signifying the change from empty to filled states of the traps.[2,3]

When traps exist in the energy gap with a single discrete energy E_s, then the trapped charge concentration is

$$n_t = n_s = \frac{N_s}{1 + \exp\left[\left(E_s - F_n/kT\right)\right]}, \qquad (3.5)$$

where F_n is the quasi-Fermi level, and N_s is the trap density, assuming the degeneracy factor is equal to one. Because the free carrier concentration is $n = N_C \exp(F_n/kT)$ with N_C the effective density of states of the lowest unoccupied molecular orbitals (LUMOs) (assumed to be at zero as an energy level reference), the following relationship between n and n_s can be obtained:

$$n_s = \frac{N_s}{1 + \theta N_s/n}, \qquad (3.6)$$

with $\theta = (N_C/N_s)\exp(E_s/kT)$. At low injection levels, e.g., $n \ll \theta N_s$, we can obtain the discrete-TCL (D-TCL) current–voltage relationship:

$$J \approx \frac{9}{8} \frac{\varepsilon\mu}{1 + 1/\theta} \frac{V^2}{L^3}. \qquad (3.7)$$

Apparently, the D-TCL current has the same $J \sim V^2$ dependence but a smaller magnitude than that of the trap-free SCL current. This magnitude difference is the origin of the abrupt increase of the current transitioning from the D-TCL to the trap-filled SCL regimes.[2,3]

The interplay among the free, discrete-trap, and exponential-trap charges is illustrated in Figure 3.6, where the charges are plotted as functions of the Fermi level. We first discuss the case where only discrete traps are present. When the Fermi level is below the trap level ($E_s = -0.5$ eV), the trap filling rate [the slope in the $\log(n_s)$–F_n plot] is the same as the free-carrier concentration increase rate (the slope = $1/kT$), even though n_s can be much larger in magnitude than n (by a factor of θ^{-1}). That is why the D-TCL current has the same voltage dependence as the trap-filled SCL current ($J \sim V^2$). When F_n approaches E_s, the discrete traps are quickly filled and saturated while n continues to increase with F_n. Beyond $F_n = -0.07$ eV, the free carrier concentration becomes larger

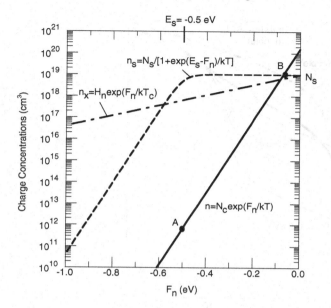

Figure 3.6 Trends of different electron charges (free electron concentration n, discrete-trap charge concentration n_s, and exponential-trap charge concentration n_x) in an OLED as functions of the electron Fermi energy F_n. The discrete trap level is assumed to be at $E_s = -0.5$ eV in reference to the LUMO. The LUMO density is $N_C = 1.6 \times 10^{20}$ cm^{-3}. Both the exponential-trap density (H_n) and discrete-trap density (N_s) are assumed to be 10^{19} cm^{-3}. (From Yang, J. and Shen, J., *J. Appl. Phys.*, 85, 2699, 1999. With permission.)

than n_s, and the conduction turns into the trap-filled or "trap-free" SCL regime. The region between $F_n = -0.5$ and -0.07 eV is a transition region when n is still smaller than n_s but n_s is no longer increasing. Within this transition region, the free carrier concentration increases by a factor of $\theta^{-1} \sim 1.7 \times 10^6$ (from point A to point B), which is basically the amount of increase in the current from the D-TCL to the trap-free SCL regions (the voltage change in the transition region is relatively small). Figure 3.6 can be very helpful because it graphically illustrates the amount of current changes, the boundary between various regions, and how the discrete trap parameters

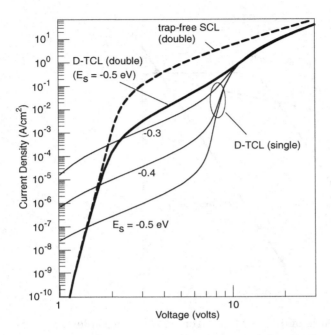

Figure 3.7 Calculated current–voltage characteristics of several space-charge-limited conduction devices without the exponential traps. The three thin solid curves correspond to the single-carrier (electron) injection case with three different discrete trap energies. The thick solid curve corresponds to the double-carrier injection case with a discrete electron trap level $E_s = -0.5$ eV. The thick dashed curve corresponds to the double-carrier trap-free SCL case. The parameters we used are $H_p = H_n = 0$ cm^{-3}, $N_s = 10^{18}$ cm^{-3}, $T = 300$ K, $\varepsilon_r = 3.4$, $n_i = 1$ cm^{-3}, $E_g = 2.4$ eV, $K_b = 10^{-4}$ cm^3/s, $\mu_n = 5 \times 10^{-5}$ cm^2/V.s, $\mu_p = 0.01\mu_n$, $n_0 = p_0 = \rho_0 = 5 \times 10^{20}$ cm^{-3}, $L = 60$ nm. (From Yang, J. and Shen, J., *J. Appl. Phys.*, 85, 2699, 1999. With permission.)

(E_s and N_s) affect the results. For example, if the discrete trap level is shallower than $E_s = -0.07$ eV (with the same N_s and N_C), then the discrete trap will have little effect on the current–voltage characteristics because the conduction is always in the trap-free SCL region. Similar interesting conclusions can be drawn from moving N_s and N_C.

Figure 3.7 shows several calculated current–voltage characteristics of a single-layer device with or without discrete traps.

The three thin curves represent the case of single-carrier (electron) injection with various discrete electron trap energies (E_s = –0.3, –0.4, and –0.5 eV). For E_s = –0.5 eV, a sharp transition occurs at V~8 V, corresponding to the trap-filling regime. As discussed above, the current magnitude increase in this region is θ^{-1} = $(N_s/N_C)\exp(-E_s/kT)$ = 1.7 × 10^6 by using the assumed parameter values (N_C = 1.6 × 10^{20} cm^{-3}, N_s = 10^{18} cm^{-3}, E_s = –0.5 eV, T = 300 K), which is in close agreement with the simulation result. This result can also be obtained graphically from Figure 3.6 as discussed above. For E_s = –0.3 eV, θ^{-1} ≈ 7 × 10^2 so the transition region is small. What is interesting is the double-carrier injection case with an electron discrete trap at E_s = –0.5 eV. The sharp transition region disappears almost completely even though the trap is rather deep. By analyzing the energy and charge profiles of the device, we found that *p*- and *n*-type regions are formed adjacent to the recombination zone. Unlike the single-carrier case, the electron discrete levels are already full in the *n*-region even at a small bias while they are always empty in the *p*-region. When the bias is increased, the recombination zone gradually shifts toward the anode, enlarging the *n*-region and shrinking the *p*-region. So there is no instantaneous transition of all trap levels from empty to full states, which is reflected in the absence of a sharp transition region in the current–voltage characteristics.

3.5 HETEROJUNCTION DEVICE

Heterojunctions play a crucial role in improving the EL efficiency in OLEDs. When Tang and VanSlyke[37,46] incorporated a bilayer structure consisting of a hole transporting layer (HTL) and an electron transporting and light emitter layer (ETLEL), the EL efficiency was improved several orders of magnitude. Since then, heterojunction structures have become almost indispensable in making high-efficiency OLEDs.[47,48] For a typical bilayer structure, TPD/Alq$_3$, for example, the hole transporting TPD layer imposes an energy barrier of about 0.7 eV to the electrons coming from Alq$_3$, and on the other hand, the Alq$_3$ layer has a barrier of about 0.3

eV to the holes in TPD. The 0.7 eV barrier effectively blocks the electrons and confines them to the TPD/Alq$_3$ interface. The 0.3 eV hole barrier, however, is relatively small and thus can still allow significant amount of hole injection into the Alq$_3$ layer. This configuration apparently improves the EL efficiency by forcing the recombination to occur in the Alq$_3$ and limiting the electron leakage current.[18,49] Furthermore, recent and past experiments have shown great influences of the hole-injecting and -transporting layers on the characteristics (current-voltage, efficiency, etc.) of bilayer OLEDs,[50–52] which challenges the notion that the single-carrier (electron-only) SCL theory is sufficient to explain the experimental observations.[6] In this section, we show quantitatively that there is an optimal value for the hole barrier height to maximize the electron–hole recombination in the Alq$_3$ layer near the heterojunction interface.

In our calculation, the coupled one-dimensional Poisson and current continuity equations are similar to Equations 3.A1 and 3.A2 (see the appendix) except for an additional term $(1/\varepsilon)(d\psi/dx)(d\varepsilon/dx)$ in the right-hand side of Equation 3.A1, which is from the position-dependent permittivity, where ψ is the electrostatic potential.[53] Expressions for the carrier concentrations and current densities also need to be modified to take into account the spatial nonuniformity. The LUMO and the highest-occupied molecular orbital (HOMO) levels can be written as:

$$E_c = q(\psi_0 - \psi) - \chi_c \qquad (3.8)$$

$$E_v = q(\psi_0 - \psi) - \chi_c - E_g , \qquad (3.9)$$

where ψ_0 is a reference potential and χ_c is the position-dependent electron affinity, i.e., the position-dependent energy gap.

Figure 3.8 shows the calculated profiles (energy, charge, and electron–hole overlap) of a bilayer OLED structure at 18.5 V. The electrons are effectively blocked by the 0.7 eV barrier and accumulate sharply at the interface. Few electrons ($\sim 10^{11}$ cm^{-2}) are injected into the hole-transporting layer (on the left). On the other hand, although holes are also

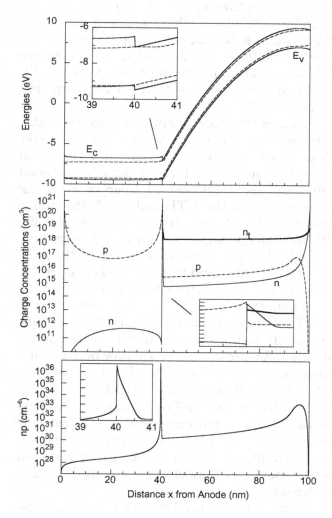

Figure 3.8 Calculated profiles of energies, charge concentrations, and the electron–hole overlap (np) in a bilayer OLED. The electron and hole barrier heights are $\Delta_e = 0.7$ eV and $\Delta_h = 0.3$ eV at the heterojunction, respectively. The current density is 26 mA cm^{-2} and the corresponding voltage is 18.5 V. The ETLEL (on the right) parameters are the same as those chosen by Burrows et al.[6] (From Yang, J. and Shen, J., *J. Phys. D Appl. Phys.*, 33, 1768, 2000. © IOP Publishing Ltd. With permission.)

blocked by the hole barrier, they enter the ETLEL in significant numbers ($\sim 10^{15}$ cm^{-2}), which allows efficient formation
of electron–hole pairs and subsequent generation of photons
at or near the interface on the emitting layer side. This picture
is consistent with the experimental observation that the EL
in the Diamine/Alq$_3$ device occurs within ~ 5 nm from the
interface on the Alq$_3$ side.[37]

Because the electron-transporting layer is also the light-
emitting layer, a larger electron barrier Δ_e is apparently better
because it confines more effectively the electrons to within
the emitting layer. As a matter of fact, under our assumed
conditions, we find that a significant number of electrons
($\sim 10^{14}$ cm^{-2}) are injected into the HTL and reach the anode
when Δ_e is lowered to 0.3 eV, which can degrade the EL
efficiency (details presented elsewhere). But a very large hole
barrier (say, $\Delta_h > 0.7$ eV) is bad because it limits the hole
injection into the emitting layer and forces the holes to recombine with electrons across the interface (which can be
nonradiative[50]) or causes relatively more electrons to be
injected into the HTL side when the total current is constant
(Figure 3.9). It is worth noting that too small a hole barrier
may *not* be good either. As a result of the smaller hole barrier,
the applied voltage obviously becomes smaller for the same
current because holes now can move into ETLEL more easily.
The increased hole injection into ETLEL may not be good
because a larger portion of the holes can leak into the cathode
or recombine near the cathode where photo quenching centers
are known to be abundant.[54,55] Such a problem is aggravated
especially in the SCL devices where charges accumulate near
the injecting electrodes.

To quantify such a trend further, we calculated the
recombination efficiency in the ETLEL as a function of the
hole barrier height (Δ_h) (Figure 3.10). A region with a variable
thickness $t = 0$ to $t = 10$ nm in ETLEL near the cathode is
intentionally excluded to discount the nonradiative part of
the recombination. When $\Delta_h > 0.7$ eV, the photon emission
from the ETLEL (for all t) is vanishing. For $t = 0$, more and
more holes are being injected into the ETLEL and recombine
with electrons there. For $t \geq 5$ nm, when $\Delta_h \leq 0.3$ eV; however,

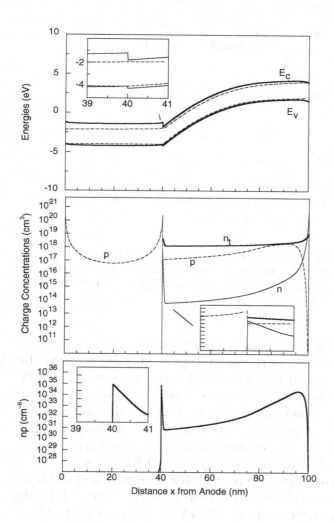

Figure 3.9 Calculated profiles of energies, charge concentrations, and the electron–hole overlap (*np*) in a bilayer OLED. The electron and hole barrier heights are Δ_e = 0.7 eV and Δ_h = 0.2 eV at the heterojunction, respectively. Other parameters are the same as in Figure 3.8. The corresponding voltage is 8 V. (From Yang, J. and Shen, J., *J. Phys. D Appl. Phys.*, 33, 1768, 2000. © IOP Publishing Ltd. With permission.)

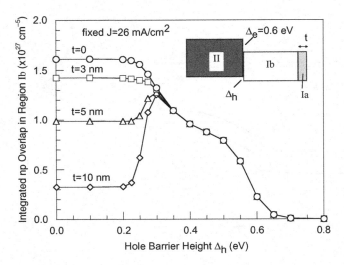

Figure 3.10 The calculated recombination efficiency in the ETLEL (excluding a region near the cathode) as functions of Δ_h. (From Yang, J. and Shen, J., *J. Phys. D Appl. Phys.*, 33, 1768, 2000. © IOP Publishing Ltd. With permission.)

a smaller electric field (because of the smaller voltage for the fixed current) can support less charge accumulation near the heterojunction, which in turn reduces the amount of recombination in the region. At the same time, the recombination or leakage in the region is increased by the same amount because the total integrated np overlap remains constant (fixed current). Our picture seems to be consistent with the experimental results of the hole-transporting layer in OLEDs.[50–52]

In the above calculations, we have assumed that the bulk carrier transport is the limiting factor by assuming relatively small injection barrier heights (0.2 eV) for both contacts. However, an unbalanced carrier injection can also drastically affect the performance of the OLEDs. To see how sensitive the device is to the injection imbalance, we calculated the recombination efficiency as a function of cathode barrier height with a fixed anode barrier height of 0.2 eV. When the cathode barrier height is smaller than 0.3 eV, the recombination rates change little because the contact is already in the ohmic regime. When the barrier height exceeds 0.3 eV, the recombination rates drop precipitately. The larger cathode barrier effectively blocks electron injection into the organic layer, causing larger portion of holes to leak out or to recombine near the cathode where photo quenching centers practically diminish their contribution to the light output. A particular metal contact can easily give rise to a barrier that is too large, or in some cases, too small. Furthermore, recent experiments show that the barrier height cannot be made smaller than ~0.6 eV with elemental metal contacts to Alq_3.[11] Contacts with LiF or its composites have been shown to improve the carrier injection.[9,10]

3.6 SUMMARY

Several issues related to the carrier transport in OLEDs with large concentrations of traps have been discussed. Current–voltage characteristics, band, charge, and electric field profiles are obtained by solving Poisson and current continuity equations numerically. The simulation methods are extended to cases with doping, discrete traps, and heterojunctions. Relevance to some experimental data is also discussed.

ACKNOWLEDGMENTS

The authors are thankful to Kodak for partial funding of this project. Stimulating discussions with Dr. C.W. Tang and Dr. J. Shi are acknowledged also. One of the authors (JS) would like to thank his former colleagues and students, especially Dr. J. Yang, who contributed to most of the joint work cited in this chapter.

APPENDIX: THEORETICAL MODELS

In studying the carrier transport in OLEDs in the SCL and TCL regimes, we solve the coupled one-dimensional Poisson and current continuity equations:

$$\frac{d^2\psi}{dx^2} = -\frac{q}{\varepsilon}\left(p + p_t - n - n_t\right) \tag{3.A1}$$

$$\frac{dJ_n}{dx} = -\frac{dJ_p}{dx} = qK_b\left(np - n_i^2\right), \tag{3.A2}$$

where ψ is the electrostatic potential (defined to be $-E_i/q$ with E_i the intrinsic Fermi level), n, p, n_t, p_t, and n_i are the free electron, free hole, trapped electron, trapped hole, and intrinsic carrier concentrations, respectively, J_n and J_p are the electron and hole current densities, respectively, K_b is the recombination coefficient, q is the electron charge, ε is equal to $\varepsilon_0\varepsilon_r$ with ε_0 the permittivity of vacuum and ε_r the relative dielectric constant. We introduce two quasi-Fermi levels (F_n and F_p) and express the carrier concentrations as follows:

$$n = n_i e^{(F_n + q\psi)/kT} \tag{3.A3}$$

$$p = n_i e^{(-q\psi - F_p)/kT}. \tag{3.A4}$$

The trap density is generally described by an exponential function:

$$h(E) = \frac{H}{kT_c} e^{-E/kT_c}, \tag{3.A5}$$

which decays exponentially from the band edge into the gap. The trapped charge concentration (n_t for electrons or p_t for holes) can be obtained by integrating the product of the trap density and the proper Fermi–Dirac probability function (note: the $T = 0$ K assumption is used):

$$n_t = H_n e^{(F_n + q\psi - \Delta_c)/kT_{cn}} \tag{3.A6}$$

$$p_t = H_p e^{(\Delta_v - q\psi - F_p)/kT_{cp}}, \tag{3.A7}$$

where H_n and H_p are the total corresponding trap concentrations (as defined in Equation 3.A5), T_{cn} and T_{cp} are the characteristic temperatures for the corresponding trap distributions. Here we adopt a labeling scheme that the sub-scripts n and p are for electrons and holes, respectively. We also introduce two parameters, $l_n = T_{cn}/T$ and $l_p = T_{cp}/T$. In this work, we have chosen $T_{cn} = T_{cp}$ so that $l = l_n = l_p$. For $T = 300$ K and $T_{cn} = T_{cp} = T_c = 2100$ K, we have $l = 7$, which is the power factor that appeared in the previously derived analytical $J \sim V^{l+1}$ expression.[1-3] In dealing with organic materials, it has been a convention to use the terms the lowest unoccupied molecular orbital (LUMO) and the highest occupied molecular orbital (HOMO) in place of the conduction and valence band edges, respectively. However, here we have used the symbols Δ_c and Δ_v conveniently to label the LUMO and HOMO in reference to E_i (intrinsic Fermi level) in an OLED. We have included both the drift and diffusion current components. By using the Einstein relationship, the current densities can be written as

$$J_n = -q\mu_n n \frac{d\psi}{dx} + qD_n \frac{dn}{dx} = n\mu_n \frac{dF_n}{dx} \qquad (3.A8)$$

$$J_p = -q\mu_p p \frac{d\psi}{dx} - qD_p \frac{dp}{dx} = p\mu_p \frac{dF_p}{dx} , \qquad (3.A9)$$

where D_n and D_p are the diffusion constants of electrons and holes, respectively. Ohmic boundary conditions are also used in this work by assuming large enough constant charge concentrations at the carrier-injecting electrodes:

$$n(L) = n_0, \text{ for the right-hand}$$
$$\text{electron injecting electrode,} \qquad (3.A10)$$

$$p(0) = p_0, \text{ for the left-hand}$$
$$\text{hole injecting electrode,} \qquad (3.A11)$$

where L is the organic layer thickness. The quasi-Fermi levels are assumed to be equal at the boundaries:

$$F_n(0) = F_p(0) \text{ and } F_n(L) = F_p(L). \qquad (3.A12)$$

In solving the equations, we have used the Gummel method.[56-61] We have studied the current voltage characteristics (trap-free and trap-limited) and their dependence on several device parameters (e.g., layer thickness, temperature, recombination constant, carrier mobility, etc.). We have also extended the model to describe doping and heterojunctions as discussed above.

REFERENCES

1. Mark, P. and Helfrich, W., *J. Appl. Phys.*, 33, 205, 1962.

2. Lampert, M.A. and Mark, P., *Current Injection in Solids,* Academic Press, New York, 1970.

3. Kao, K.C. and Hwang, W., *Electrical Transport in Solids, with Particular Reference to Organic Semiconductors*, Pergamon Press, New York, 1981.

4. Schöbel, J., Ammermann, D., Böhler, A., Dirr, S., and Kowalsky, W., *SID97 Digest*, 782, 1997.

5. Bak, G.W., *J. Phys. Condens. Matter,* 8, 4145, 1996.

6. Burrows, P.E., Shen, Z., Bulovic, V., McCarty, D.M., Forrest, S.R., Cronin, J.A., and Thompson, M.E., *J. Appl. Phys.*, 79, 7991, 1996.

7. Blom, P.W.M., de Jong, M.J.M., and Breedijk, S., *Appl. Phys. Lett.*, 71, 930, 1997.

8. Helfrich, W., Space-charge-limited and volume-controlled currents in organic solids, in *Physics and Chemistry of the Organic Solid State*, D. Fox et al., Ed., John Wiley & Sons, New York, 1967).

9. Hung, L.S., Tang, C.W., and Mason, M.G., *Appl. Phys. Lett.,* 70, 152, 1997.

10. Jabbour, G.E., Kippelen, B., Armstrong, N.R., and Peyghambarian, N., *Appl. Phys. Lett.,* 73, 1185, 1998.

11. Campbell, I.H. and Smith, D.L., *Appl. Phys. Lett.,* 74, 561, 1999.

12. Campbell, I.H., Hagler, T.W., Smith, D.L., and Ferraris, J.P., *Phys. Rev. Lett.,* 76, 1900, 1996.

13. Kalinowski, J., *Synth. Met.,* 64, 123, 1994.

14. Schottky, W., *Z. Phys.,* 118, 539, 1942.

15. Bethe, H.A., *MIT Radiat. Lab. Rep.*, 43-12, 1942.

16. Fowler, R.H. and Nordheim, L., *Proc. R. Soc. Lond.*, 119, 173, 1928.

17. Parker, I.D., *J. Appl. Phys.*, 75, 1659, 1994.

18. Khramtchenkov, D.V., Bässler, H., and Arkhipov, V.I., *J. Appl. Phys.*, 79, 9283, 1996.

19. Davids, P.S., Kogan, S.M., Parker, I.D., and Smith, D.L., *Appl. Phys. Lett.*, 69, 2270, 1996.

20. Davids, P.S., Campbell, I.H., and Smith, D.L., *J. Appl. Phys.*, 82, 6319, 1997.

21. Scott, J.C. and Malliaras, G.G., *Chem. Phys. Lett.*, 299, 115, 1999.

22. Heeger, A.J., Nature of the primary photoexcitations in poly(arylene-vinylenes), in *Primary Photoexcitations in Conjugated Polymers*, N.S. Sariciftci, Ed., World Scientific, London, 1997.

23. Abkowitz, M.A., Mizes, H.A., and Facci, J.S., *Appl. Phys. Lett.*, 66, 1288, 1995.

24. Gartstein, Y.N. and Conwell, E.M., *Chem. Phys. Lett.*, 255, 93, 1996.

25. Wolf, U. and Bässler, H., *Appl. Phys. Lett.*, 74, 3848, 1999.

26. Langevin, P., *Am. Chem. Phys.*, 28, 289, 1903.

27. Albrecht, U. and Bässler, H., *Phys. Status Solidi b*, 191, 455, 1995.

28. Bässler, H., *Phys. Status Solidi b*, 175, 15, 1993.

29. Onsager, L., *Phys. Rev.*, 54, 554, 1938.

30. Onsager, L., *J. Chem. Phys.*, 2, 599, 1934.

31. Kalinowski, J., *J. Phys. D Appl. Phys.*, 32, R179, 1999.

32. Kalinowski, J. et al., *Appl. Phys. Lett.*, 72, 513, 1998.

33. Godlewski, J. et al., *Solid State Commun.*, 25, 473, 1978.

34. Shen, J. and Yang, J., *J. Appl. Phys.*, 83, 7706, 1998.

35. Kalinowski, J. et al., *Int. J. Electron.*, 81, 377, 1996.

36. Kalinowski, J. et al., *J. Appl. Phys.*, 89, 1866, 2001.

37. Tang, C.W., VanSlyke, S.A., and Chen, C.H., *J. Appl. Phys.*, 65, 3610, 1989.

38. Adachi, C., Tsutsui, T., and Saito, S., *Optoelectronics*, 6, 25, 1991.

39. Mori, T., Miyachi, K., and Mizutani, T., *J. Phys. D Appl. Phys.*, 28, 1461, 1995.

40. Utsugi, K. and Takano, S., *J. Electrochem. Soc.*, 139, 3610, 1992.

41. Aminaka, E., Tsutsui, T., and Saito, S., *J. Appl. Phys.*, 79, 8808, 1996.

42. Suzuki, H. and Hoshino, S., *J. Appl. Phys.*, 79, 8816, 1996.

43. Shi, J. and Tang, C.W., *Appl. Phys. Lett.*, 70, 1665, 1997.

44. Dunlap, D.H., Parris, P.E., and Kenkre, V.M., *Phys. Rev. Lett.*, 77, 542, 1996.

45. Pope, M. and Swenberg, C.E., *Electronic Processes in Organic Crystals*, Clarendon, Oxford, 1982, 202.

46. Tang, C.W. and VanSlyke, S.A., *Appl. Phys. Lett.*, 51, 913, 1987.

47. Sheats, J.R., Antoniadis, H., Hueschen, M.R., Leonard, W., Miller, J., Moon, R., Roitman, D., and Stocking, A., *Science*, 273, 884, 1996.

48. Rothberg, L.J. and Lovinger, A.J., *J. Mater. Res.*, 11, 3174, 1996.

49. Staudigel, J., Stobel, M., Steuber, F., and Simmerer, J., *J. Appl. Phys.*, 86, 3895, 1999.

50. Giebeler, C., Antoniadis, H., Bradley, D.D.C., and Shirota, Y., *J. Appl. Phys.*, 85, 608, 1999.

51. Roitman, D.B., Antoniadis, H., Hueschen, M., Moon, R., and Sheats, J.R., *IEEE J. Sel. Top. Quantum Electron*, 4, 58, 1998.

52. Tamoto, N., Adachi, C., and Nagai, K., *Chem. Mater.*, 9, 1077, 1997.

53. Yang, J. and Shen, J., *J. Phys. D Appl. Phys.*, 33, 1768, 2000.

54. Choong, V.-E., Park, Y., Shivaparan, N., Tang, C.W., and Gao, Y., *Appl. Phys. Lett.*, 71, 1005, 1997.

55. Choong, V.-E., Park, Y., Gao, Y., Mason, M.G., and Tang, C.W., *J. Vac. Technol.*, A16, 1838, 1998.

56. Gummel, H.K., *IEEE Trans. Electron. Dev.*, ED-11, 455, 1964.

57. Scharfetter, D.L. and Gummel, H.K., *IEEE Trans. Electron. Dev.*, ED-16, 64, 1969.

58. Gokhale, B.V., *IEEE Trans. Electron. Dev.*, ED-17, 594, 1970.

59. Rafferty, C.S., Pinto, M.R., and Dutton, R.W., *IEEE Trans. Electron. Dev.*, ED-32, 2018, 1985.

60. Torpey, P.A., *J. Appl. Phys.*, 56, 2284, 1984.

61. Hurm, V., Hornung, J.C.R., and Manck, O., *J. Appl. Phys.*, 60, 3214, 1986.

62. Yang, J. and Shen, J., *J. Appl. Phys.*, 84, 2105, 1998.

63. Yang, J. and Shen, J., *J. Appl. Phys.*, 85, 2699, 1999.

4

Amorphous Molecular Materials for Carrier Injection and Transport

YASUHIKO SHIROTA

CONTENTS

4.1 INTRODUCTION

Organic light-emitting diodes (OLEDs) have attracted a great deal of attention, both academic interest and interest in their practical applications for full-color, flat-panel displays and lighting.[1-5] The operation of OLEDs involves charge injection from the electrodes, transport of charge carriers, recombination of holes and electrons to generate electronically excited states, followed by the emission of either fluorescence or phosphorescence. The main factors that determine luminous and quantum efficiencies are the following: efficiency of charge carrier injection from electrodes, charge balance, spin multiplicity of the luminescent state, emission quantum yield, and light output coupling factor. To attain high quantum efficiency for electroluminescence (EL), it is necessary to attain high charge injection efficiency and good charge balance, and to confine charge carriers within the emitting layer to lead to an enhanced recombination probability of the charge carriers. Generally, layered devices consisting of charge-transporting and charge-emitting layers can achieve higher charge injection efficiency and better charge balance than can single-layer devices using emitting materials alone. This is because a suitable combination of charge-transporting and emitting materials in layered devices reduces the energy barrier for the injection of charge carriers from the electrodes into the organic layer and blocking charge carriers from escaping from the emitting layer, leading to better balance in the number of injected holes and electrons, as shown in Figure 4.1.

The performance of OLEDs depends on materials functioning in various specialized roles, including emitting, charge-transporting, and charge-blocking materials. It is therefore of crucial importance to develop high-performance materials for the fabrication of high-performance OLEDs. Generally, materials for use in OLEDs including both emitting and charge-transporting materials should meet the following requirements:[6] (1) materials should possess suitable ionization potentials and electron affinities for energy level matching for the injection of charge carriers at interfaces between the electrode/organic material and the organic material/organic material; (2) materials should permit the formation of uniform films without pinholes; (3) materials should be morphologically stable; and (4) materials should be thermally stable. Both small organic molecules and polymers have been studied for use as materials in OLEDs. Small organic

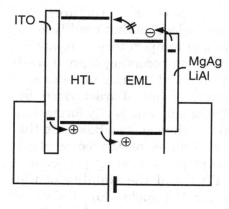

HTL: Hole-Transport Layer
EML: Emitting Layer

Figure 4.1 Structure of a layered OLED.

molecules that readily form stable amorphous glasses, i.e., amorphous molecular materials, function well as materials for use in OLEDs.

This chapter deals with amorphous molecular materials for carrier injection and transport. First, amorphous molecular materials are briefly overviewed and then discussion focuses on amorphous molecular materials for carrier injection and transport for use in OLEDs. Molecular design concepts for such amorphous molecular materials are presented. Charge transport in amorphous molecular materials is also discussed.

4.2 AMORPHOUS MOLECULAR MATERIALS

Generally, low-molecular-weight organic compounds tend to crystallize very readily, and hence, little attention had been paid to the amorphous glasses of small organic molecules. Since the late 1980s, intensive studies on the creation of small organic molecules that readily form stable amorphous glasses have been performed, and it has now been recognized that not only polymers but also low-molecular-weight organic compounds can form amorphous glasses if their molecular structures are properly designed.[4] Small organic molecules that readily form stable amorphous glasses at ambient temperatures, i.e., amorphous molecular materials, have been receiving a great deal of attention as a new class of functional organic materials.

Amorphous molecular materials are characterized by well-defined glass-transition phenomena and ready formation of uniform amorphous thin films by themselves. Based on the information obtained from the studies of the correlation between molecular structures and glass-forming properties, glass-transition temperatures (Tg), and the stability of the glassy state, several concepts for the molecular design of amorphous molecular materials have been proposed.[4,7–15] Nonplanar molecular structures together with the existence of different conformers are responsible for the ready formation of the amorphous glassy state.

The following π-electron starburst molecules have been designed and synthesized, which include the families of 4,4′,4″-tris(diphenylamino)triphenylamine (TDATA),[7,8] 1,3,5-tris(diphenylamino)benzene (TDAB),[9,10] 1,3,5-tris[4-(diphenylamino)phenyl]benzene (TDAPB),[13] and others. These compounds except for TDAB have been found to readily form amorphous glasses when their melt samples are cooled either on standing in air or by rapid cooling with liquid nitrogen. The formation of the amorphous glassy state is evidenced by polarizing optical microscopy, differential scanning calorimetry (DSC), and X-ray diffraction (XRD).

A number of triarylamine-containing compounds have also been found to form amorphous glasses, which include oligoarylenylamines such as tri(p-terphenyl-4-yl)amine (p-TTA),[12] oligothiophenes such as α,ω-bis{4-[bis(4-methylphenyl)amino]phenyl}oligothiophenes (BMA-nT),[16–19] azo-compounds such as 4,4′-bis[bis(4′-tert-butylbiphenyl-4-yl)amino]azobenzene (t-BuBBAB),[19] and others. π-Electron starburst molecules have been extended to dendrimer-type molecules such as 1,3,5-tris[4-bis(4-methylphenyl)aminophenyl-4-diphenylaminophenylamino]benzene (TDAB-G1)[20] and tris(bis{4-bis[4-bis(4-methylphenyl)aminophenyl]amino-phenyl})amine (**1**).[21] In addition, other nonplanar molecules

Table 4.1 Glass-Transition Temperatures (Tgs) of Amorphous Molecular Materials

Material	Tg/°C	Ref.	Material	Tg/°C	Ref.
TDATA	89	7	BMA-4T	98	16–18
m-MTDATA[a]	75	7	t-BuBBAB	177	19
p-MTDAB[b]	55	9,10	TDAB-G1	134	20
TDAPB	121	13	1	169	21
m-MTDAPB[c]	105	13	spiro-6Φ	212	22
p-TTA	132	12	C(^{t}BuSSB)$_4$	175	23
BMA-3T	93	16–18			

[a] 4,4′,4″-tris[3-methylphenyl(phenyl)amino]triphenylamine.
[b] 1,3,5-tris[4-methylphenyl(phenyl)amino]benzene.
[c] 1,3,5-tris{4-[3-methylphenyl(phenyl)amino]phenyl}benzene.

such as spiro compounds and tetraphenylmethane derivatives have also been reported to form amorphous glasses readily.[22,23] The Tgs of these amorphous molecular materials are summarized in Table 4.1, and their molecular structures are shown below.

Amorphous Molecular Materials

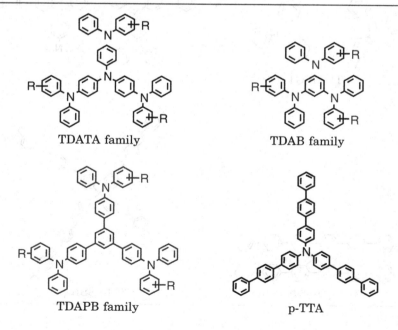

TDATA family

TDAB family

TDAPB family

p-TTA

Amorphous Molecular Materials (continued)

BMA-nT family

t-BuBBAB

TDAB-G1

1

spiro-6Φ

C(t-BuSSB)$_4$

Table 4.2 Amorphous Molecular Materials with High Glass-Transition Temperatures (Tgs)

Material	Tg/°C	Ref.	Material	Tg/°C	Ref.
TPTTA	141	8	TFATA	131	26
TCTA	151	24	FFD	165	26
t-Bu-TBATA	203	25	TBFAPB	189	27

Thermal stability is required for technological applications of such amorphous molecular materials. The Tg is a measure of thermal stability of amorphous molecular materials. One of the guidelines for raising Tg is the incorporation of rigid moieties such as phenothiazine, carbazole, biphenyl, and fluorene groups to form nonplanar molecular structures.[8,19,24–27] Based on this concept, amorphous molecular materials with high Tgs, e.g., 4,4′,4″tri(N-phenothiazinyl) triphenylamine (TPTTA),[8] 4,4′,4″-tri(N-carbazolyl)triphenylamine (TCTA),[24] 4,4′,4″-tris[bis(4′-*tert*-butylbiphenyl-4-yl) amino]triphenylamine (t-Bu-TBATA),[25] 4,4′,4″-tris[9,9-dimethylfluoren-2-yl(phenyl)amino]triphenylamine (TFATA),[26] N,N,N′,N″tetrakis(9,9-dimethylfluoren-2-yl)-[1,1′-biphenyl]-4,4′-diamine (FFD),[26] and 4,4′,4″-tris[bis(9,9-dimethylfluoren-2-yl)amino]triphenylbenzene (TBFAPB),[27] have been developed. The Tgs of these amorphous molecular materials are listed in Table 4.2. They function as hole-transporting materials for OLEDs, the molecular structures of which are shown in Section 4.4.

The studies on the creation of amorphous molecular materials have merged with the studies on OLEDs, and amorphous molecular materials have demonstrated their suitability and versatility as materials for OLEDs.[4,28]

4.3 HOLE-TRANSPORTING AMORPHOUS MOLECULAR MATERIALS

The hole-transport layer in layered OLEDs generally plays the roles of facilitating hole injection from the anode into the organic layer, accepting holes, and transporting injected holes to the emitting layer. Usually, an indium-tin-oxide (ITO)-coated glass substrate has been used as the anode for OLEDs.

The hole-transport layer also functions to block electrons from escaping from the emitting layer to the anode. Therefore, hole-transporting materials should fulfill the requirements of energy level matching for the injection of holes from the anode. They should possess electron-donating properties, and their anodic oxidation processes should be reversible to form stable cation radicals. The hole mobilities of hole transporters should be desirably high. They should form homogenous thin films with both morphological and thermal stability.

N,N'-Bis(3-methylphenyl)-N,N'-diphenyl-[1,1'-biphenyl]-4,4'-diamine (TPD),[29] which has been used by being dispersed in polycarbonate as a charge carrier transport layer for photoreceptors in electrophotography, has also been used widely as a hole transporter in OLEDs;[30] however, TPD lacks thermal and morphological stability.[31] TPD forms an amorphous glass with a Tg of 63°C, but it tends to crystallize. Other families of hole-transporting amorphous molecular materials include 1,3,5-tris[N-(4-diphenylaminophenyl)phenylamino]benzene (p-DPA-TDAB),[32] TDAPB and its methyl-substituted derivatives,[13] and TCTA.[24] These hole-transporting amorphous molecular materials are thermally much more stable than TPD. Thermally stable OLEDs have been fabricated using these hole-transporting amorphous molecular materials in combination with tris(8-quinolinolato)aluminum (Alq_3) as an emitting material.[24,25,33,34]

Recently, other hole-transporting amorphous molecular materials (Table 4.3) with relatively high Tgs have also been reported; these include triphenylamine oligomers such as TPTE,[35] triarylamine-containing spiro-compounds such as spiro-TAD,[36] a triaryl-amine-containing fluorene (**2**),[37] a triarylamine-containing carbazole derivative (**3**),[38] and fluorene-containing TDAPB derivatives, e.g., 4,4',4''-tris[9,9-dimethylfluoren-2-yl(phenyl)amino]triphenylbenzene (TFAPB), 4,4',4''-tris[9,9-dimethylfluoren-2-yl(4-methylphenyl)amino]triphenylbenzene (MTFAPB), and 4,4',4''-tris[bis(9,9-dimethylfluoren-2-yl)amino]triphenylbenzene (TBFAPB).[27] New TPD derivatives with higher Tgs have also been developed, which include N,N'-di(1-naphthyl)-N,N'-diphenyl-[1,1'-biphenyl]-4,4'- diamine (α-NPD),[39] N,N'-di(9-phenanthryl)-N,N'-diphenyl-[1,1'-biphenyl]-4,4'-diamine (PPD),[40] 4,4'-di(N-carbazolyl)bi-phenyl

Table 4.3   Glass-Transition Temperatures (Tgs) and Oxidation Potentials ($E_{1/2}^{ox}$) of Hole-Transporting Amorphous Molecular Materials

Material	Tg/°C	$E_{1/2}^{ox}$ [a]	Ref.	Material	Tg/°C	$E_{1/2}^{ox}$ [a]	Ref.
m-MTDATA	75	0.06	7	TBFAPB	189	0.51	27
1-TNATA	113	0.08	45	TPTE	140	—	35
2-TNATA	110	0.11	45	spiro-TAD	133	—	36
t-Bu-TBATA	203	0.09	25	**2**	167	—	37
o-PTDATA	93	0.06	42	**3**	163	—	38
m-PTDATA	91	0.10	42	TPD	60	0.48	41
p-PMTDATA	110	0.08	42	α-NPD	100	0.51	41
TFATA	131	0.08	26	CBP	—	0.72	41
TCTA	151	0.69	24	SBB	125	0.44	41
p-DPA-TDAB	108	0.23	32	p-BPD	102	0.50	42
p-MTDAPB	110	0.64	13	PFFA	135	0.32	43
TFAPB	150	0.61	27	FFD	165	0.40	26
MTFAPB	154	0.54	27				

[a] vs. Ag/Ag+ (0.01 mol dm⁻³).

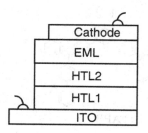

Figure 4.2   Structure of OLEDs consisting of double hole-transport layers.

(CBP),[41] 4-(iminodibenzyl)-4′-(iminostilbenyl)biphenyl (SBB),[41] *N*, *N*′-di(biphenyl-2-yl)-*N*, *N*′-diphenyl-[1,1′-biphenyl]-4,4′-diamine (o-BPD),[42] *N*, *N*′-di(biphenyl-3-yl)-*N*, *N*′-diphenyl-[1,1′-biphenyl]-4,4′-diamine (m-BPD),[42] *N*, *N*′-di(biphenyl-4-yl)-*N*, *N*′-diphenyl-[1,1′-biphenyl]-4,4′-diamine (p-BPD),[42] *N*, *N*′-bis (9,9-dimethylfluoren-2-yl)-*N*, *N*′-diphenyl-9,9-dimethylfluorene-2,7- diamine (PFFA),[43] and FFD.[26] These hole-transporting materials are used either for the single hole-transport layer or for the second hole-transport layer in double hole-transport layers in OLEDs (Figure 4.2), as described later.

Hole-Transporting Amorphous Molecular Materials

p-DPA-TDAB

p-MTDAPB

TCTA

TFAPB

MTFAPB

TBFAPB

Hole-Transporting Amorphous Molecular Materials (continued)

TPTE

spiro-TAD

2

TPD

α-NPD

PPD

3

CBP

Hole-Transporting Amorphous Molecular Materials (continued)

SBB

p-BPD

PFFA

FFD

Hole-transporting amorphous molecular materials with very low solid-state ionization potentials facilitate hole injection from the ITO electrode, functioning as a hole-injection layer in layered OLEDs consisting of double hole-transport layers such as shown in Figure 4.2.[44] The materials include the family of TDATA, e.g., 4,4′,4″-tris[3-methylphenyl (phenyl)amino]triphenylamine (m-MTDATA), 4,4′,4″-tris[1-naphthyl(phenyl)amino]triphenylamine (1-TNATA),[45] 4,4′,4″-tris[2-naphthyl(phenyl)amino]triphenylamine (2-TNATA),[45] and TFATA.[26] These compounds readily form stable amorphous glasses with well-defined Tgs. These materials are characterized by very low solid-state ionization potentials of ~5.0 to 5.1 eV and by the good quality of their amorphous films. The amorphous films of these materials, in particular, m-MTDATA, are very stable without undergoing crystallization in an ambient atmosphere. They form smooth, uniform amorphous films by vacuum deposition and spin coating form solution. They are transparent in the wavelength region of visible light.

Hole-Transporting Amorphous Molecular Materials for Use as the
Hole-Injection Layer in OLEDs

m-MTDATA

1-TNATA

2-TNATA

p-PMTDATA

t-Bu-TBATA

The Tgs and oxidation potentials of hole-transporting amorphous molecular materials are listed in Table 4.3.

Multilayer OLEDs consisting of an Alq_3 emitting layer and double hole-transport layers of the materials of the TDATA family (HTL1) and materials with higher ionization potentials, i.e., TPD and TDAPB families (HTL2), exhibited higher luminous efficiency and significantly enhanced operational stability compared with the double-layer device consisting of only the emitting layer of Alq_3 and the single hole-transport layer of hole transporters.[33,44–47] A thermally stable OLED using 1-TNATA and α-NPD as the double hole-transport layers, N,N'-diethylquinacridone (DEQ)-doped Alq_3 as the emitting layer, and Alq_3 as the electron-transport layer exhibited high performance with luminous efficiencies of 4.6 lm W^{-1} at room temperature and 7.5 lm W^{-1} at 90°C for obtaining 100 cd m^{-2} and an external quantum efficiency of 2.2%. The half-decay time of the initial luminance (~800 cd m^{-2}) of the device was 3200 h at room temperature at constant DC current.[47] m-MTDATA has been widely used as the hole-injection layer in OLEDs. Copper phthalocyanine has also been used as the hole-injection layer in OLEDs.

4.4   ELECTRON-TRANSPORTING AMORPHOUS MOLECULAR MATERIALS

As compared with a number of hole-transporting materials, there have been reported fewer electron-transporting materials. The electron-transport layer in OLEDs plays the roles of facilitating electron injection from the cathode into the organic layer, accepting electrons, and transporting injected electrons to the emitting layer. The electron-transport layer also plays a role of blocking holes from escaping from the emitting layer to the cathode. Therefore, electron-transporting materials should fulfill the requirements of energy level matching for electron injection from the cathode. The cathodic reduction processes of electron-transporting materials should be reversible. They should possess electron-accepting properties to permit the ready formation of stable anion radicals. The electron mobilities of electron transporters should be desirably high. They should form homogeneous thin films with both morphological and thermal stability.

Alq$_3$ has been widely used not only as a green emitter but also as an electron transporter in OLEDs. It has been reported that Alq$_3$ takes up different polymorphs and that the deposited thin films are of amorphous nature.[48] Likewise, beryllium and zinc complexes have been reported to function as electron transporters in OLED,[46,49] but their morphologies have not been reported. Other electron transporters that have been developed include oxadiazole derivatives such as 2-(biphenyl-4-yl)-5-(4-*tert*-butylphenyl)-1,3,4-oxadiazole (t-Bu-PBD)[30] and 1,3-bis[5-(4-*tert*-butylphenyl)-1,3,4-oxadiazol-2-yl]benzene (OXD-7),[50] 3-(biphenyl-4-yl)-4-phenyl-5-(4-*tert*-butylphenyl)-1,2,4-triazole (TAZ),[51] and 1,1-dimethyl-2,5-di(2-pyridyl)silole (PYSPY),[52] although their glass-forming properties and morphological changes have not been reported in detail. 1,3,5-Tris(4-*tert*-butylphenyl-1,3,4-oxadizolyl)benzene (TPOB),[45,53,54] dendrimer-type oxadiazole,[55] and oxadiazole derivative containing a spiro center[56] readily form amorphous glasses with relatively high Tgs. Boron-containing oligothiophenes such as 5,5′-bis(dimesitylboryl)-2,5-thiophene (BMB-1T), 5,5′-bis(dimesitylboryl)-2,2′-bithiophene (BMB-2T), and 5,5′-bis(dimesitylboryl)- 2,2′:5′,2″-terthiophene (BMB-3T), which readily form stable amorphous glasses, function as electron-transporting materials.[57–59] These compounds undergo reversible cathodic reductions, with BMB-2T and BMB-3T exhibiting two sequential cathodic and the corresponding anodic waves to generate the radical anion and dianion species. BMB-nT (n = 0, 1, 2, and 3) have stronger electron-accepting properties than Alq$_3$.[57–59] 1,3,5-Tris[5-(dimesitylboryl)thiophen-2-yl]benzene (TMB-TB) also functions as an electron transporter with better hole-blocking ability than Alq$_3$.[60] TMB-TB also fulfills the requirement of reversible cathodic reduction for electron-transporting materials. It is of interest to note that TMB-TB has multiple redox properties, exhibiting three sequential cathodic and the corresponding anodic waves in the cyclic voltammogram. The reduction potential of TMB-TB ($E_{1/2}^{red}$ = −1.98 V vs. Ag/Ag$^+$ (0.01 mol dm^{-3})) is more or less the same as that of Alq$_3$ ($E_{1/2}^{red}$ = −2.01

Table 4.4   Glass-Transition Temperatures (Tgs) and Reduction Potentials ($E_{1/2}^{red}$) of Electron-Transporting Amorphous Molecular Materials

Material	Tg/°C	$E_{1/2}^{red}$a	Ref.	Material	Tg/°C	$E_{1/2}^{red}$a	Ref.
OXD-7	—	—	50	TPQ	147	−2.67	61
TAZ	—	—	51	TPBI	—	—	62
PYSPY	—	—	52	BMB-1T	71	−1.76	59
TPOB	137	−2.10	45	BMB-2T	107	−1.76	57
Dendrimer oxadizaole	248	—	55	BMB-3T	115	−1.76	57
spiro-PBD	163	—	56	TMB-TB	160	−1.98	60

a vs. Ag/Ag⁺ (0.01 mol dm⁻³).

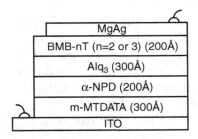

Figure 4.3 Side view of the ITO/m-MTDATA (300 Å)/α-NPD (200 Å)/Alq₃ (300 Å)/BMB-nT (n = 2 and 3) (200 Å)/MgAg device.

V vs. Ag/Ag⁺ (0.01 mol dm⁻³)), but the optical band gap of TMB-TB (3.2 eV) estimated from the absorption threshold is 0.5 eV wider than that of Alq₃ (2,7 eV). Therefore, TMB-TB is expected to function as an electron transporter with a better hole-blocking ability than Alq₃.

The TDAB analogues, tris(phenylquinoxaline) (TPQ)[61] and a benzimidazole derivative (TPBI)[62] have also been reported to function as electron transporters. Table 4.4 lists the Tgs and reduction potentials of electron transporters.

Multilayer OLEDs using BMB-2T or BMB-3T as an electron-transporting material, Alq₃ as an emitting material with electron-transporting properties, and m-MTDATA and α-NPD as hole-transporting materials, ITO/m-MTDATA (300 Å)/α-NPD (200 Å)/Alq₃ (300 Å)/BMB-nT (n = 2 and 3) (200 Å)/MgAg (Figure 4.3), emitted bright green light originating from Alq₃,

Electron-Transporting Materials

Alq₃

t-Bu-PBD

OXD-7

TAZ

TPOB

spiro-PBD

PYSPY

dendrimer oxadiazole

Electron-Transporting Materials (continued)

TPQ

$R_1 = R_2 = CF_3 / H$

TPBI

BMB-1T

BMB-2T

BMB-3T

TMB-TB

exhibiting approximately 10 to 20% higher luminous and quantum efficiencies and 1.6 to 1.8 times higher maximum luminance than the OLED without the BMB-nT layer, ITO/m-MTDATA (300 Å)/α-NPD (200 Å)/Alq$_3$ (500 Å)/MgAg. These results show that both BMB-nT function as excellent electron-transporting materials, facilitating electron injection from the Mg electrode in OLEDs.[57]

The better hole-blocking ability of TMB-TB as the electron transporter permitted the blue emission from p-TTA in OLEDs.[60] That is, the device, ITO/m-MTDATA (300 Å)/p-TTA (400 Å)/TMB-TB (300 Å)/MgAg, emitted bright blue light resulting from p-TTA. Contrastingly, the device using Alq_3 as the electron-transport layer, ITO/m-MTDATA (300 Å)/p-TTA (400 Å)/Alq_3 (300 Å)/MgAg, emitted green light originating from Alq_3 instead of the desired blue emission from p-TTA. At higher drive voltage, the green light from Alq_3 still prevailed, although blue light from p-TTA also contributed to the overall emission. This result indicates that holes injected from the ITO electrode into the p-TTA (HOMO level: 5.6 eV)[61] layer via the m-MTDATA layer enter the Alq_3 layer due to the lack of efficient hole blocking by Alq_3 and that TMB-TB serves both as the electron transporter and the hole blocker in blue-emitting OLEDs using p-TTA as the emitter.

Likewise, TPOB with better electron-blocking ability than Alq_3 enabled the blue emission from BMA-1T in OLEDs.[18]

4.5 HOLE-BLOCKING AMORPHOUS MOLECULAR MATERIALS

Generally, hole and electron transporters in OLEDs play a role of blocking electrons and holes, respectively, from escaping from the emitting layer as well as facilitating charge injection from the electrodes. When materials with hole-transporting properties are used as emitters in OLEDs, the presence of the electron-transport layer with an effective hole-blocking ability is required because of the large energy barrier for electron injection from the cathode. However, there have been few electron transporters that function as effective hole blockers; a well-known electron transporter Alq_3 does not necessarily function well as an effective hole blocker for emitters with hole-transporting properties as described in Section 10.4. A promising approach is the use of hole blockers inserted between the emitting layer and the electron-transport layer in OLEDs, where the hole-blocking layer and the electron-transport layer play each role of blocking holes from escaping from the emitting layer to confine holes within the emitting layer, and facilitating electron injection from the cathode, respectively. Such hole blockers are expected to enable the fabrication of OLEDs using emitters with hole-transporting properties.

Hole-blocking materials for use in OLEDs should fulfill several requirements. They should possess proper energy levels of the highest occupied molecular orbital (HOMO) and the lowest unoccupied molecular orbital (LUMO) to be able to block holes from escaping from the emitting layer into the electron-transport layer but to pass on electrons from the electron-transport layer to the emitting layer. In other words, the difference in the HOMO energy levels between the emitting material and the hole-blocking material should be much larger than that in their LUMO energy levels. Their cathodic reduction processes should be reversible to form stable anion radicals. They should not form any exciplexes with emitting materials having hole-transporting properties. In addition, they should form thermally and morphologically stable, uniform amorphous thin films.

It has recently been reported that bathocuproine serves as a hole blocker in blue-emitting OLEDs using α-NPD as the emitter;[64] however, its morphological and thermal stabilities are poor. In addition, it tends to form exciplexes with a number of materials with hole-transporting properties such as TPD, m-MTDATA, and others, stronger exciplex emissions taking place in the longer wavelength region.[65]

Two families of hole-blocking amorphous molecular materials have recently been developed. One is a family of triarylbenzenes, e.g., 1,3,5-tri(biphenyl-4-yl)benzene (TBB),[65,66] 1,3,5-tris(4-fluorobiphenyl-4´-yl)benzene (F-TBB),[67,68] 1,3,5-tris (9,9-dimethylfluoren-2-yl)benzene (TFB),[68] and 1,3,5-tris[4-(9,9-dimethylfluoren-2-yl)phenyl]benzene (TFPB).[68] Another class of compounds are boron-containing compounds, e.g., tris(2,3,5,6-tetramethylphenyl)borane (TPhB),[69] tris(2,3,5,6-tetramethyl-1,1´;4´,1″-terphenyl-4-yl)borane (TTPhB),[69] and tris[2,3,5,6-tetramethyl-4-(1,1´;3´,1″-terphenyl-5´-yl)phenyl] borane (TTPhPhB).[69]

Table 4.5 Glass-Transition Temperatures (Tgs) and Reduction Potentials ($E_{1/2}^{red}$) of Hole-Blocking Amorphous Molecular Materials

Material	Tg/°C	Ref.	Material	Tg/°C	$E_{1/2}^{red}$ [a]	Ref.
TBB	88	68	TPhB	127	−2.51	69
F-TBB	87	67	TTPhB	163	−2.48	69
TFB	133	68	TTPhPhB	183	−2.49	69
TFPB	149	68				

[a] vs. Ag/Ag+ (0.01 mol dm−3).

Hole-Blocking Amorphous Molecular Materials

TBB   F-TBB   TFB   TFPB

TPhB   TTPhB   TTPhPhB

These compounds readily form amorphous glasses with well-defined Tgs and possess weaker electron-accepting properties than usual electron transporters, as shown in Table 4.5.

High-performance blue- and blue-violet-emitting OLEDs have been developed by using these hole blockers and p-TTA, α-NPD, and TPD as emitters.[67–69]

4.6 CHARGE TRANSPORT IN AMORPHOUS MOLECULAR MATERIALS

As charge transport is involved in the operation processes of OLEDs, it is important to understand charge transport in amorphous molecular materials. Charge transport in organic disordered systems has been studied extensively with regard to polymers and molecularly doped polymer systems, where low-molecular-weight organic materials are dispersed in binder polymers. It is known that charge transport in organic disordered systems such as amorphous polymers and molecularly doped polymer systems generally show the following characteristic features: (1) the drift mobility of charge carriers is very small, on the order from ~10^{-8} to 10^{-4} cm^{2}V^{-1}s^{-1}, (2) the drift mobility of charge carriers is electric-field dependent, and (3) charge transport is a thermally activated process. It has generally been accepted that charge transport in organic disordered systems takes place by a hopping process. A few models have been proposed to explain the temperature and electric-field dependencies of drift mobilities of disordered systems, which include the Poole–Frenkel model,[70] the small-polaron model,[71-73] and the disorder formalism.[74,75]

It has been revealed that charge transport in molecularly doped polymer systems is greatly dependent on the binder polymer. In fact, the hole drift mobility of molecularly doped polymer systems has been reported to vary by two orders of magnitude depending on the binder polymer.[76-78] Charge-dipole interactions between a charge on the transport molecule and the binder polymer are thought to cause the fluctuation of both the hopping site energy and the overlap integral. To clarify the intrinsic charge-transport properties of low-molecular-weight organic compounds in the disordered system, charge transport in the amorphous glassy state of low-molecular-weight organic materials in the absence of a binder polymer needs to be investigated.

Creation of amorphous molecular materials has enabled the investigation of charge transport in the glassy state of small organic molecules. The charge carrier drift mobilities of a variety of amorphous molecular materials have been

Table 4.6 Hole Drift Mobilities of Some Molecular Glasses[a]

Material	μ_h cm^{2}V^{-1}s^{-1}	Ref.	Material	μ_h cm^{2}V^{-1}s^{-1}	Ref.
m-MTDATA	3.0×10^{-5}	80	TPD	1.0×10^{-3}	29
o-PTDATA	7.2×10^{-5}	42	α-NPD	8.8×10^{-4}	84
m-PTDATA	1.7×10^{-5}	42	o-BPD	6.5×10^{-4}	85
p-PMTDATA	3.0×10^{-5}	42	m-BPD	5.3×10^{-5}	85
TFATA	1.7×10^{-5}	26	p-BPD	1.0×10^{-3}	85
p-DPA-TDAB	1.4×10^{-5}	79	FFD	4.1×10^{-3}	26
m-MTDAPB	1.6×10^{-5}	13	DPH	2.2×10^{-4}	11,86–89
TBA	1.5×10^{-4}	12,81	DPMH	4.5×10^{-5}	87,88,90
o-TTA	7.9×10^{-4}	82	M-DPH	6.4×10^{-5}	86–88
m-TTA	2.3×10^{-5}	79	ECH	3.7×10^{-5}	89,91
p-TTA	8.8×10^{-4}	12,79	ECMH	2.6×10^{-6}	91
TTPA	4.8×10^{-3}	83	M-ECH	4.4×10^{-6}	79
TPTPA	1.0×10^{-2}	4	spiro-TAD	2.5×10^{-4}	92

[a] Measured at an electric field of ~10^5 V cm^{-1} at room temperature.

determined, and their electric-field and temperature dependencies have been analyzed in terms of the disorder formalism:

$$\mu(\sigma,\Sigma,E) = \mu_0 \exp\left[-\left(\frac{2\sigma}{3kT}\right)^2\right]\exp\left\{C\left[\left(\frac{\sigma}{kT}\right)^2 - \Sigma^2\right]E^{1/2}\right\}, \quad (4.1)$$

where σ and Σ are the parameters that characterize the degree of energetic disorder and positional disorder, respectively, μ_0 is a hypothetical mobility in the absence of energetic disorder, C is a constant, k is the Boltzmann constant, E is the electric field, and T is the temperature.

Table 4.6 lists the hole drift mobilities of the molecular glasses of m-MTDATA, 4,4′,4″-tris[biphenyl-2-yl(phenyl) amino]triphenylamine (o-PTDATA), 4,4′,4″-tris[biphenyl-3-yl (phenyl)amino]triphenylamine (m-PTDATA), 4,4′,4″-tris[biphenyl-4-yl(3′-methylphenyl)amino]triphenylamine (p-PMTDATA), TFATA, p-DPA-TDAB, m-MTDAPB, tri(biphenyl-4-yl)amine (TBA), o-TTA, m-TTA, p-TTA, tris[4-(2-thienyl)phenyl]amine (TTPA), tris[4-(5-phenylthiophen-2-yl)phenyl]amine (TPTPA),

TPD, α-NPD, o-BPD, m-BPD, p-BPD, FFD, 4-diphenylamino-benzaldehyde diphenylhydrazone (DPH), 4-diphenylamino-benzaldehyde methylphenylhydrazone (DPMH), 4-diphenyl-aminoacetophenone diphenylhydrazone (M-DPH), 9-ethyl-carbazole-3-carbaldehyde diphenylhydrazone (ECH), 9-ethyl-carbazole-3-carbaldehyde methylphenylhydrazone (ECMH), 3-acethyl-9-ethylcarbazole diphenylhydrazone (M-ECH), and spiro-TAD. It is shown that the hole drift mobility of the molecular glass is strongly dependent on the molecular structure and varies greatly from 10^{-6} to 10^{-2} cm^2 V^{-1} s^{-1} at an electric field of 1.0×10^5 V cm^{-1} at room temperature.[65,79]

Very few studies have been made of the relationship between molecular structure and charge-transport properties. It has been reported that the hole drift mobilities of the o- and p-isomers of TTA and BPD are more than one order of magnitude larger than those of the corresponding m-isomers (Table 4.6). The analysis of the electric-field and temperature dependencies of the drift mobilities of these materials in terms of the disorder formalism shows that the difference in the energetic disorder is responsible for the difference in the hole drift mobility among o-, m-, and p- isomers: the σ value significantly increases in the order o-TTA (0.059 eV) < p-TTA (0.071eV) < m-TTA (0.093eV) and o-BPD (0.071 eV) < p-BPD (0.075 eV) < m-BPD (0.105 eV).

It is understood that the energetic disorder σ, namely, the fluctuation of hopping site energy, is the superposition of both intramolecular and intermolecular contributions. The intramolecular contribution arises from the variation of molecular geometry caused by bond rotation, and the inter-molecular contribution arises from the fluctuation of polar-ization energy due to charge-dipole and van der Waals interactions. The difference in the variation of molecular geometry may also lead to the difference in the fluctuation of the polarization energy. It is thought that the variation of molecular geometry caused by the rotation along the C–C and C–N bonds is responsible for the difference in the σ value among the o-, m-, and p-isomers. That is, the variation of the molecular geometry of o-isomer may be much smaller than those of m- and p-isomer because of the restricted internal rotation for o-isomer due to the large steric constraint. The number of conformers formed by rotation is suggested to be smaller for p-isomer than for m-isomer; this decreases the variation of molecular geometry in p-isomer relative to m-isomer. This result suggests that decreasing the variation of the molecular geometry by restricting internal rotation or by increasing the symmetry of the molecule results in the decrease of σ value.

The drift mobility of charge carriers generally increases with increasing electric field, following the relation $\mu \propto \exp(SE^{1/2})$, where S is a coefficient; however, a phenomenon such as the decrease of the mobility with increasing electric field has been observed for a few molecularly doped polymer systems at low concentrations $(5 \sim 25 \text{ wt\%})$[93,94] and for the molecular glass of o-TTA.[82] The hole drift mobility of the o-TTA glass increased with increasing electric field in a temperature region below 285 K, following the electric-field dependence of $\exp(SE^{1/2})$; however, the hole drift mobility began to decrease with increasing electric field in a temperature region above 285 K.

The negative electric-field dependence of charge-carrier drift mobility has been explained as due to the presence of a large positional disorder in terms of the disorder formalism and as arising from the contribution of the diffusion process of charge carriers against the direction of electric field due to the presence of a positional disorder Σ caused by the fluctuation of the intermolecular π-electron overlap. That is, faster detour routes against the electric field exist due to the presence of the positional disorder Σ and the rate of the diffusion of charge carriers through the detour routes, which increases with increasing temperature because the mean energy of charge carriers increases with increasing temperature, gradually decreases with increasing electric field; this leads to the negative electric-field dependence of charge-carrier drift mobility. Equation 4.1 shows that the slope of the plots of ln μ_h vs. $E^{1/2}$ is proportional to $C[(\sigma/kT)^2 - \Sigma^2]$ and that both the magnitude and the sign of the electric-field dependence are determined by the balance between the two disorder parameters, energetic disorder (σ) and positional disorder (Σ). When

Table 4.7 Electron Drift Mobilities of Several Electron-Transporting Materials

Material	μ_e/cm^{2}V^{-1}s^{-1}	Ref.	Material	μ_e/cm^{2}V^{-1}s^{-1}	Ref.
Alq$_3$	1.4×10^{-6}	95	TPQ	1×10^{-4}	98
Alq$_3$	4×10^{-6}	96	BPhen[b]	4.2×10^{-4}	99
PyPySPyPy[a]	2.0×10^{-4}	97	TPBI	7.5×10^{-6}	100

[a] 2,5-Bis[6'-(2',2''-bipyridyl)]-1,1-dimethyl-3,4-diphenylsilole.
[b] 4,7-Diphenyl-1,10-phenanthroline.

$(\sigma/kT)^2$ becomes smaller than Σ^2, the slope of the electric-field dependence of the charge-carrier drift mobility should become negative.

It has been reported that a large Σ value of ~5 is responsible for the phenomenon of the negative electric-field dependence of the hole drift mobility for the lightly doped polymer systems.[93,94] The analysis of the electric-field and temperature dependencies of the drift mobilities of o-TTA in terms of the disorder formalism gave the following hole-transport parameters in Equation 4.1 for the o-TTA molecular glass: $\mu_0 = 9.7 \times 10^{-3}$ cm^{2}V^{-1}s^{-1}, $\sigma = 0.059$ eV, $\Sigma = 2.4$, and C $= 4.4 \times 10^{-4}$ (cm V^{-1})$^{1/2}$.[82] The molecular glass of o-TTA is characterized by much smaller values of both σ and Σ. The small σ value, as well as the presence of the positional disorder Σ, is responsible for the negative electric-field dependence of the hole drift mobility observed for the molecular glass of o-TTA.

The electron mobilities of electron transporters have been reported to be in the range from 10^{-6} to 10^{-4} cm^{2}V^{-1}s^{-1} at room temperature.[95–100] Table 4.7 lists the electron mobilities of several electron-transporting materials for use in OLEDs.

REFERENCES

1. C.W. Tang and S.A. VanSlyke, *Appl. Phys. Lett.*, 51, 913, 1987.

2. R.H. Friend, R.W. Gymer, A.B. Holmes, J.H. Burroughes, R.N. Marks, C.Taliani, D.D.C. Bradley, D.A. Dos Santos, J.L. Brédas, M. Lögdlund, and W.R. Salaneck, *Nature*, 397, 121, 1999.

3. A. Kraft, A.C. Grimsdale, and A.B. Holmes, *Angew. Chem. Int. Ed. Engl.*, 37, 403, 1998.

4. Y. Shirota, *J. Mater. Chem.*, 10, 1, 2000 and references cited therein.

5. U. Mitschke and P. Bäuerle, *J. Mater. Chem.*, 10, 1471, 2000.

6. Y. Shirota, *SPIE-Int. Soc. Opt. Eng.*, 3148, 186, 1997.

7. Y. Shirota, T. Kobata, and N. Noma, *Chem. Lett.*, 1145, 1989.

8. A. Higuchi, H. Inada, T. Kobata, and Y. Shirota, *Adv. Mater.*, 3, 549, 1991.

9. W. Ishikawa, H. Inada, H. Nakano, and Y. Shirota, *Chem. Lett.*, 1731, 1991.

10. W. Ishikawa, H. Inada, H. Nakano, and Y. Shirota, *Mol. Cryst. Liq. Cryst.*, 211, 431, 1992.

11. K. Nishimura, T. Kobata, H. Inada, and Y. Shirota, *J. Mater. Chem.*, 1, 897, 1991.

12. A. Higuchi, K. Ohnishi, S. Nomura, H. Inada, and Y. Shirota, *J. Mater. Chem.*, 2, 1109, 1992.

13. H. Inada and Y. Shirota, *J. Mater. Chem.*, 3, 319, 1993.

14. E. Ueta, H. Nakano, and Y. Shirota, *Chem. Lett.*, 2397, 1994.

15. H. Kageyama, K. Itano, W. Ishikawa, and Y. Shirota, *J. Mater. Chem.*, 6, 675, 1996.

16. T. Noda, H. Ogawa, N. Noma, and Y. Shirota, *Adv. Mater.*, 9, 239, 1997.

17. T. Noda, H. Ogawa, N. Noma, and Y. Shirota, *Adv. Mater.*, 9, 720, 1997.

18. T. Noda, H. Ogawa, N. Noma, and Y. Shirota, *J. Mater. Chem.*, 9, 2177, 1999.

19. Y. Shirota, K. Moriwaki, S. Yoshikawa, T. Ujike, and H. Nakano, *J. Mater. Chem.*, 8, 2579, 1998.

20. K. Katsuma and Y. Shirota, *Adv. Mater.*, 10, 223, 1998.

21. J. Louie and J.F. Hartwig, *J. Am. Chem. Soc.*, 119, 11695, 1997.

22. J. Salbeck, N. Yu, J. Bauer, F. Weissörtel, and H. Bestgen, *Synth. Met.*, 91, 209, 1997.

23. W.J. Oldham, Jr., R.J. Lachicotte, and G.C. Bazan, *J. Am. Chem. Soc.*, 120, 2987, 1998.

24. Y. Kuwabara, H. Ogawa, H. Inada, N. Noma, and Y. Shirota, *Adv. Mater.*, 6, 677, 1994.

25. H. Ogawa, H. Inada, and Y. Shirota, *Macromol. Symp.*, 125, 171, 1997.

26. K. Okumoto and Y. Shirota, *Chem. Lett.*, 1034, 2000.

27. K. Okumoto, H. Doi, and Y. Shirota, *J. Photopolym. Sci. Technol.*, 15, 239, 2002.

28. M.R. Robinson, S. Wang, A. J. Heeger, and G.C. Bazan, *Adv. Funct. Mater.*, 11, 413, 2001.

29. M. Stolka, J.F. Yanus, and D.M. Pai, *J. Phys. Chem.*, 88, 4707, 1984.

30. C. Adachi, T. Tsutsui, and S. Saito, *Appl. Phys. Lett.*, 55, 1489, 1989.

31. E.-M. Han, L.-M. Do, Y. Nidome, and M. Fujihira, *Chem. Lett.*, 969, 1994.

32. W. Ishikawa, K. Noguchi, Y. Kuwabara, and Y. Shirota, *Adv. Mater.*, 5, 559, 1993.

33. K. Itano, T. Tsuzuki, H. Ogawa, S. Appleyard, M.R. Willis, and Y. Shirota, *IEEE Trans. Electron Devices*, 44, 1218, 1997.

34. H. Inada, Y. Yonemoto, T. Wakimoto, K. Imai, and Y. Shirota, *Mol. Cryst. Liq. Cryst.*, 280, 331, 1996.

35. H. Tanaka, S. Tokito, Y. Taga, and A. Okada, *Chem. Commun.*, 2175, 1996.

36. J. Salbeck, N. Yu, J. Bauer, F. Weissörtel, and H. Bestgen, *Synth. Met.*, 91, 209, 1997.

37. C.-W. Ko and Y.-T. Tao, *Synth. Met.*, 126, 37, 2002.

38. N.-X. Hu, S. Xie, Z.D. Popovic, B. Ong, and A.-M. Hor, *Synth. Met.*, 111–112, 421, 2000.

39. S.A. VanSlyke, C.H. Chen, and C.W. Tang, *Appl. Phys. Lett.*, 69, 2160, 1996.

40. D.F. O'Brien, P.E. Burrows, S.R. Forrest, B.E. Koene, D.E. Loy, and M.E. Thompson, *Adv. Mater.*, 10, 1108, 1998.

41. B.E. Koene, D.E. Loy, and M.E. Thompson, *Chem. Mater.*, 10, 2235, 1998.

42. Y. Shirota, K. Okumoto, and H. Inada, *Synth. Met.*, 111–112, 387, 2000.

43. K. Okumoto and Y. Shirota, *Mater. Sci. Eng.*, B85, 135, 2001.

44. Y. Shirota, Y. Kuwabara, H. Inada, T. Wakimoto, H. Nakada, Y. Yonemoto, S. Kawami, and K. Imai, *Appl. Phys. Lett.*, 65, 807, 1994.

45. Y. Shirota, Y. Kuwabara, D. Okuda, R. Okuda, H. Ogawa, H. Inada, T. Wakimoto, H. Nakada, Y. Yonemoto, S. Kawami, and K. Imai, *J. Luminescence*, 72–74, 985, 1997.

46. Y. Hamada, T. Sano, K. Shibata, and K. Kuroki, *Jpn. J. Appl. Phys.*, 34, L824, 1995.

47. H. Murata, C.D. Merritt, H. Inada, Y. Shirota, and Z.H. Kafafi, *Appl. Phys. Lett.*, 75, 3252, 1999.

48. M. Brinkmann, G. Gadret, M. Muccini, C. Taliani, N. Masciocchi, and A. Sironi, *J. Am. Chem. Soc.*, 122, 5147, 2000.

49. N. Donzé, P. Péchy, M. Gräzel, M. Schaer, and L. Zuppiroli, *Chem. Phys. Lett.*, 315, 405, 1999.

50. D. O'Brien, A. Bleyer, D.G. Lidzey, and D.D.C. Bradley, *J. Appl. Phys.*, 82, 2662, 1997.

51. J. Kido, C. Ohtaki, K. Hongawa, K. Okuyama, and K. Nagai, *Jpn. J. Appl. Phys.*, 32, L917, 1993.

52. K. Tamao, M. Uchida, T. Izumizawa, K. Furukawa, and S. Yamaguchi, *J. Am. Chem. Soc.*, 118, 11974, 1996.

53. J. Bettenhausen and P. Strohriegl, *Adv. Mater.*, 8, 507, 1996.

54. H. Ogawa, R. Okuda, and Y. Shirota, *Mol. Cryst. Liq. Cryst.*, 315, 187, 1998.

55. J. Bettenhausen, M. Greczmiel, M. Jandke, and P. Strohriegl, *Synth. Met.*, 91, 223, 1997.

56. J. Salbeck and F. Weissörtel, *Macromol. Symp.*, 125, 121, 1997.

57. T. Noda and Y. Shirota, *J. Am. Chem. Soc.*, 120, 9714, 1998.

58. A.J. Mäkinen, I.G. Hill, T. Noda, Y. Shirota, and Z.H. Kafafi, *Appl. Phys. Lett.*, 78, 670, 2001.

59. A.J. Mäkinen, I. G. Hill, M. Kinoshita, T. Noda, Y. Shirota, and Z.H. Kafafi, *J. Appl. Phys.*, 91, 5456, 2002.

60. M. Kinoshita and Y. Shirota, *Chem. Lett.*, 614, 2001.

61. M. Jandke, P. Strohriegl, S. Berleb, E. Werner, and W. Brütting, *Macromolecules*, 31, 6434, 1998.

62. Z. Gao, C.S. Lee, I. Bello, S.T. Lee, R.-M. Chen, T.-Y. Luh, J. Shi, and C.W. Tang, *Appl. Phys. Lett.*, 74, 865, 1999.

63. H. Ogawa, K. Ohnishi, and Y. Shirota, *Synth. Met.*, 91, 243, 1997.

64. Y. Kijima, N. Asai, and S. Tamura, *Jpn. J. Appl. Phys.*, 38, 5274, 1999.

65. Y. Shirota, M. Kinoshita, and K. Okumoto, *SPIE Int. Soc. Opt. Eng.*, 4464, 203, 2002.

66. T. Noda, H. Ogawa, and Y. Shirota, *Adv. Mater.*, 11, 283, 1999.

67. K. Okumoto and Y. Shirota, *Appl. Phys. Lett.*, 79, 1231, 2001.

68. K. Okumoto and Y. Shirota, *Chem. Mater.*, 15, 699, 2003.

69. M. Kinoshita, H. Kita, and Y. Shirota, *Adv. Funct. Mater.*, 12, 780, 2002.

70. W.D. Gill, *J. Appl. Phys.*, 43, 5033, 1972.

71. D. Emin, in P.G. Le Comber and J. Mort, Eds., *Electronic and Structural Properties of Amorphous Semiconductors*, Academic Press, New York, 1973, chap. 7.

72. L.B. Schein and J.X. Mack, *Chem. Phys. Lett.*, 149, 109, 1988.

73. A. Peled, L.B. Schein, and D. Glatz, *Phys. Rev. B*, 41, 10835, 1990.

74. H. Bässler, *Phys. Status Solidi b,* 107, 9, 1981.

75. H. Bässler, *Phys. Status Solidi b*, 175, 15, 1993.

76. T. Sasakawa, T. Ikeda, and S. Tazuke, *J. Appl. Phys.*, 65, 2750, 1989.

77. P.M. Borsenberger, *J. Appl. Phys.*, 68, 5188, 1990.

78. P.M. Borsenberger, E.H. Magin, and J.J. Fitzgerald, *J. Phys. Chem.*, 97, 8250, 1993.

79. Y. Shirota, S. Nomura, and H. Kageyama, *SPIE Int. Soc. Opt. Eng.*, 3476, 132, 1998.

80. C. Geibeler, H. Antoniadis, D.D.C. Bradley, and Y. Shirota, *Appl. Phys. Lett.*, 72, 2448, 1998.

81. H. Inada, K. Ohnishi, S. Nomura, A. Higuchi, H. Nakano, and Y. Shirota, *J. Mater. Chem.*, 4, 171, 1994.

82. H. Kageyama, K. Ohnishi, S. Nomura, and Y. Shirota, *Chem. Phys. Lett.*, 277, 173, 1997.

83. J. Sakai, H. Kageyama, S. Nomura, H. Nakano, and Y. Shirota, *Mol. Cryst. Liq. Cryst.*, 296, 445, 1997.

84. Z. Deng, S.T. Lee., D.P. Webb, Y.C. Chen, and W.A. Gambling, *Synth. Met.*, 107, 107, 1999.

85. K. Okumoto, K. Wayaku, T. Noda, H. Kageyama, and Y. Shirota, *Synth. Met.*, 111–112, 473, 2000.

86. K. Nishimura, H. Inada, T. Kobata, Y. Matsui, and Y. Shirota, *Mol. Cryst. Liq. Cryst.*, 217, 235, 1992.

87. S. Nomura, K. Nishimura, and Y. Shirota, *Mol. Cryst. Liq. Cryst.*, 253, 79, 1994.

88. S. Nomura, K. Nishimura, and Y. Shirota, *Thin Solid Films*, 273, 27, 1996.

89. S. Nomura and Y. Shirota, *Chem. Phys. Lett.*, 268, 461, 1997.

90. S. Nomura, K. Nishimura, and Y. Shirota, *Mol. Cryst. Liq. Cryst.*, 313, 247, 1998.

91. S. Nomura and Y. Shirota, *Mol. Cryst. Liq. Cryst.*, 315, 217, 1998.

92. U. Bach, K. De Cloedt, H. Spreitzer, and M. Grätzel, *Adv. Mater.*, 12, 1060, 2000.

93. N. Novo, M. Van der Auweraer, F.C. De Schryver, P. Borsenberger, and H. Bässler, *Phys. Status Solidi b*, 177, 223, 1993.

94. R.H. Young, J.A. Sinicropi, and J.J. Fitzgerald, *J. Phys. Chem.*, 99, 9497, 1995.

95. R.G. Kepler, P.M. Beeson, S.J. Jacobs, R.A. Anderson, M.B. Sinclair, V.S. Valencia, and P.A. Cahill, *Appl. Phys. Lett.*, 66, 3618, 1995.

96. G.G. Malliaras, Y. Shen, D.H. Dunlap, H. Murata, and Z.H. Kafafi, *Appl. Phys. Lett.*, 79, 2582, 2001.

97. H. Murata, G.G. Malliaras, M. Uchida, Y. Shen, and Z.H. Kafafi, *Chem. Phys. Lett.*, 339, 161, 2001.

98. M. Redecker, D.D.C. Bradley, M. Jandke, and P. Strohriegl, *Appl. Phys. Lett.*, 75, 109, 1999.

99. S. Naka, H. Okada, H. Onnagawa, and T. Tsutsui, *Appl. Phys. Lett.*, 76, 197, 2000.

100. T.C. Wong, J. Kovac, C.S. Lee, L.S. Hung, and S.T. Lee, *Chem. Phys. Lett.*, 334, 61, 2001.

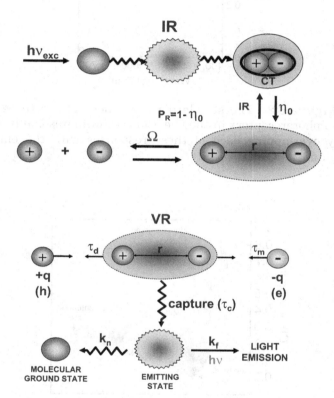

Color Figure 2.11 Initial (IR) and volume-controlled (VR) recombination (for explanation, see text).

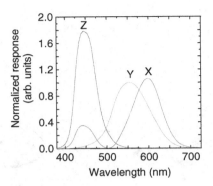

Color Figure 6.2 The standardized response of the three-color-sensitive photoreceptors in the eye. Color coordinates can be calculated by overlapping each of these responses with the emission spectrum of the OLED.

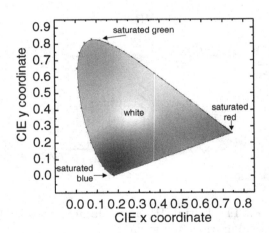

Color Figure 6.3 The CIE chart used to quantify the sensation of color. A full-color display should possess red, green, and blue pixels. Positioning these on the chart above forms a triangle that circumscribes all the colors available to the display as a combination of the red, green, and blue elements. Thus, the individual colored pixels should be as saturated as possible (i.e., their emission should contain the minimum amount of white) and they should be located on the perimeter of the chart.

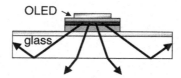

Color Figure 6.4 The fraction of photons emitted in the forward, or viewing, direction is reduced by absorption losses and waveguiding within the device and its substrate.

A Dipole-dipole coupled (Förster) energy transfer

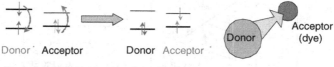

Donor ˙ Acceptor   Donor Acceptor ˙   Acceptor (dye)

Exciton non-radiatively transferred by dipole-dipole coupling if transitions are allowed.   up to ~ 100 Å

B Electron exchange energy transfer

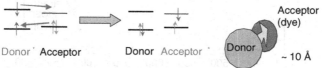

Donor ˙ Acceptor   Donor Acceptor ˙   Acceptor (dye)

Exciton hops from donor to acceptor.   ~ 10 Å

Color Figure 6.6 (A) A schematic representation of Förster energy transfer, a mechanism for the rapid transfer of energy between molecules. If both transitions on the donor and acceptor are allowed, then the range of transfer may extend to 100 Å. This is the dominant method for the transfer of singlet excitons. (B) If one of the transitions is disallowed, then Förster energy transfer is not possible; for example, we may not excite the triplet state of the acceptor. However, it is possible to transfer triplets by exciton hopping from one molecule to the next. The rate constant of this process is determined by the rate of Marcus electron and hole transfer; see Closs et al.[16]

Mg:Ag cathode

100 Å Alq$_3$

400 Å PtOEP in Alq$_3$

350 Å α-NPD

60 Å CuPc

Indium tin oxide

Color Figure 6.8 The device structure used to demonstrate PtOEP electrophosphorescence.

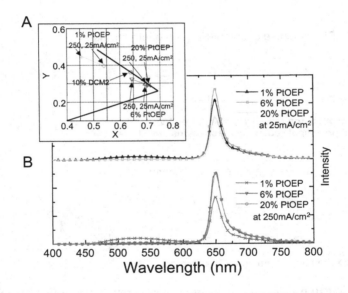

Color Figure 6.9 Spectra of the OLEDs with different molar concentrations of PtOEP at different current densities. (A) 1, 6, and 20% PtOEP in Alq$_3$ OLEDs at 25 mA/cm^2 (B) 1, 6, and 20% PtOEP in Alq$_3$ OLEDs at 250 mA/cm^2. Note the increased Alq$_3$ emission at 530 nm in the 1% PtOEP OLED. Inset: CIE chromaticity coordinates for the devices in Figure 6.8 at the specified current densities. Only the red corner of the CIE diagram is shown. Note the trend from saturated red to orange with increasing current. The 6% PtOEP in the Alq$_3$ device at 25 mA/cm^2 has a luminance of 100 cd/m^2. The DCM2 result is from Bulović et al.[15]

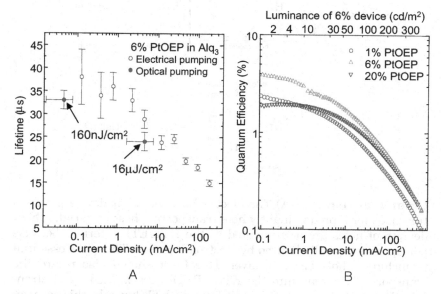

Color Figure 6.10 (A) Phosphorescent lifetime of the 6% PtOEP in Alq$_3$ OLED as function of current density. Both electrical and optical pumping show a decrease in PtOEP lifetime with increasing excitation strength, indicative of a bimolecular quenching process such as triplet–triplet annihilation. (B) Quantum efficiency of PtOEP emission as a function of doping concentration and current density. The top axis shows the luminance of the 6% PtOEP in Alq$_3$ device.

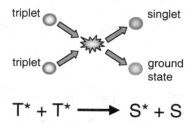

$$T^* + T^* \longrightarrow S^* + S$$

Color Figure 6.11 Total spin is conserved if two excited triplets combine and form an excited singlet and ground-state singlet exciton. The process destroys at least one triplet exciton and is known as triplet–triplet annihilation.

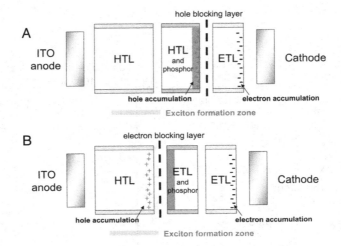

Color Figure 6.12 (A) The electrophosphorescent device architecture used for a predominantly hole-transporting host material. A high density of holes are accumulated at the HTL/ETL interface because transport is typically limited by electron-injection. Thus, it is essential to include a hole blocking layer. This layer should also retard the transport of excitons into the ETL. (B) Devices limited by electron-injection are better suited to emit from an ETL layer. In this case, an electron blocking layer is not as important since the majority of electron charge is located near the cathode. Less charge in the exciton formation zone should also reduce losses from exciton-charge quenching.

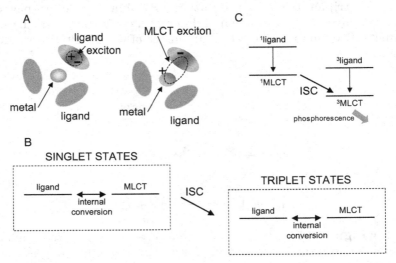

Color Figure 6.14 In metal–organic complexes the emissive state is generally a mixture of a ligand-centered exciton and a metal–ligand charge transfer (MLCT) exciton. The MLCT state has superior singlet-triplet mixing because it overlaps with the heavy-metal atom.

fac tris(2-phenylpyridine) iridium

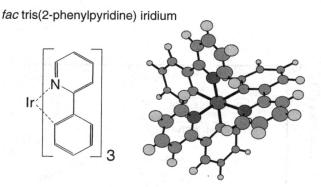

Color Figure 6.15 The chemical structure of *fac* tris(2-phenylpy-ridine) iridium.

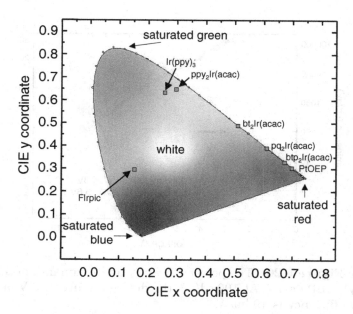

Color Figure 6.18 The chromaticity coordinates of six Ir-based phosphors together with the coordinates of PtOEP.

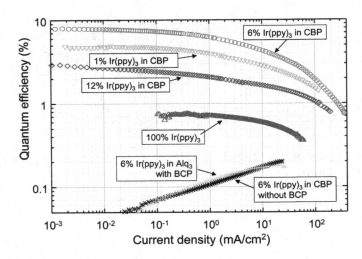

Color Figure 6.19 The external quantum efficiency of OLEDs using Ir(ppy)$_3$:CBP luminescent layers. Peak efficiencies are observed for a mass ratio of 6% Ir(ppy)$_3$:CBP. The efficiency of a Ir(ppy)$_3$:CBP device grown without a blocking layer is also shown.

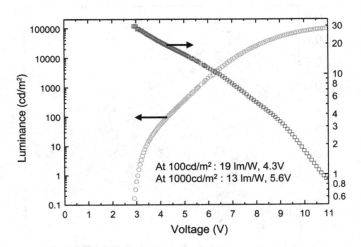

Color Figure 6.20 The power efficiency and luminance of the 6% Ir(ppy)$_3$:CBP device. At 100 cd/m^2, the device requires 4.3 V and its power efficiency is 19 lm/W.

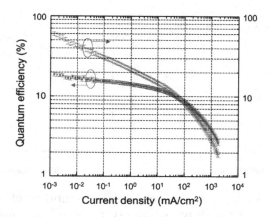

Color Figure 6.21 The quantum and power efficiency of 12% ppy$_2$Ir(acac) in TAZ. Inset: The chemical structure of the electron-transporting host material TAZ.

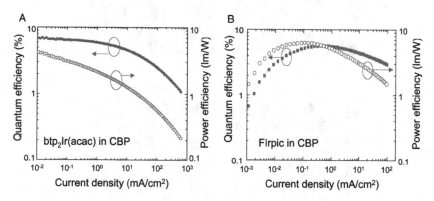

Color Figure 6.22 The quantum and power efficiency of 6% FIrpic in CBP.

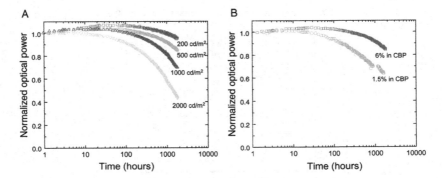

Color Figure 6.24 (A) The operational stability of electrophosphorescent devices with an active layer of 6% Ir(ppy)$_3$ in CBP and a BAlq hole-blocking layer. The projected life to 50% of an initial brightness of 500 cd/m^2 is 10,000 h. (B) A comparison of the operational stability of devices with 6 and 1.5% Ir(ppy)$_3$ in CBP at an initial brightness of 500 cd/m^2. The difference in lifetime reflects the quantum efficiency of the two devices. (After Ray Kwong et al.[52])

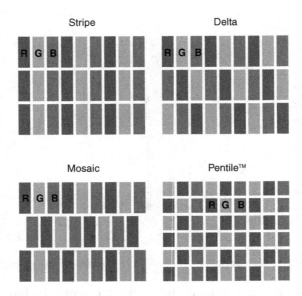

Color Figure 7.1 Common pixel array patterns.

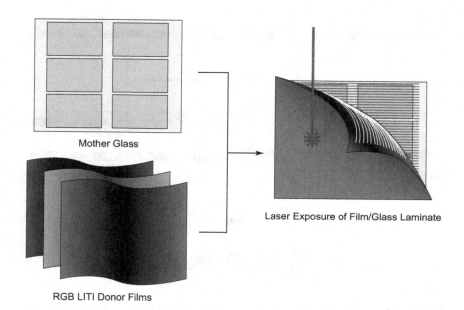

Mother Glass

RGB LITI Donor Films

Laser Exposure of Film/Glass Laminate

Color Figure 7.6 The LITI process.

Color Figure 7.11 Samsung SDI 17-in. D LITI AMOLED.

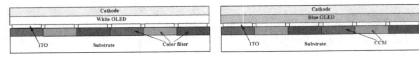

White OLED with color filter array

Blue OLED with CCM array

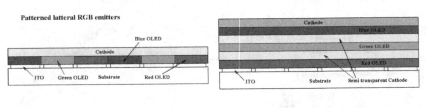

Patterned latteral RGB emitters

Stacked RGB emitters

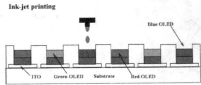

Ink-jet printing

Color Figure 9.14 Methods used to pattern OLEDs for RGB emission and full-color displays.

Color Figure 9.21 A micrograph picture of the pixel arrangements.

5

Chemistry of Electroluminescent Conjugated Polymers

BIN LIU AND GUILLERMO C. BAZAN

CONTENTS

5.1   INTRODUCTION

Although organic electroluminescence (EL) was reported in the 1950s,[1] it was not until Tang's report in 1987 that it would be seriously considered of commercial importance. Tang successfully developed organic light-emitting diodes (OLEDs) with a luminance of over 1000 cd/m^2 at a voltage of 10 V.[2] Interest further intensified with the report of EL from devices based on poly(*p*-phenylenevinylene) (PPV, **1**).[3] The manufacturing simplicity of polymer light-emitting diodes (PLEDs) opened the opportunity for LED fabrication by casting the electroluminescent layer from solution.[4] With solution methods, one can envision the manufacture of thin, flexible, large-area full-color displays.

Motivated by the new science and engineering of PLEDs and OLEDs, substantial worldwide efforts in both industry and academia have focused on the synthesis and design of electroluminescent polymeric and low-molecular-weight organic materials. In the last decade, several types of conjugated polymer structures have been developed, including PPVs (**1**), polyphenylenes (PPPs, **2**), polyfluorenes (PFs, **3**), polythiophenes (PTs, **4**), among others (shown in Scheme 5.1). Out of this range of materials, the PPVs and the PFs have

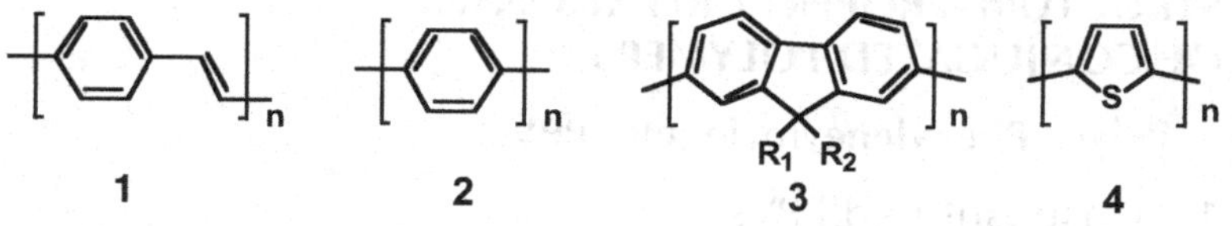

Scheme 5.1   Core structures of widely used emissive conjugated polymers.

emerged as the leading candidates for PLED applications. Through modification of chemical structure, both polymer families have achieved useful processibility to complement a wide range of emission frequencies. Many high-efficiency, high-brightness, and long-lived devices have been demonstrated using these materials.

The focus of this chapter is to survey how the molecular structure of conjugated polymers influences their performance in PLEDs and to provide a summary of the most commonly used synthetic methods for their preparation. For polymer design, one must consider not only the optoelectronic properties of the polymers, such as electron affinity, ionization potential, charge mobility, charge injection balance, emission frequency and efficiency, but also the number of defects in the molecular structure and the solid state order of the polymer chains. It is somewhat frustrating to make comparisons between PLED reports in the literature to evaluate the "ultimate" performance of a specific polymer or polymer class. This limitation results from different fabrication techniques and device architectures used by different research groups and from the levels of purity and molecular weights, which vary from sample to sample. Furthermore, device fabrication and evaluation techniques have improved considerably with time and many structure–function properties were established during the early days of PLED development. An absolute evaluation of materials performance therefore must be made with caution. Consequently, comparison between PLEDs of different conjugated polymers will be done only when they are from the same research group and only under very specific circumstances.

5.2 STRUCTURE-PROPERTY RELATIONSHIP OF CONJUGATED POLYMERS

5.2.1 Poly(*p*-Phenylenevinylene)s (PPVs)

5.2.1.1 Unsubstituted PPVs

PPV is an insoluble, intractable, and infusible yellow material with green-yellow emission. Any direct synthesis will therefore

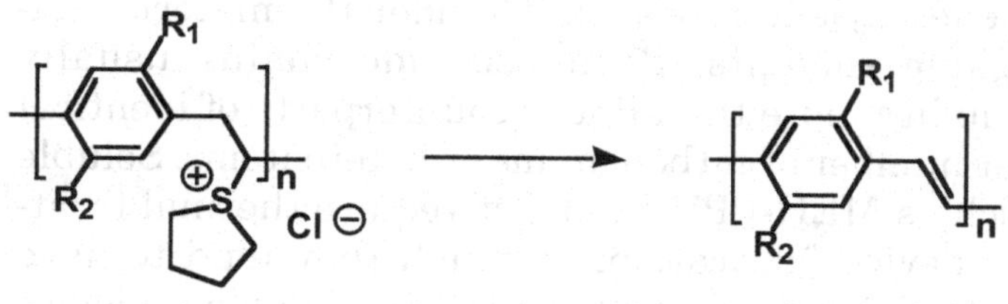

Scheme 5.2 Synthesis of PPV through a precursor polymer.

produce a solid that cannot be processed. This problem is conveniently resolved by use of a solution-processible precursor polymer, such as **5**,[5] which is first cast from solution and is subsequently converted into PPV by thermal treatment (as shown in Scheme 5.2).[6] The conversion temperature can be reduced to 100°C, from 180 to 300°C, by using the bromide salt of the precursor polymer, instead of the chloride counterpart. PPV PLEDs have been fabricated on flexible polymer substrates, such as ITO-coated poly(ethylene terephthalate).[7]

Precursor routes provide materials that differ considerably in molecular weight, molecular weight distribution, film quality, and the number of defects and impurities. Consequently, PLED efficiencies and lifetimes vary considerably (as reviewed by A.B. Holmes).[8] Undesirable side reactions with the processing solvent can result in structural defects.[9] These defects influence the average polymer conjugation length and ultimately the photoluminescence (PL) and EL properties of the film. A study of the elimination temperature on the emission properties of PPV has also been reported.[10] As will be shown later, the ring-opening metathesis polymerization offers the opportunity for precise control of PPV polydispersity and microstructure.

5.2.1.2 Alkyl- and Alkoxy-Substituted PPVs

In 1991, the EL of a soluble PPV derivative, poly[2-methoxy-5-(2'-ethylhexyloxy)-1,4-phenylenevinylene] (MEH-PPV, **7**) was reported.[11] For MEH-PPV, the two alkoxy side chains on each phenylenevinylene repeat unit render the material soluble in organic solvents. Branched side chains usually enhance solubility more than linear counterparts of identical length and/or number of carbon atoms or heteroatoms. Soluble polymers such as MEH-PPV no longer require thermal treatment during device fabrication, although they tend to have lower glass transition temperatures. A single-layer device with the structure ITO/**7**/Ca displayed an external EL efficiency (photos emitted per electron injected) of 1% with a driving voltage of 3 to 4 V.[11] A double-layer device with structure ITO/polyaniline (PANI):($\pm$)-10-camphor sulfonic acid (CSA)/**7**/Ca turns on below 2 V and reaches 100 cd/m^2 at 2.4 V with an external efficiency of 2 to 2.5%, and a luminescence efficiency of 3 to 4.5 lm/W.[12]

Several other soluble PPV derivatives with alkyl and alkoxy groups have been reported which produce films of varying emission colors. As shown in Scheme 5.3, examples include poly[2-butyl-5-(2′-ethylhexyl)-1,4-phenylenevinylene] (BuEH-PPV, **8**), poly[bis(2-ethylhexyloxy)-1,4-phenylenevinylene] (BEH-PPV, **9**), poly[bis(epi-cholestanoxy)-1,4-phenylenevinylene] (BCHA-PPV, **10**), and poly[2-methoxy-5-(3′,7′-dimethyloctyloxy)-1,4-phenylenevinylene] (C$_1$C$_{10}$-PPV, **11**).[13] Alkyl substitution provides polymers with emission that is blue-shifted relative to alkoxy counterparts of similar dimensions. The blue-shifted emission for **8**, relative to **9**, can be explained by the steric hindrance from the methylene groups attached to the phenylene rings. The use of two different side chains gives a less-ordered and more-soluble polymer. For **9**, due to its symmetric substitution pattern, more efficient packing of the polymer chains reduces the solid state PL efficiency (12%, as compared to 21% for **7** and 62% for **8**). For BCHA-PPV, the bulky side groups effectively separate the conjugated chains and the films show high PL efficiencies (53%). As the size of the alkoxy groups increases, the EL efficiency passes through a maximum and decreases again with larger side chains.[14] Very large substituents, for example cholestanoxy in **10**, dilute the semiconducting properties of the conjugated polymer. The low EL efficiency of **10** has been attributed to poor charge mobility due to the large interchain distances.[13] For **11**, the

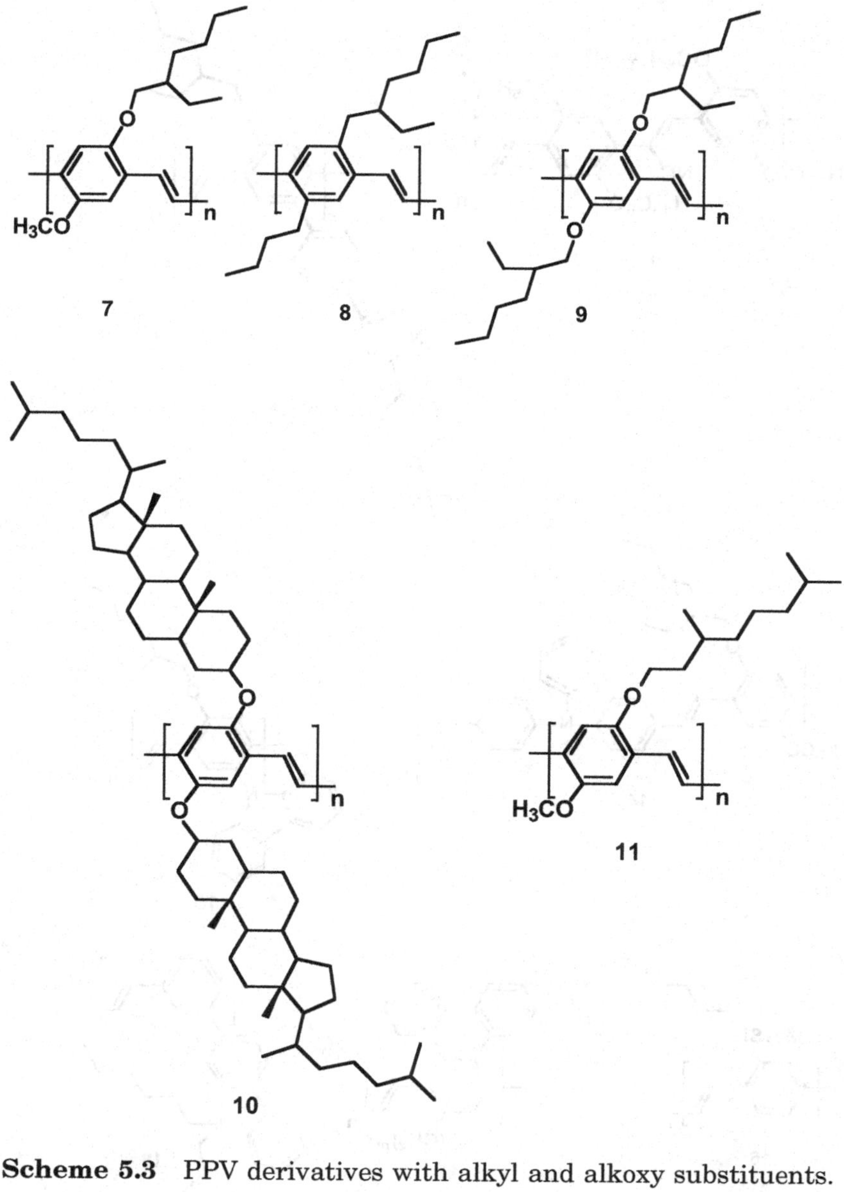

Scheme 5.3 PPV derivatives with alkyl and alkoxy substituents.

size of side groups is well balanced, giving rise to an EL efficiency of 2.1% for a single-layer device.[15]

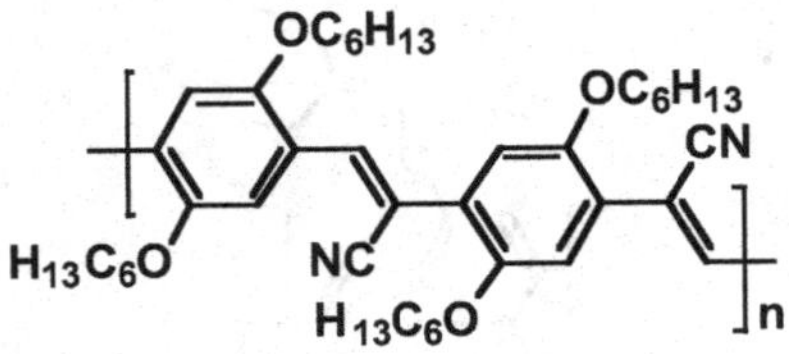

Scheme 5.4 Structural diversity in PPV derivatives.

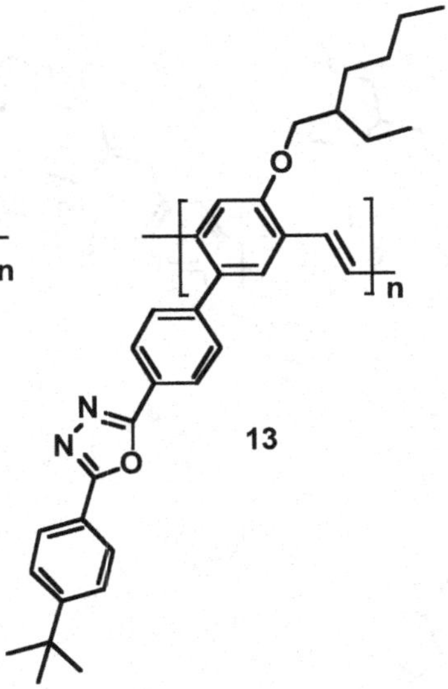

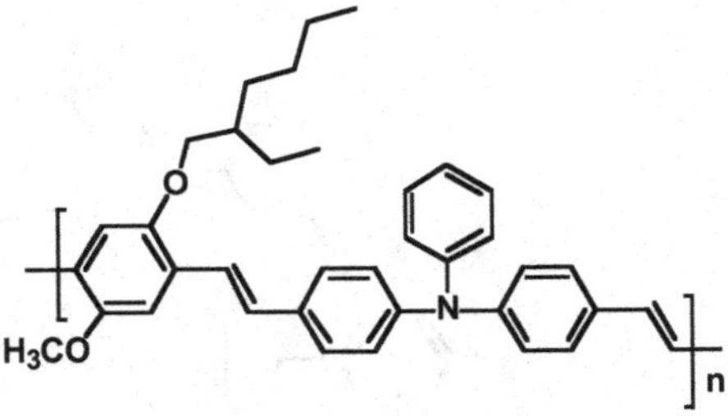

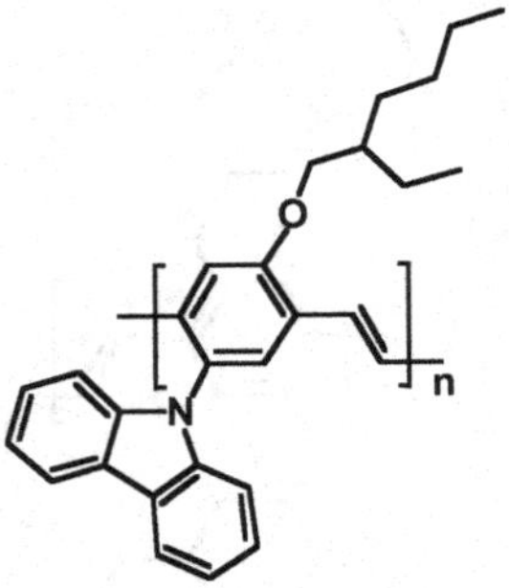

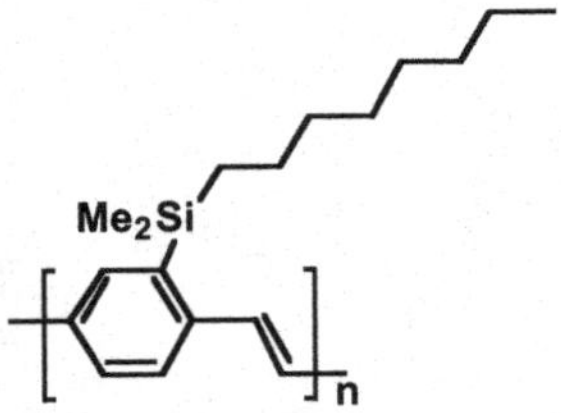

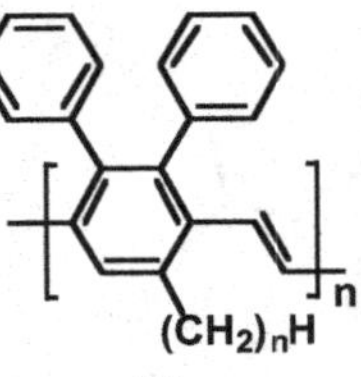

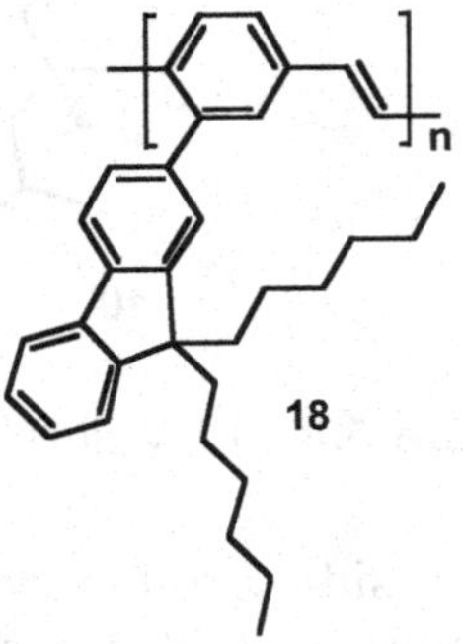

5.2.1.3 PPV Backbone Modifications

Following the observation that device properties can be modified by the choice of backbone side groups, several soluble PPV structures bearing other functionalities have been developed. With respect to improving electron transport, cyano (**12**) or oxadiazole derivatives (**13**) were introduced as side groups or within the main chain (Scheme 5.4).[16,17] To enhance hole-transport, molecular fragments such as triphenylamine (**14**) or carbazole (**15**) have been included.[18] Attaching silyl-groups (**16**) provides green-emitting materials.[19] Bulky substituents, such as phenyl (**17**) and fluorenyl groups (**18**), have been substituted at the 2- or 2,5-position of the PPV backbone to suppress intermolecular contacts, which lead to PL self-quenching.

5.2.1.3.1 Cyano-Substituted PPVs

Cyano-substituted PPVs aroused special interest since the first example was reported in 1993. It was demonstrated that CN-PPV (**12**) exhibited similar EL efficiency with either calcium or aluminum cathodes.[20] Cyclic voltammetry showed that the electron withdrawing groups shift the onset of reduction by approximately 0.6 V, as compared to MEH-PPV (**7**).[21] Additionally, the HOMO–LUMO band gap of 2.1 eV indicates that the cyano groups do not introduce steric torsion along the PPV backbone. Internal efficiencies of the red-emitting PLED ITO/**12**/Ca or Al were reported at 0.2%, and ITO/ PPV **1**/CN-PPV **12**/cathode metal devices operate with internal efficiencies of 4%.[16,22] Among the CN-PPV derivatives, MEH-CN-PPV (**19** in Scheme 5.5) exhibited high EL device efficiencies. A double-layer device of ITO/PPV**1**/**19**/Al produces red-orange light (600 nm) with a power efficiency of 5 cd/A, corresponding to a luminous intensity of over 1000 cd/m^2 at a driving voltage of 6 V.[23] A wide range of color variation can be obtained through modification of phenylene ring substituents. Bulky isopropyl substituents, as in **20**, cause a blue shift in emission and green light is observed.[8]

Simple resonance considerations make *meta*-linkages in phenylene fragments along a conjugated chain less effective

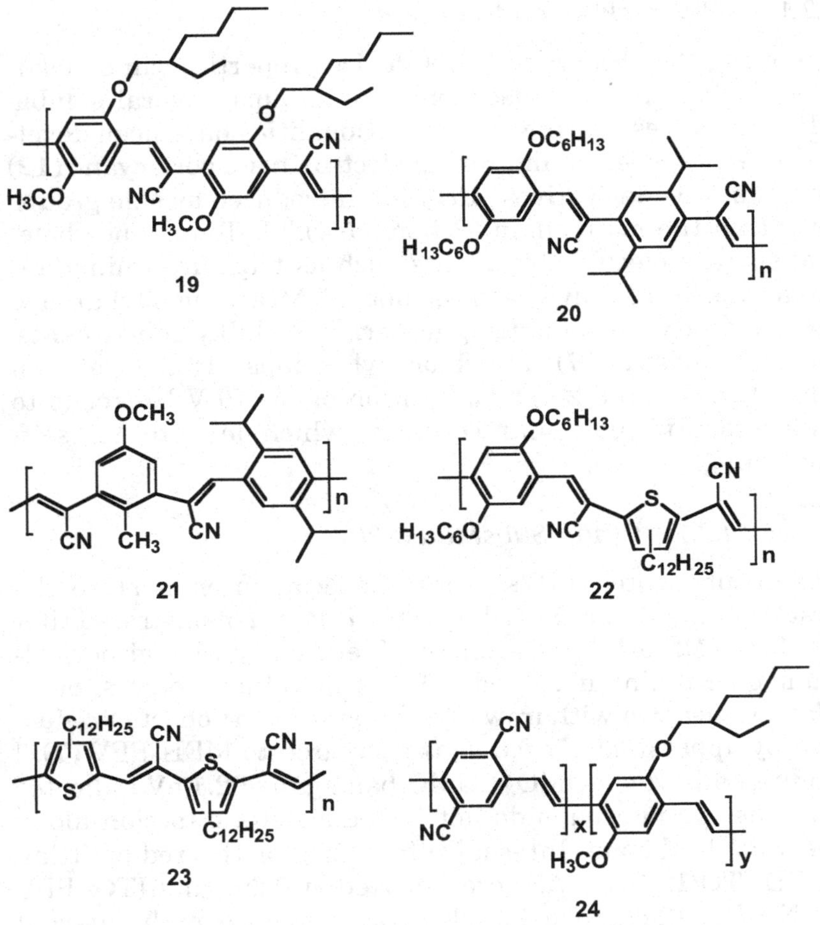

Scheme 5.5 CN-PPV derivatives with varying emission colors.

for electron delocalization, than the corresponding *para* counterparts. As a result, the introduction of these linkages within
a polymer chain results in a blue shift in emission, when
compared to *para*-substituted analogues. For example, **21** is
a blue emitter. Replacing thiophene for phenylene causes a
red shift in emission. In **22** the HOMO and LUMO energy
gap is 1.75 eV and the polymer shows substantial emission

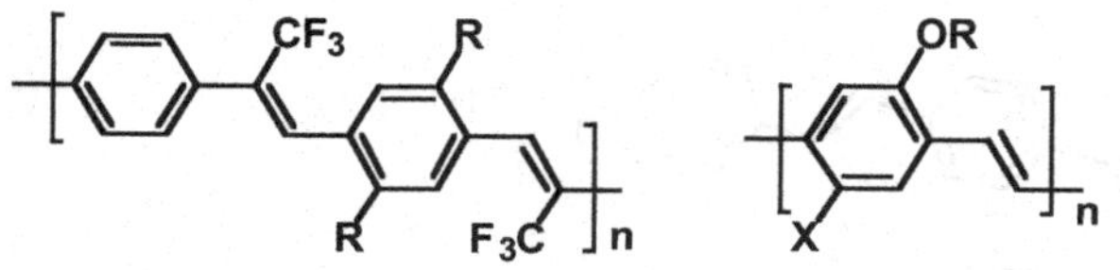

Scheme 5.6 PPV derivatives with electron-deficient groups.

in the near-infrared (IR) region. Indeed, the device ITO/PPV**1**/**22**/Al is a near-IR-emitting LED.[24] Replacement of both arylene rings with thiophene groups, as in **23**, further reduces the energy gap to 1.56 eV.[23] Copolymers containing 2,5-dicyano-1,4-phenylenevinylene and 2-methoxy-5-(2′-ethylhexyloxy)-1,4-phenylenevinylene (**24**) have also been reported, in which the HOMO and LUMO energy levels can be tuned within a range of ±0.7 V.[25] The authors claim that these copolymers can be changed from a hole-transport material to an electron-transport material by increasing the fraction of cyano-containing comonomers.[25]

In addition to CN-PPV, PPV derivatives with other electron withdrawing substituents on the olefinic bond, such as CF_3 (**25**)[26] or SO_2CF_3[27] groups, have been investigated for EL applications. Oligomers with triflyl groups on the vinylene units show a larger shift to longer wavelengths in the PL emission maxima than cyano-substituted analogs.[27,28] As shown in Scheme 5.6, PPV derivatives containing electron-withdrawing cyano,[29] halide (**26** for Cl, and **27** for Br),[30] methylsulfinyl,[31] or trifluoromethyl (**28**)[32] groups on the phenylene rings have also been reported. The halogen-substituted PPVs (**26** and **27**) emit red light (λ_{max} = 620 to 630 nm), while the trifluoromethyl derivative **28** emits yellow to yellow-orange light (λ_{max} = 540 to 570 nm). The PL quantum yield of **28** is lower than for PPV, and efficient single-layer devices were not reported. In view of their poor PL quantum yield, these materials may be suitable for use as electron-transporting layers.[8]

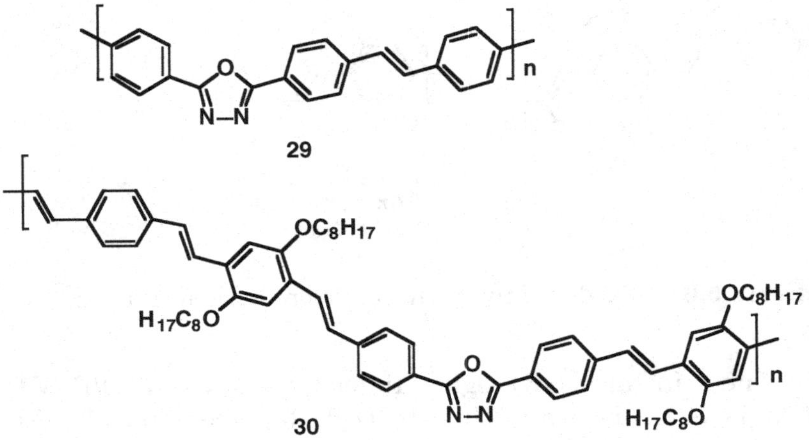

Scheme 5.7 PPV derivatives with oxadiazole functionalities.

5.2.1.3.2   PPV Derivatives with Oxadiazole Functionalities

The oxadiazole fragment is a functionality widely used to increase the electron affinity of conjugated polymers. Examples of oxadiazole-containing copolymers are shown in Scheme 5.7.[33] A polymer with the phenylenestilbeneoxadiazole repeat unit (**29**) shows a PL maximum in chloroform at 440 nm, while as a film one observes emission at 470 nm. The spectrum redshifts into the green region when the phenylene units in the polymer are disubstituted by octyloxy groups (**30**).[34]

Polymers with oxadiazole pendant groups also exist,[35–37] and their solubility can be improved by copolymerization with other structures or by introducing more substantial side chains. The copolymer **31** shows an emission maximum at 620 nm and has been used to fabricate single-layer devices with Al cathodes.[36a] With increasing amounts of oxadiazole content, the electron injection properties improve. Polymer **32** shows that it is possible to introduce oxadiazole groups in the main chain and as side groups within a single polymer structure. The single-layer device of ITO/**32**/Al has shown external efficiencies of 0.07%, and that of ITO/**32**/Ca has shown external efficiencies of 0.15%. By introducing a long alkyl spacer

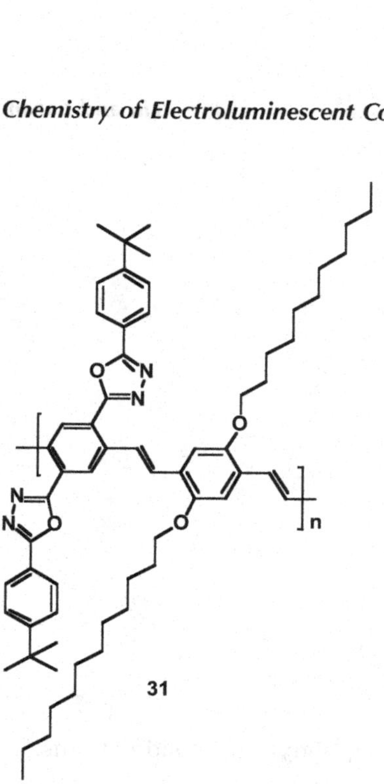

Scheme 5.7 (continued)

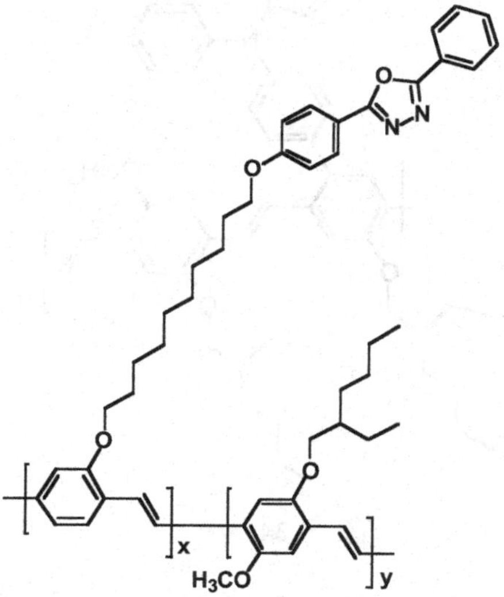

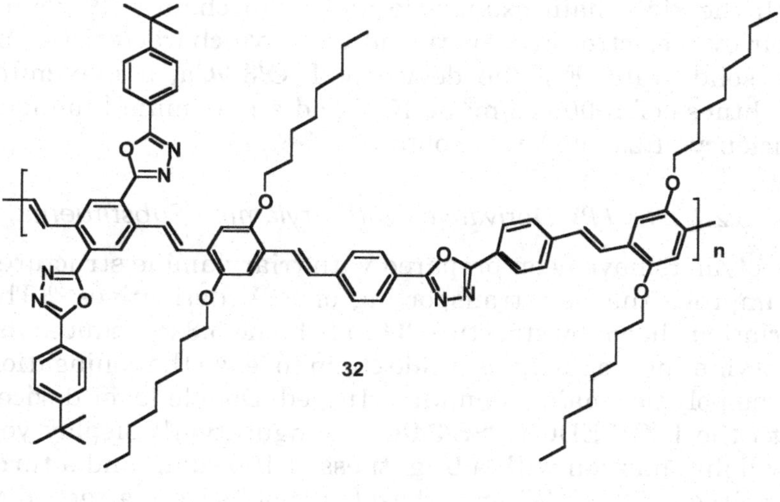

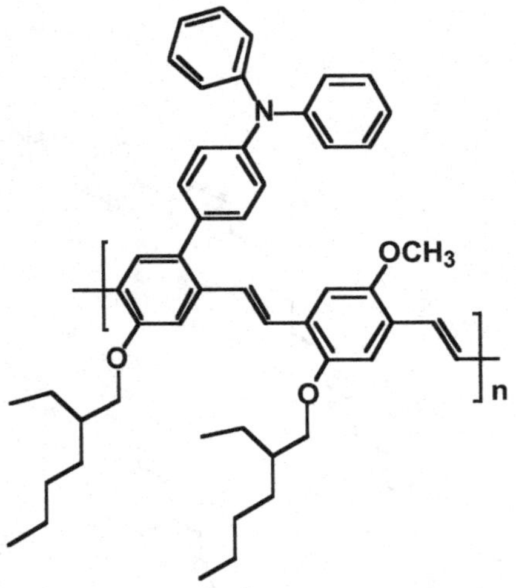

34

Scheme 5.8 PPV derivatives with triphenylamine side chains.

between oxadiazole and the PPV backbone, as in polymer **33**, both the side chain oxadiazole and main chain PPV retain their own electron-transport and emissive characteristics in the solid state. For the device of ITO/**33**/Ca, a maximum brightness of 15000 cd/m^2 at 15 V and a maximum luminous efficiency of 2.3 cd/A were obtained.[38,39]

5.2.1.3.3 PPV Derivatives with Arylamine Substituents

Copolymers have been prepared with triarylamine structures to improve the hole transporting of PPV derivatives.[40] The variation shown by structure **34** in Scheme 5.8 introduces the triarylamine fragment as a side chain to leave the conjugation of the polymer main chain unperturbed. Double-layer devices with the ITO/PEDOT:PSS/**34**/Ca configuration[41] display yellow light emission with a brightness of 100 cd/m^2 and a turn-on voltage of 3 V.[42] When carbazole moieties were attached to the polymer main chain, devices with the configuration ITO/PEDOT/**15**/Ca/Al were reported with a turn on electric field of 0.31 MV cm^{-1}.[43]

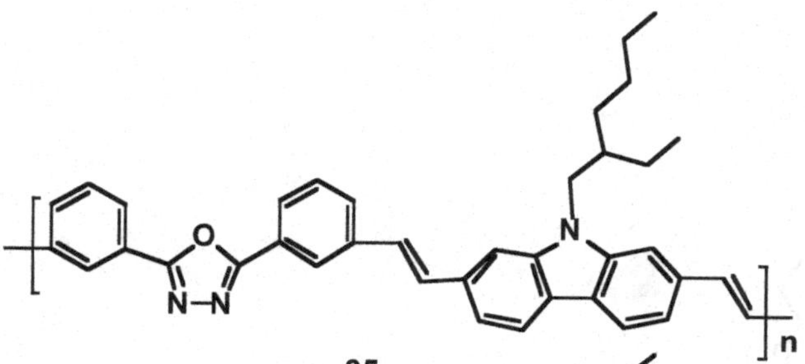

35

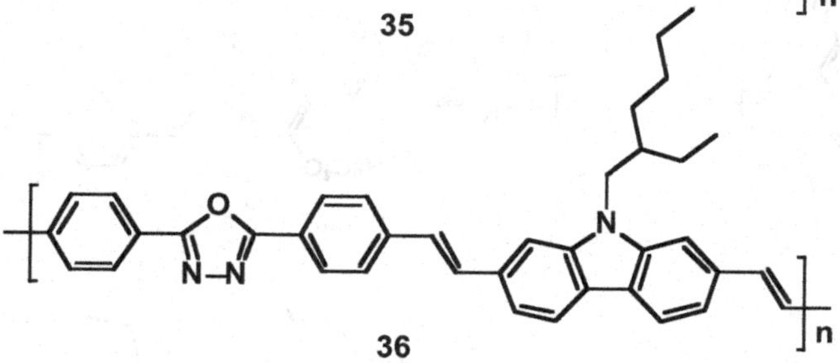

36

Scheme 5.9 PPV derivatives bearing electron-donating and electron-accepting groups.

5.2.1.3.4 PPV Derivatives with Oxadiazole and Carbazole Units

Fully conjugated alternating polymers with distyryleneoxadiazole and carbazole units show emission that reflects the regiochemistry of the polymer backbone. For example, the emission maxima of **35** and **36** (Scheme 5.9) are 450 and 495 nm, respectively.[37] As mentioned previously, when the oxadiazole unit is *m*-linked, the effective conjugation is less than that of the *p*-linked chain. The EL quantum yields for LEDs with the *p*-linked structure are 18 times higher than those prepared the *m*-linked structures.[37]

5.2.1.3.5 Silyl-Substituted PPVs

Poly[2-(3-epi-cholestanoxy)-5-dimethylthexylsilyl-1,4-phenylenevinylene] (**37**) contains silyl groups directly attached to the phenylene fragments.[19] Thin films exhibit green EL with efficiencies as high as 0.3% for the device ITO/**37**/ETL/Al. The

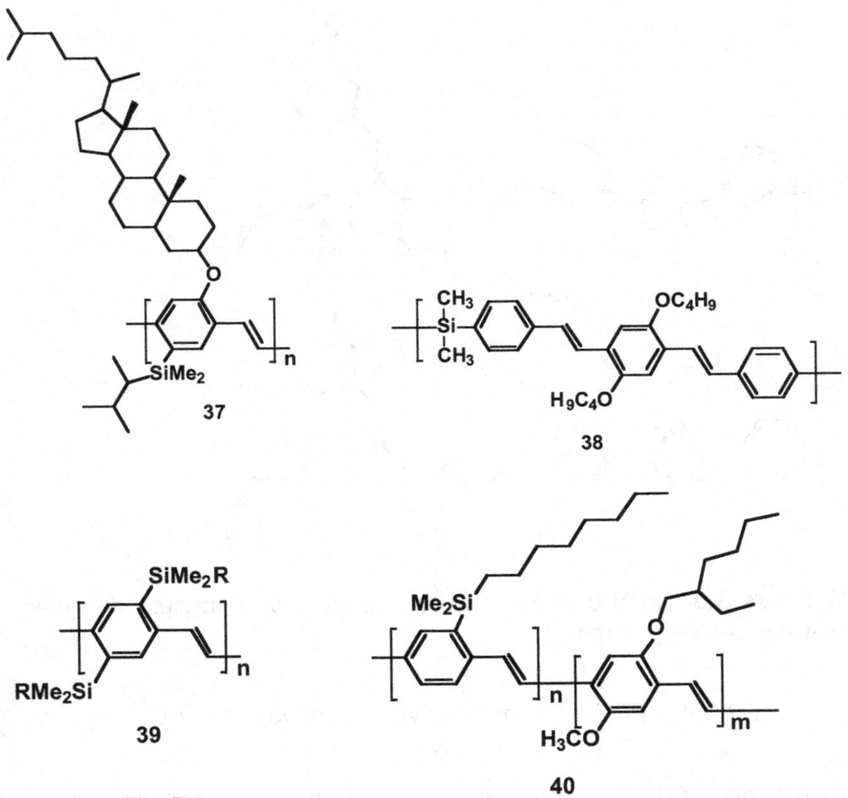

39a R = CH_3
39b R = C_4H_9
39c R = $C_{10}H_{21}$
39d R = $C_{12}H_{25}$
39e R = $C_{18}H_{37}$

Scheme 5.10 Examples of silyl-substituted PPVs.

green emission of **37** contrasts with the orange emission of **10**, indicating that an increase in the HOMO–LUMO energy difference takes place upon silyl substitution. Attractive properties of silyl-substituted PPVs are their high PL quantum efficiency, good solubility, and uniform film morphology.[45,46]

Incorporating silylene or oligosilylene segments into a polymer backbone has the effect of confining electron delocalization over the polymer chain by alternating σ-bonds and π-bonds.[47-49] Poly[(dimethylsilylene-*p*-phenylenevinylene-(2,5-dibutoxy-*p*-phenylene)-vinylene-*p*-phenylene] (**38** in Scheme 5.10) shows a PL maximum in THF at 452 nm.[50] The alkoxy group on the phenylene ring may be butyloxy or octyloxy, with no effect on the PL frequency. However, when alkoxy groups are replaced by alkyl groups, such as an octyl group, the PL blue-shifts to 412 nm.[51] Molecular simulations for the polymer structure reveals that conjugation is disrupted at the silane–phenylene linkage and coplanarity is retained only with the *trans* configuration of stilbene units.[51]

In the PPV copolymers represented by **40**, which contain silyl-bearing and alkoxy-substituted monomers, the PL and EL efficiencies are determined by the monomer feed ratio. Higher ratios of the silicon-containing monomer give rise to higher efficiencies.[52] Bis-silyl-substituted homopolymers such as **39** have shown more than 85% PL quantum yield in solution, and the chemical, electrical stability and charge injection or transporting ability for the polymers with shorter side chains appear to be improved, relative to those with longer ones.[53] A double-layer device with ITO/PEDOT:PSS/**39c**/Mg:Ag configuration displayed a maximum current efficiency of 2.3 cd/A and a power efficiency of 0.65 lm/W, with a turn-on voltage of 4 V.[53]

The absorption of the poly[*o* (*m,p*)-phenylenevinylene-*alt*-2,5-bis(trimethylsilyl)-*p*-phenylenevinylene (*o,m,p*-PBTMS-PPV) series (Scheme 5.11) varies according to stereochemistry.[54] The maximum absorption wavelengths of *o,m*-PBTMS are blue shifted, as compared to those of *p*-linkages. Due to the

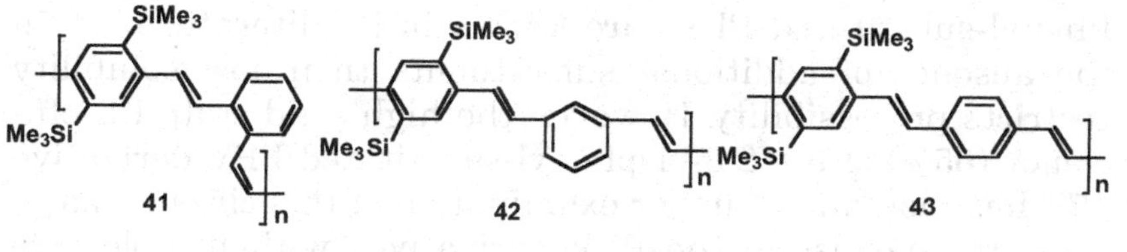

Scheme 5.11 Silyl-substituted PPVs with different regiochemistries.

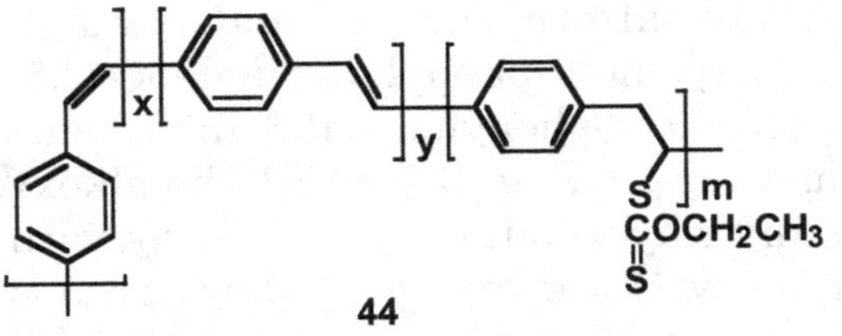

44

Scheme 5.12 An example of a "kinked" structure to decrease interchain contacts in PPV.

larger steric encumbrance of the *o*-linkage and the less π-electron delocalization of the *m*-linkage, polymers **42**, **41**, and **43** have emission maxima at 450, 475, and 495 nm, respectively. *p*-PBTMS-PPV (**43**) exhibits the highest EL power efficiency at lower operating voltages. Other studies exist that show a similar impact of arene regiochemistry on the optical properties of the conjugated polymers.[55,56]

It should also be noted at this stage that *cis* linkages within a PPV structure decrease the effective conjugation between adjacent units. Because of less-pronounced steric interference, the *trans*-isomer can more effectively maintain a coplanar arrangement and, therefore, the coexistence of *cis*- and *trans*-linkage units results in less-effective chain packing. An LED containing PPV with a mixture of *cis*- and *trans*-alkene units (**44**, Scheme 5.12) shows an internal EL efficiency of 0.22% for a single-layer device, and 2% for a double-layer device.[57]

5.2.1.3.6   Aryl-Substituted PPVs

Phenyl-substituted PPVs are known in the literature.[58,59] In the absence of additional substituents their low solubility restricts processibility. However, the high solid state PL efficiency (65%) of the 2,3-bisphenyl-substituted PPV derivative **17**[60] has motivated further examination of this class of materials. In contrast to the PPV derivatives with flexible side chains, such as **7** and **8**, attaching an additional phenyl ring with two 2-methylhexyloxy groups (such as **45** in Scheme 5.13)

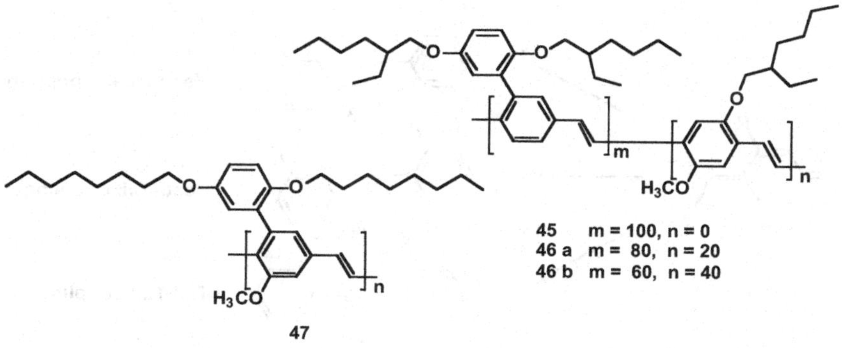

45   m = 100, n = 0
46 a   m = 80, n = 20
46 b   m = 60, n = 40

47

Scheme 5.13   Phenyl-substituted PPVs.

leads to a new series of polymers, which are more stable toward photooxidation than alkoxy-substituted PPVs.[61]

Poly{2-[2′,5′-bis(2″-ethylhexyloxy)-phenyl]-1,4-phenylenevinylene} (**45**), has good solubility in organic solvents, high stability against photooxidation, and efficient PL, as a result of the well-separated conjugated polymer chains.[62] Changing to copolymers **46**, which have an increased amount of MEH-PPV content, one observes a decrease in the PL efficiency and a red shift to yellow emission.[62] Single-layer devices (ITO/**46b**/Ca) and double-layer devices of (ITO/PVK/**46b**/Ca) having external efficiencies of up to 0.68 and 1.15%, respectively, were reported.[62,63] However, as shown in Scheme 5.14, defects from tolane-bisbenzyl moieties (TBB) **49**, which are explained by head-to-head coupling instead of the more common head-to-tail reaction,[64] were present at higher levels for the phenyl-substituted PPVs[65] (5 to 6%) when compared to dialkoxy PPVs, such as MEH-PPV (1.5 to 2.2%).[66]

For **45**, the PL efficiencies in the solid state increased from 28% (polymerized at 144°C) to 60% when the polymerization was carried out at 0°C. The higher PL efficiency was attributed to the lower concentration of HH (head-to-head) defects.[67] Double-layer devices (ITO/PEDOT/**45**/Ca) show an EL quantum efficiency of 0.94%.[67] Further improvement in the EL was obtained with poly(2-(2′,5′-bis(octyloxy)benzene)-5-methoxy-1,4-phenylenevinylene) (**47**). Decreasing the HH

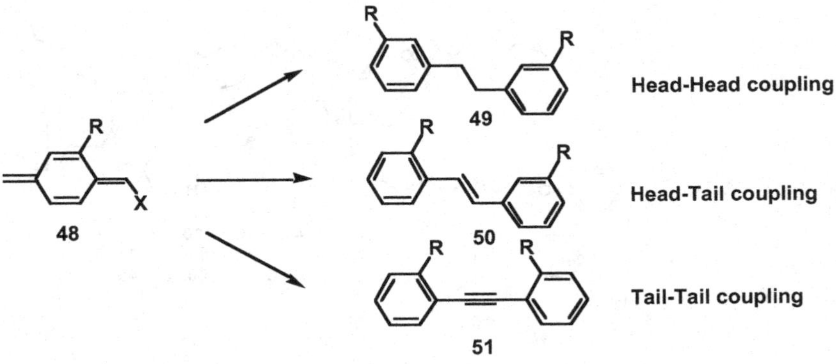

Scheme 5.14   Regiochemistry of coupling in PPV synthesis.

content by a factor of two leads to a 30-fold increase in device lifetime.[66] An external EL efficiency of 1.74% (HH < 1) was reported for **47**.[64]

Incorporating the 9,9-dialkylfluorenyl side groups into the parent PPV backbone yields polymers **52** and **53**, as shown in Scheme 5.15.[68] In these structures there is substantial steric hindrance at the inter-ring linkages between the fluorene fragments and the PPV backbone. The alkyl chains are out of plane, relative to the fluorene plane, and separate the chains in the solid. Polymers such as **52** and **53** have PL efficiencies in the range of 70% in chloroform.[68] The glass transition temperatures range from 113 to 148°C. For poly[2-(9,9-dihexylfluorenyl)-1,4-phenylenevinylene] (**18**) LEDs with the architecture ITO/PEDOT:PSS/**18**/Ca have a low turn-on voltage and a maximum brightness of 12,000 cd/m^2.[68] Improved performance was observed by the introduction of comonomers corresponding to the MEH-PPV structure. The copolymer (92.5 DHF (dihexylfluorene)/7.5 MEH-PPV) (**53a**) exhibits a higher light output per current density (2.4 cd/A), when compared with MEH-PPV **7**, DHF-PPV **18**, and another copolymer (**53b**) with a higher (50 mol%) MEH-PPV content. It has been proposed that the inclusion of a relatively low level of the alkoxy-containing comonomer unit along the main chain of the polymer provides sites for exciton confinement.

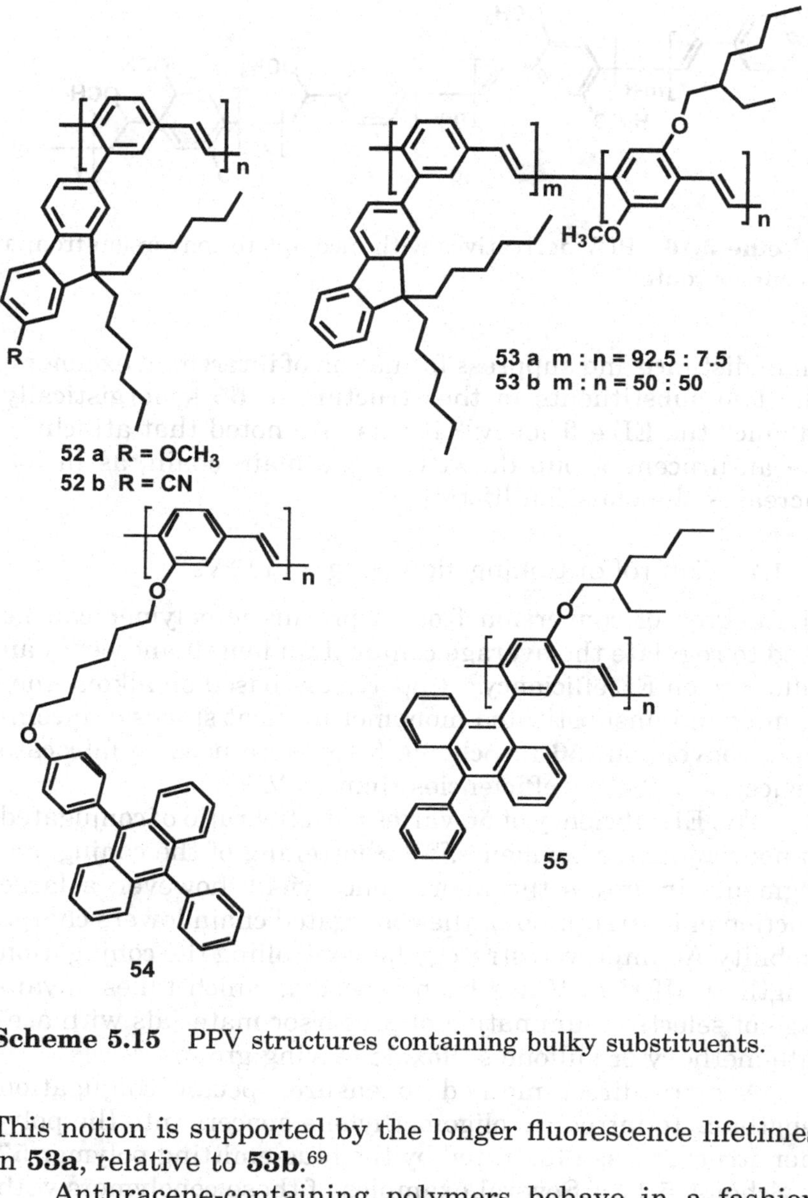

52 a R = OCH_3
52 b R = CN

53 a m : n = 92.5 : 7.5
53 b m : n = 50 : 50

54

55

Scheme 5.15 PPV structures containing bulky substituents.

This notion is supported by the longer fluorescence lifetimes in **53a**, relative to **53b**.[69]

Anthracene-containing polymers behave in a fashion similar to **53a** and **53b**. In the case of the polymer **54**, the phenylanthracene hexyloxy substituents increase the inter-

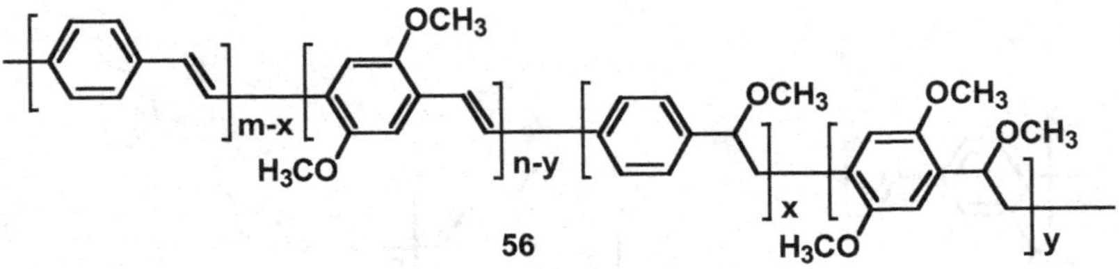

Scheme 5.16 PPV derivatives with incomplete conversion from a precursor route.

chain distance and suppress formation of interchain excimers. The two substituents in the structure of **55** synergistically enhance the EL efficiency.[70] It was also noted that attaching the anthracene group directly to the main chain, as in **55**, increases the emission lifetime.[70]

5.2.1.4 Control of Conjugation Length in PPVs

The degree of conversion from a precursor polymer can be used to regulate the average conjugation length and bears an influence on EL efficiency.[71] Copolymers based on alkoxy-containing and unsubstituted monomer units at stages of incomplete conversion (**56** in Scheme 5.16) were used to fabricate devices with better efficiencies than PPV.[72]

The EL efficiency of **56** varies with the ratio of conjugated to nonconjugated segments.[72,73] Shortening of the conjugated segments improves the fluorescence yield; however, a large fraction of interruption in the conjugated chain lowers charge mobility. An improved strategy for controlling the conjugation length of MEH-PPV has been reported, which takes advantage of selective elimination of precursor materials with acetate-methoxy or sulfone-sulfoxide leaving groups.[74,75]

A more direct method to ensure specific conjugation lengths is to introduce oligomethylene spacers into the polymer structure, as illustrated by the blue-emitting polymer **57** in Scheme 5.17.[76] Several examples of these copolymers with well-defined alternating segments exist.[76] However, the EL

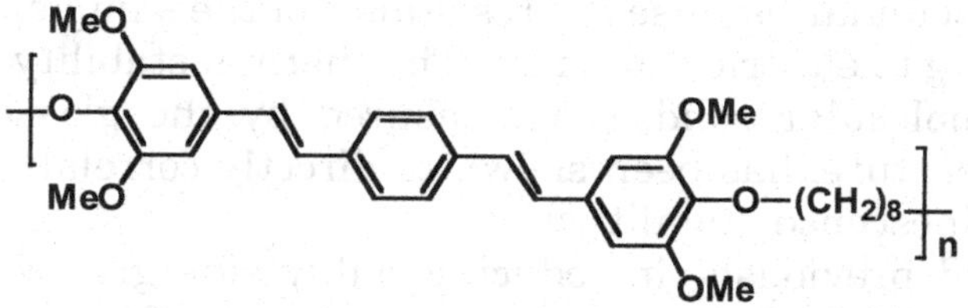

57

Scheme 5.17 An example of a blue-emitting copolymer with well-defined isolated chromophores.

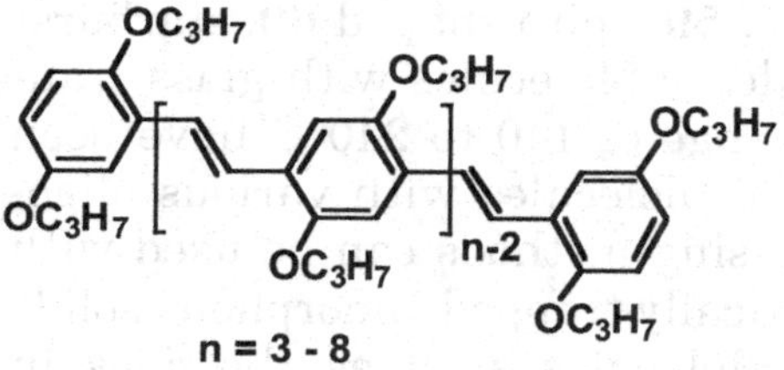

58 a n = 3
58 b n = 8

Scheme 5.18 Examples of PPV oligomers with different conjugation lengths.

performance of this family of materials has not matched the best performance of fully conjugated polymers.

5.2.1.5 Oligomer Approach

Oligomers with a well-defined number of repeat units can be used to model the corresponding polymers, and several important structure/property relationships have been established in this manner.[77] With an increase in repeat units, the emission of the oligomers changes from blue-green **58a** ($n = 3$) to yellow ($n = 8$) **58b**, as shown in Scheme 5.18.[78] However, the general tendency of small molecules to spontaneously crystallize[79] presents a limitation for their use in LED fabrication. Crystal formation destroys film homogeneity and the ensuing crystal boundaries raise the resistance of the sample, eventually leading to electric shorting.[80] The thermal stability of amorphous molecular solids, as measured by the glass transition temperature, has been shown to directly correlate with electroluminescence stability.[81]

As described previously, introducing bulky side groups generally leads to limited charge mobility and lower EL efficiencies.[14] In response to these limitations, considerable efforts have been dedicated to developing molecules of intermediate molecular weight that minimize the aliphatic content and at the same time possess geometries that resist crystallization. Examples of molecules based on different molecular shape include the tetrahedral **59**, spiro-shaped **60**, starburst **61**, and dendritic **62** molecules.[82] Molecules with glass transition temperatures in the range of 140 to 240°C have been reported for some tetrahedral molecules with various oligomeric arms.[82] Solution processing methods can be used with these materials to yield kinetically trapped, amorphous solids that resist crystallization. Molecules such as the ones in Scheme 5.19 embody the beneficial properties of small molecules, namely, purity and well-defined structure, with the ability to cast films directly from solution, a property characteristic of polymeric materials.

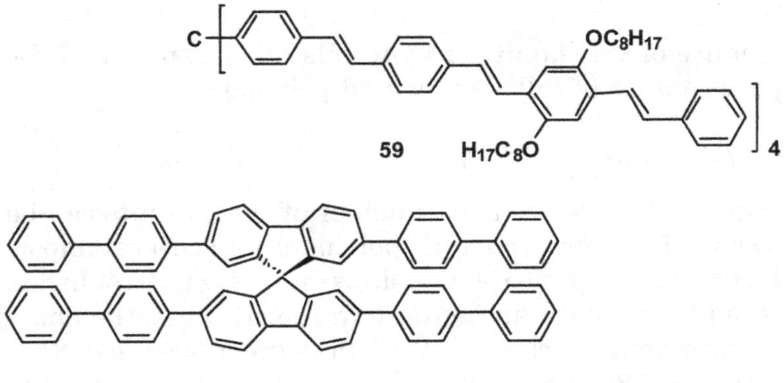

Scheme 5.19 Examples of molecular shapes that resist crystallization.

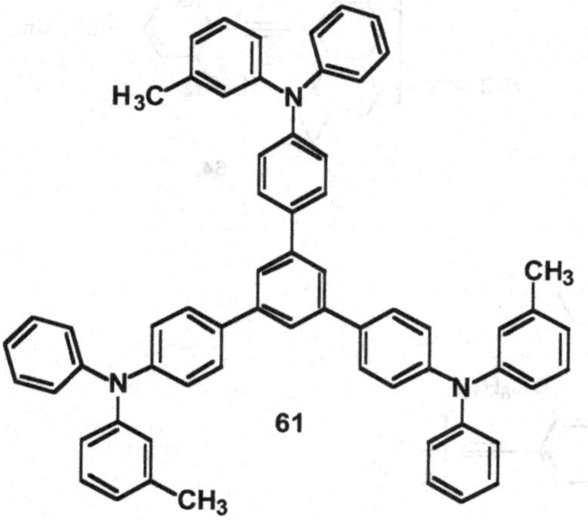

61

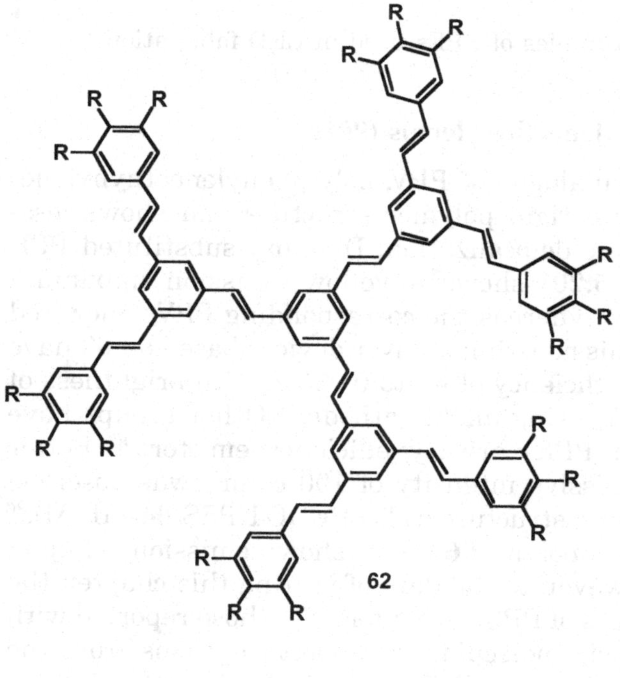

62

Scheme 5.19 (continued)

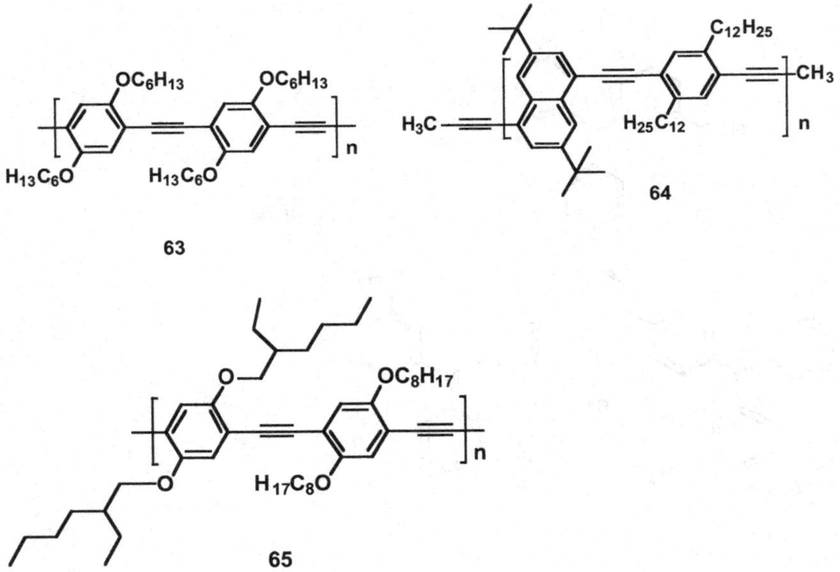

63

64

65

Scheme 5.20   Examples of PPEs used in LED fabrication.

5.2.2   Poly(Phenyleneethynylene)s (PPEs)

As the dehydro-analogue of PPV, poly(phenyleneethynylene) (PPE) has a more rigid polymer structure and shows less-effective electronic delocalization. Dialkoxy-substituted PPE (**63** in Scheme 5.20) shows a yellow emission maximum around 600 nm,[83] whereas the corresponding PPVs show red or red-orange emission. Single-layer devices based on **63** have shown external efficiency of up to 0.035%, with brightness of 80 cd/m^2 at 22 V, using an Al cathode.[84] Other groups have also shown that PPEs are low-efficiency emitters.[85] Better performance (emissive intensity of 100 cd/m^2) was observed from a multilayer structure (ITO/PEDOT:PSS/**64**/LiF/Al).[86] PLEDs with the copolymer **65** have shown emissions of up to 300 cd/m^2.[87] However, to the date of writing this chapter, the LED performances of PPEs are inferior to those reported with PPVs. An intrinsic molecular disadvantage stems from the *sp*-hybridized carbons in PPEs, which render the polymer

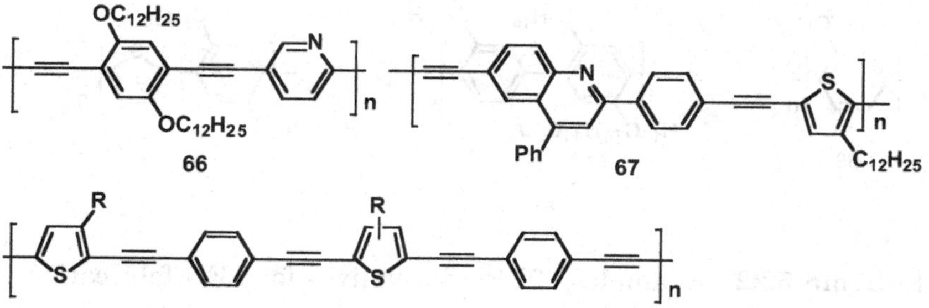

Scheme 5.21 Examples of PPEs with heterocyclic structures.

more electron poor than the sp^2-hybridized PPVs. Hole injection is thus more difficult. PPE derivatives containing pyridine **66**, quinoline **67**, or thiophene **68** structures have also been studied (Scheme 5.21).[88]

5.2.3 Poly(p-Phenylene)s (PPPs)

5.2.3.1 Substituted PPPs and Copolymers

PPP has a HOMO–LUMO gap between 2.8 and 3.5 eV, and has received considerable attraction for use in blue-emitting PLEDs. Like PPV, the unsubstituted PPP is insoluble and cannot be processed. Although films of oligomeric poly(phenylene)s can be obtained by vacuum deposition,[89] the preparation of the polymer requires suitable precursor routes.[90] In analogy to the modifications in the PPV structure, soluble PPPs can be designed by incorporation of alkyl-, aryl-, alkoxy-, or perfluoroalkyl[91]-solubilizing groups.

PPPs show excellent thermal stability. They are also interesting because one can tune the emission frequency by the torsion angle between consecutive rings along the main chain. The influence of this angular relationship on the HOMO–LUMO energy difference has been examined via molecular orbital calculations (such as VEH band structure calculations) and it was found to be hyperlinear.[92] Alkyl chains on the PPP skeleton lead to polymers with band gaps of ~3.5

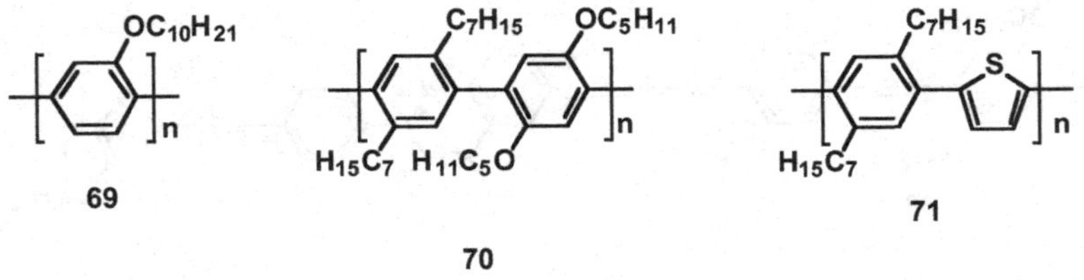

Scheme 5.22   Examples of PPP derivatives for LED fabrication.

eV, larger than PPP itself.[93] For alkoxy-substituted PPPs, the HOMO–LUMO energy differences are larger than that of PPP, but are smaller than those of alkyl-substituted analogs. Modeling studies show that the steric bulk of alkoxy groups is less demanding than an equivalent alkyl substituent and causes a smaller twist angle along the PPP skeleton. In addition, the stronger electron-donating properties (in a π sense) of the alkoxy groups decrease the HOMO–LUMO energy gap.

Although alkyl- and alkoxy-PPPs have high thermal and oxidative stability, they exhibit relatively low EL quantum efficiencies in single-layer LEDs. Devices with the decyloxy derivatives **69** show external efficiencies of up to 3% with the double-layer structure ITO/PVK/**69**/Ca.[94] EL efficiencies have also been reported from polymer blends.[95] For blends of poly(2,5-diheptyl-2′,5′-dipentoxybiphenylene) (**70**, Scheme 5.22) with a copolymer of poly(2,5-diheptyl-1,4-phenylene-*alt*-2,5-thienylene) (**71**), the EL quantum efficiency was in the order of 2%.[95b]

5.2.3.2   Copolymers Containing Oligophenylene and Olefin Units

Conjugated copolymers with phenylene units and systematic variations of benzene, stilbene, and phenyleneacetylene fragments have been synthesized. Some of these structures are shown in Scheme 5.23.[96] A triple-layer device of ITO/PVK/**72**/2-(4-biphenylyl)-5-(4-tert-butylphenyl)-1,3,4-oxadiazole (PBD)/Ca displayed an internal EL efficiency of 4%.[96] The other structures in Scheme 5.23 showed less efficient EL performance,

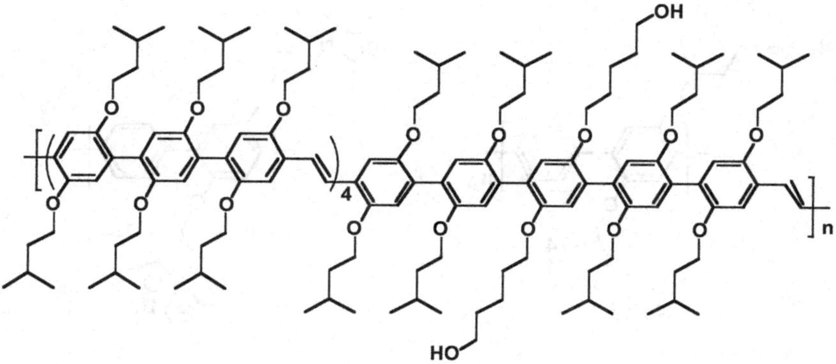

72

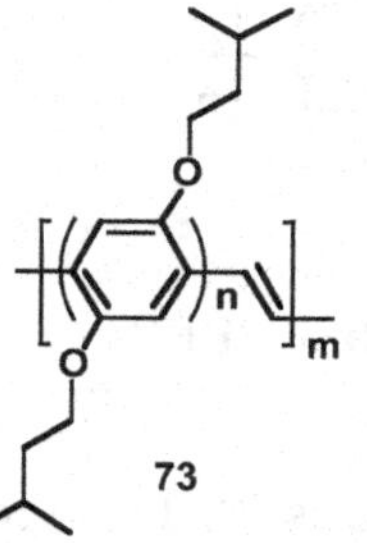

73 a   n = 2
73 b   n = 3
73 c   n = 5

73

Scheme 5.23 Oligophenylene derivatives connected with olefin links.

relative to **72**.[96] Poly[(2,5-dimethylbutyloxy)-*p*-biphenylen-evinylene] (**73a**), poly[(2,5-dimethylbutyloxy)-*p*-triphenylen-evinylene] (**73b**), and poly[(2,5-dimethylbutyloxy)-*p*-penta-phenylenevinylene] (**73c**) show emission maxima at 470, 445, and 448 nm, respectively.[98]

5.2.3.3   Water-Soluble PPPs

PPP derivatives with sulfonate (**74**), ammonium (**75**), or car-boxy functional groups (**76**) have been synthesized (as reviewed by Pinto and Schanze);[97] see Scheme 5.24. Because of their charged nature, these polymers are soluble in aqueous solutions and can be deposited onto a surface by layer-by-layer

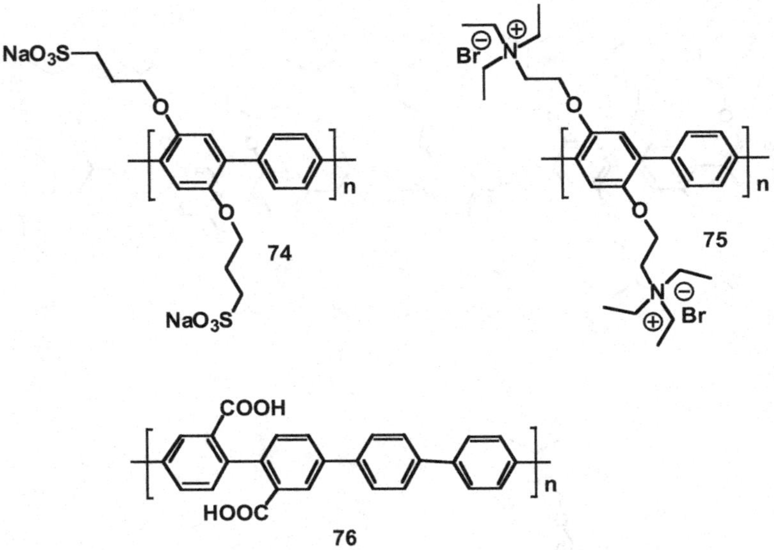

Scheme 5.24 Examples of water-soluble PPPs.

sequential adsorption techniques.[98] Devices based on ITO/(–)PPP **74**/(+)PPP **75**/Al with a bilayer thickness of 210 nm were reported with a turn-on voltage of 7 V.[98a]

5.2.3.4 Ladder-Structure PPPs

Side groups on the PPP backbone increase solubility but twist the phenylene rings relative to each other. It is possible to circumvent this problem by locking the repeat units into a coplanar arrangement within a ladder structure (Scheme 5.25). Ladder formation can be accomplished with high regularity and no defects, such as cross-linking or incomplete cyclization.[99]

Polymer **77** exhibits good solubility in organic solvents and a PL maximum at 450 nm in toluene.[100] The Stokes shift is small (150 cm^{-1}), a consequence of the structural rigidity of the ladder structure. Films of **77** show red-shifted emission, with a maximum at 600 nm, which was attributed to excimer formation and which suggests efficient interchain packing in

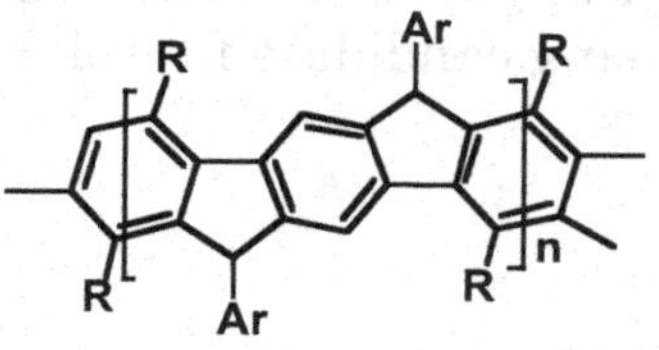

77 Ar = p-$C_{10}H_{21}C_6H_4$
    R = C_6H_{13}

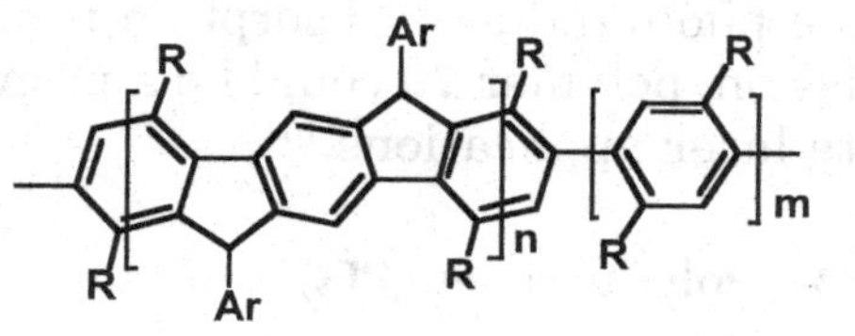

78 Ar = p-$C_{10}H_{21}C_6H_4$
    R = C_6H_{13}

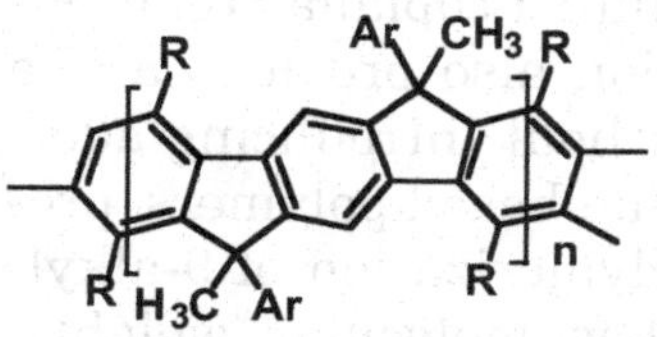

Ar = p-$C_{10}H_{21}C_6H_4$
79 R = C_6H_{13}

Scheme 5.25 Examples of ladder-type PPPs.

the solid state.[101] An LED device with ITO/**77**/Al exhibits an average quantum efficiency of 0.5%.[102] By diluting ladder polymers in a host polymer matrix, such as polystyrene or polyvinylcarbazole, there is suppression of the yellow emission.[102,103] A different strategy to suppress aggregate emission is the use of the stepladder structure in **78**, in which the phenylene comonomers twist out of the ladder plane and discourage interchain contacts.[100]

The additional methyl group on the methylene carbon of Me-LPPP (**79**) serves as an interchain spacer.[104] The solid state PL of the Me-LPPP show only weak aggregate emission and there is little difference between solution and solid state spectra.[105] Suppression of aggregate formation is accompanied by a substantial increase of PL quantum efficiency (90% in solution and up to 60% in solid state). By using Me-LPPP as the electroluminescent layer, homogeneous bright light is obtained in the blue-green spectral range (λ_{max} = 461 nm) with EL efficiencies of 4%.[106,107] It has also been suggested that, because of the low defect absorption and the spectral separation of the photo-induced absorption from the photoluminescence emission, polymer **79** could be a promising candidate for solid state laser applications.[108]

5.2.4 Polyfluorenes (PFs)

5.2.4.1 PF Homopolymers

Poly(fluorene-2,7-diyl) may be regarded as a derivative of PPP, with pairs of phenylene rings locked into a coplanar relationship by the C-9 atom. The C-9 location also provides a site where substituents may be placed without introducing additional torsional strain. The first fluorene-based polymers, prepared via ferric chloride oxidative polymerization of 9-alkyl- and 9,9'-dialkylfluorenes **3**,[109] had low molecular weights, some degree of branching, and linkages through sites other than the 2,7-positions.[110] Substantially improved structures with exclusively 2,7-linkages have been obtained through improved methods, such as the Yamamoto method[111] and the Suzuki coupling reaction.[112]

PFs and their copolymers have evolved as a major class of emitting materials for LEDs.[113] Solid state PL efficiencies of up to 80% have been reported, and LEDs fabricated from PF homopolymers or copolymers exhibit blue emission.[114] A second important property of PFs is their thermotropic liquid crystallinity.[115] Using an aligned PF emitting layer, an LED was fabricated which emitted with a polarization ratio of 15.[116]

The first paper on electroluminescence from PF featured a single-layer LED, with a turn-on voltage of about 10 V.[117] Later reports included PL and EL from poly(9,9-bis(3',6'-dioxaheptyl)-fluorene-2,7-diyl) (**80** in Scheme 5.26), a homopolymer containing oligo(ethylene oxide) side groups.[118] This polymer exhibited PL efficiencies of 77 and 73% in solution and in the solid state, respectively. Single-layered LEDs with an ITO anode, and a Ca cathode turned on at 9 V, reached 100 cd/m^2 at 30 V, with an EL efficiency of 0.3%. By adding lithium triflate, the device properties could be improved with EL efficiency up to 4%, 1000 cd/m^2, and 8.6 lm/W at 3.5 V. However, during operation, the emission color changed from sky blue to white, a phenomenon that is widely attributed to

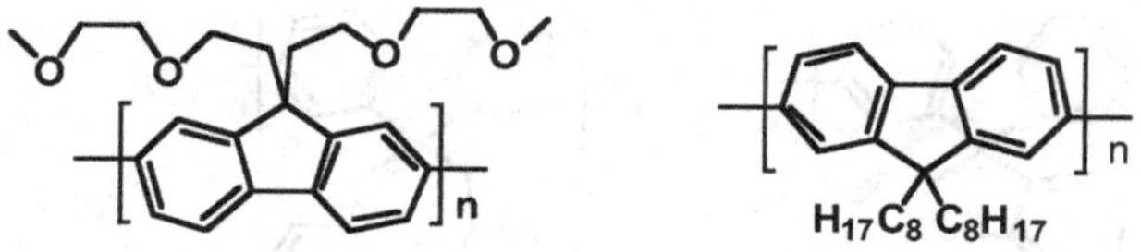

Scheme 5.26 Examples of PF homopolymers.

excimer formation in the solid state, although other mechanisms may also be involved.[119]

5.2.4.2 Strategies to Minimize Excimer Emission

5.2.4.2.1 Side Chain Modification

A substantial fraction of PF research has been centered on the elimination of the emission component in their solid state PL and EL. Many PF derivatives have been designed and synthesized with bulky groups, in the hope of minimizing interchain delocalization (Scheme 5.27). This "excimer dogma" prompted much molecular design and the synthesis of novel structures. Substitution at the 9-position minimizes excimer formation.[120] As an example, consider the spiro-functionalized PF derivative **82**.[121] No significant excimer formation was observed after annealing at 100°C in air for 2 h, but a significant green contribution was noticed when annealing at 150°C.[122] It was shown that for **83**, the dendritic side groups completely suppress the green emission contribution in PL, even after annealing at 100°C for 24 h. PLEDs fabricated on ITO with a PEDOT:PSS hole injection layer and a Ca/Al top electrode exhibited an emission onset at 3 to 4 V, and a brightness of 400 cd/m² at 10 V. Similar strategies can be observed in the synthesis homopolymers and copolymers containing benzyl ether dendrons (**84**).[123]

Other strategies toward suppression of excimer formation involve the synthesis of statistical polyfluorene copolymers, or end-capping (reviewed by D. Neher).[124] Copolymers containing low band gap repeat units, such as anthracene,

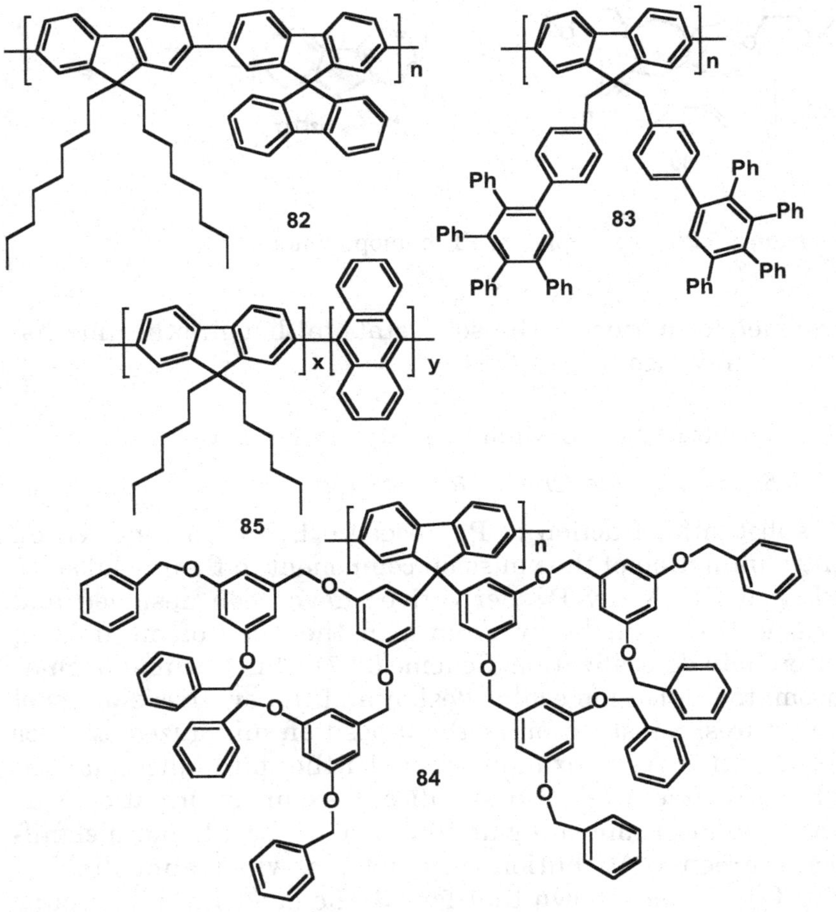

Scheme 5.27 Polyfluorene with side groups to minimize excimer formation.

display little or no change in their emission bands, even after annealing at 200°C for 4 h.[125] Relatively small amounts of anthracene units (5 to 15%) within the PF main chain in **85** suppress excimer formation.[126] One possible explanation is that the plane of the anthracene units is orthogonal, relative to the adjacent fluorenyl groups, and discourages interchain interactions in the solid.

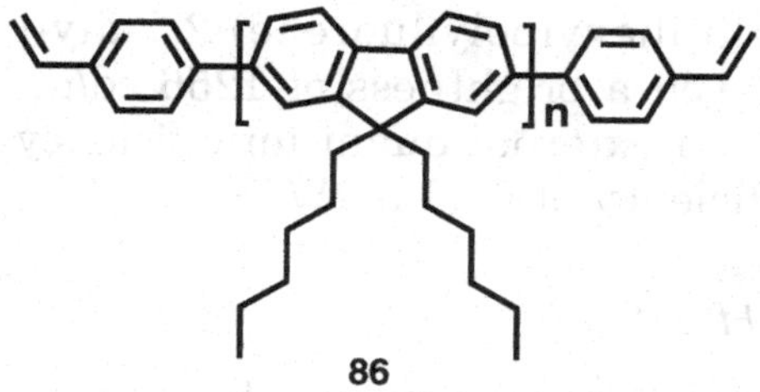

86

Scheme 5.28   An example of PF with end groups suitable for cross-linking.

5.2.4.2.2   Molecular Weight Control

Excimer formation appears to be molecular weight dependent. Lower-molecular-weight samples show a larger contribution from excimer emission, probably as a result of increased chain mobility.[127] Comparison of the performance of batches of **81** as-prepared, and after size fractionation, was also reported.[128] The as-prepared batches are characterized by a broad molecular weight distribution (M_w = 53,000 g/mol, PDI = 2.1) and show excimer emission in the PL of annealed films, while the PL and EL prepared from the higher-molecular-weight fraction (M_w > 10,000 g/mol) consisted mainly of emission from non-interacting chains. The lower content of aggregate sites in the high-molecular-weight PF layers was attributed to the lower concentration of the more mobile chain ends.

5.2.4.2.3   Cross-Linkable PFs

Oligomers end-capped with 2-bromofluorene are more prone to oxidation due to the existence of benzylic hydrogens in the 9-position of the fluorene end groups.[127,129,130] Color stability and blue EL was demonstrated for cross-linked PF layers.[131,132] After the styrene end-capped dihexylfluorene homopolymers **86** (Scheme 5.28) were thermally cross-linked at 200°C to yield a rigid amorphous network, the PL showed no excimer emission, even after annealing at 200°C in nitrogen for 12 h, indicating a reduction in chain mobility. A triple-layer device of ITO/cross-linkable poly[(4-n-hexyltriphenyl)amine] (HTPA)/cross-linkable poly[(dihexylpolyfluorene)-2,7-diyl] (DHF)/3-oxadiazole/Ca displayed a brightness of 1256 cd/m^2 at a driving voltage at 10 V, an external quantum efficiency of 1.25%, and a luminous efficiency of 0.23 lm/W.[132]

5.2.4.2.4   End-Capping Effects

The color stability and the EL efficiency of PFs can be improved with addition of low-molecular-weight hole-transporting molecules with a low ionization potential.[133] Based on this information, fluorene-triarylamine copolymers were designed and have shown improved hole mobilities.[134] In addition, triarylamine- or triarylamine-containing end groups[135] improve color stability of PF, without altering the electronic property of the backbone.[136] Examples include poly(9,9-diethylhexylfluorene)s end-capped with N,N-bis(4-methyl)-N-phenylamine (**87**) and N-(biphenyl-4-yl)-N-(naphth-2-yl)-N-phenylamine (**88**) (Scheme 5.29). Different monomer/end-group feed ratios were used to control the molecular weight of the resulting polymers and to vary the molar concentration of the end groups in the final product. When these materials were used in an LED device, the green emission component decreased with increased end group concentration, and was completely suppressed in the emission from the device containing **87** (corresponding to a monomer to end-capper feed ratio is 91 to 9). It is not clear at this stage whether the improved color stability is solely due to a competition for carrier capture between the end groups and the excimer sites, or if the end groups suppress excimer formation.[135]

Polymer **89** (Scheme 5.30) was developed in which triphenylamine units are introduced as substituents at the 9-position of fluorene. These groups simultaneously improve solubility, suppress aggregation, and improve hole injection.[137]

5.2.4.3   PF Copolymers

Many copolymers (Scheme 5.31) containing fluorene and other conjugated monomers have been reported. As is usual in polymer chemistry, the preparation of random copolymers and

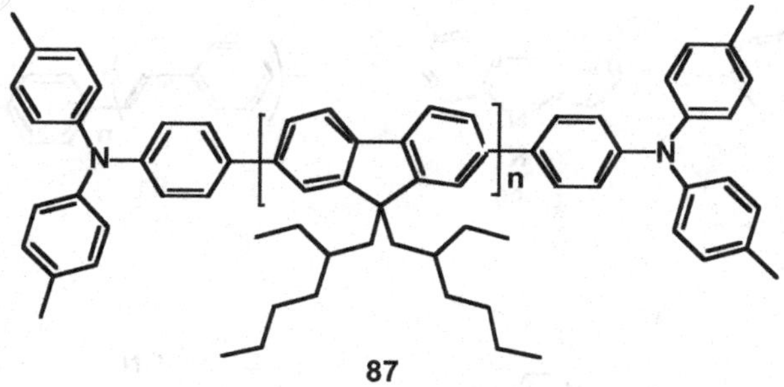

87

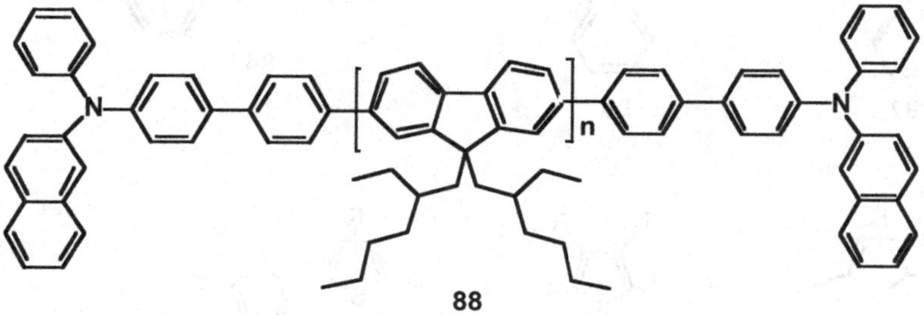

88

Scheme 5.29 PFs with aryl amine end groups.

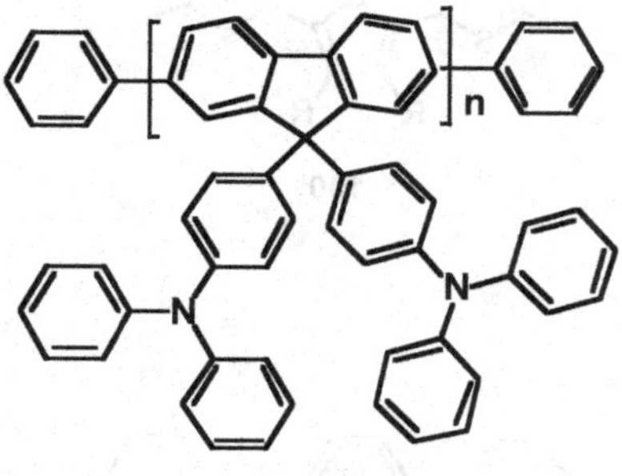

89

Scheme 5.30 A PF homopolymer with triphenylamine side groups.

alternating copolymers perturbs the short- and long-range order in the material. Examples of comonomers include the phenylene, thiophene, bithiophene, carbazole, stilbene, oxadiazole,

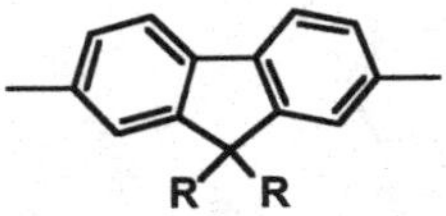

3 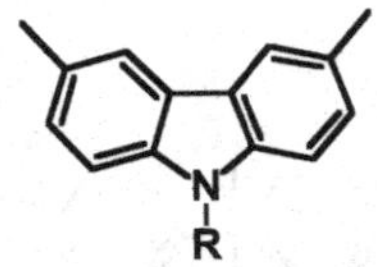   **90** 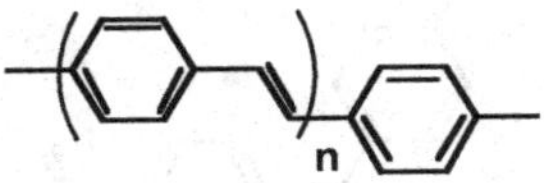   **91**

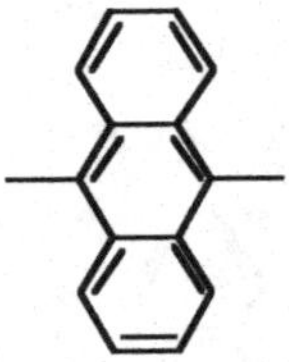

92 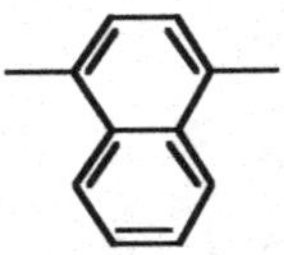   **93** 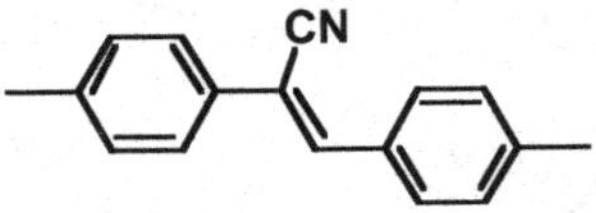   **94**

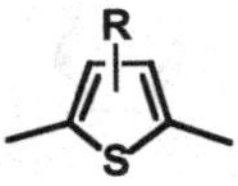

95 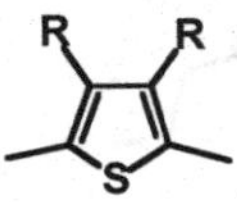   **96** 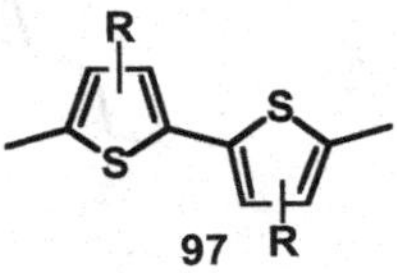   **97**

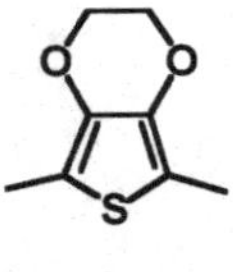

98 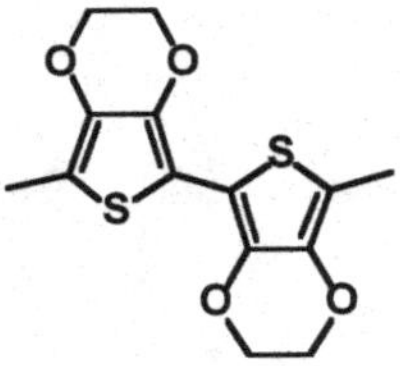   **99** 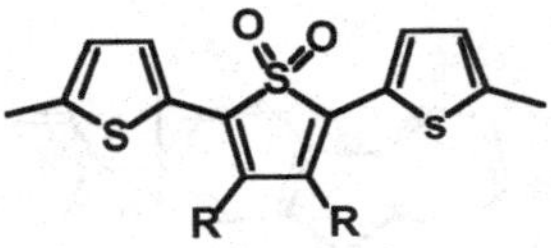   **100**

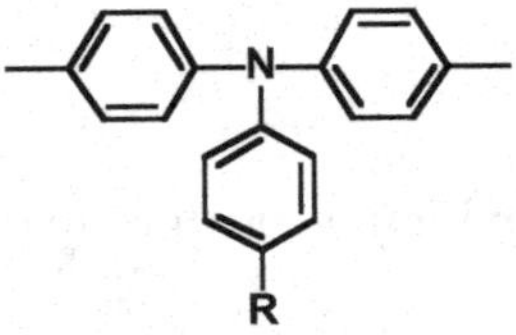

101 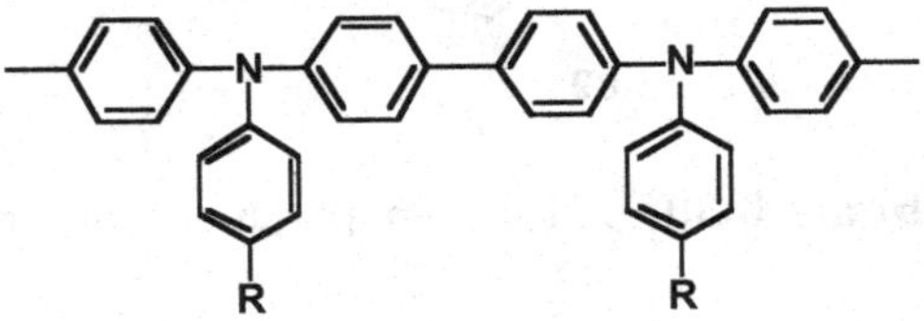   **102**

Scheme 5.31   Examples of PF copolymers.

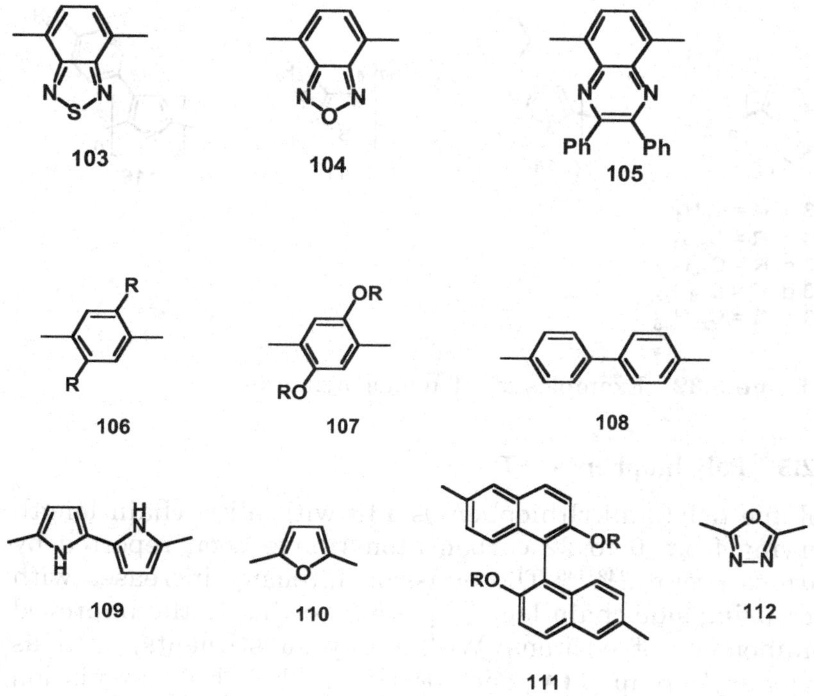

103 **104** **105**

106 **107** **108**

109 **110** **111** **112**

Scheme 5.31 (continued)

furan, quinoline derivatives and others (as reviewed by M. Leclerc).[138] The thiophene polymer (**95**) emits blue-green light, the cyanostilbene copolymer (**94**) emits green light, the bithiophene polymer emits yellow light (**99**), and the benzothiadiazole polymer (**103**) shows green emission. Comonomers can be chosen for designing polymers with well-balanced hole- and electron-transporting properties.[114a,139] The time-of-flight (TOF) mobilities of the fluorene-co-triphenylamine polymers (**101** and **102**) are relatively high for a non-aligned polymeric material.[134,140] Overall, PF copolymers offer the full range of color with high efficiency, low operating voltages, and long lifetime when applied in a device configuration and are under serious consideration for commercialization.[127,128,130,141]

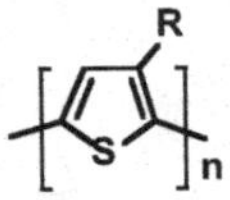

113 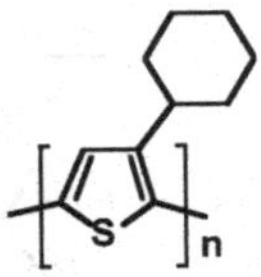   **114** 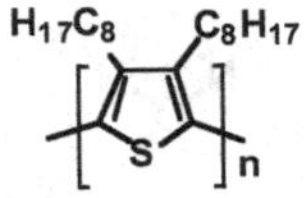   **115** 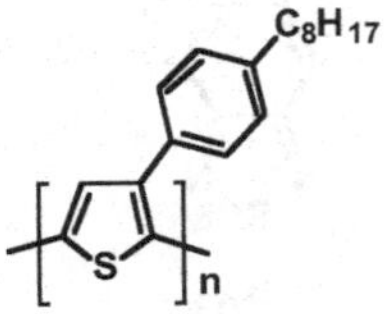   **116**

113 a R = C_6H_{13}
113 b R = C_8H_{17}
113 c R = $C_{12}H_{25}$
113 d R = $C_{18}H_{37}$
113 e R = $C_{22}H_{45}$

Scheme 5.32 Examples of PT homopolymers.

5.2.5 Polythiophenes (PTs)

Soluble poly(3-alkylthiophene)s **113** with alkyl chain length varying from 6 to 22 carbon atoms have been reported by various groups.[142–145] The emission intensity increases with increasing side chain length, possibly owing to the improved confinement of excitons. With bulky substituents, such as cyclohexyl group (**114**), one observes a blue shift in emission (555 nm).[146] Blue emission can be obtained by substitution at both the 3- and 4-positions, as in **115**;[146,147] see Scheme 5.32. When thiophenes are polymerized by use of ferric chloride or Grignard coupling reactions, the resulting polymers are regiorandom, with a preponderance toward head-to-tail coupling, varying from 52 to 80% according to the method used.[144] Examples of polythiophenes with regioregular structures are shown in Scheme 5.33.

Although the coupling of oligomeric alkylthiophenes using ferric chloride results in a random stereoregularity,[148] under carefully controlled conditions, the method can give high-molecular-weight, regioregular (94% head to tail) polymers, such as **116** (Scheme 5.32). Regiochemical control can be achieved because the aryl group preferentially stabilizes the cationic charge at the 2-position in the reaction intermediates.[149] More general routes for the synthesis of regioregular poly(3-alkylthiophene)s have been developed by use of the nickel-catalyzed coupling of 2-halo-5-metallothiophenes.[150–152]

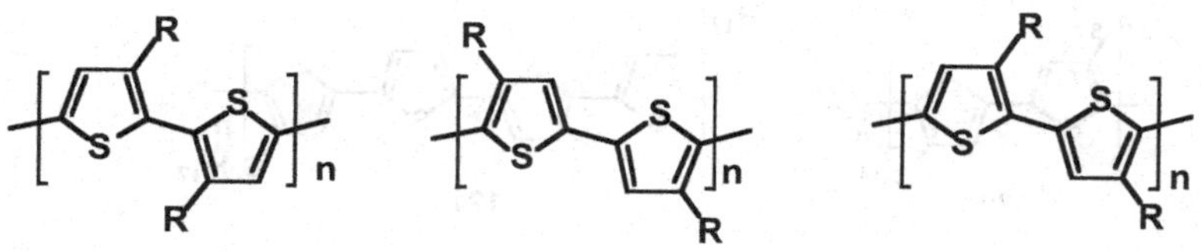

head to head   **tail to tail**   **head to tail**

117   **118**   **119**

Scheme 5.33 Regiochemistry of poly(alkylthiophene)s.

The emission frequency of PTs is controlled by the side chains and the stereoregularity of the backbone. For the alkylated PTs **120** and **121** in Scheme 5.34, there is a red shift with an increase of thiophenes between two consecutive HH (2,2′-coupled) dyads. Polymers **120** and **121** show emission maxima at 530 and 550 nm, respectively. Similarly, the emission maximum for poly(3-hexylthiophene) **113a** blue shifts with increasing amounts of HH linkages.[153,154] The influence of regioregularity on PL efficiencies was also demonstrated. For straight alkyl side chain–substituted PTs, regioregular structures with HH linkages show higher PL efficiencies than head-to-tail linkages.[155]

Despite good film forming properties and high thermal stability, PTs exhibit low PL quantum efficiency in the solid state, typically less than 5%, such as for poly(2-methyl-3-octylthiophene) **122**[156] and poly(3-hexylthiophene) **113a**,[157] which have solid state PL efficiencies of ~2%. PL values for oligothiophenes are even lower than those of thiophene polymers.[158,159]

In poly[3-(2,5-dioctylphenyl)thiophene] (**123** in Scheme 5.34), the 2,5-dioctyl-phenyl groups act as molecular "bumpers" to separate the conjugated chains.[160] The interchain separation in **123** increases the film PL quantum efficiency to 16%. A single-layer device of ITO/**123**/Ca/Al exhibits a maximum external efficiency of 0.1%, and a double-layer device of ITO/**123**/PBD/Ca shows an external efficiency of 0.7%.[160] Higher PL efficiencies are obtained for polythiophenes in solution and blending with inert polymer matrix.[161] Use of

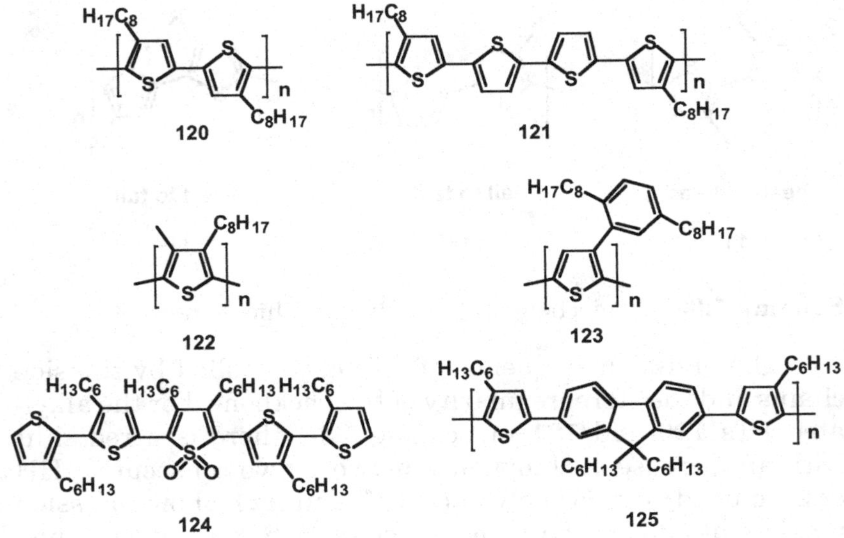

Scheme 5.34   Examples of PT derivatives for PLEDs.

charge carrier confinement layers[144] or doping with fluorescence dye materials[162] markedly increases device efficiency.

Oligothiophenes featuring sulfone substitution, as in **124**, show improved emission characteristics. A solid state efficiency of 37% has been reported for this oligomer, and devices of ITO/PVK/**124**/Ca display an external efficiency of 1%.[163] Encouraged by this report, polythiophene copolymers based on main chain modifications have been reported by different groups.[164] Polythiophene copolymers with PL efficiency of 37% have been reported for **125**. The double-layer device ITO/PVK/**125**/Ca shows an external efficiency of 0.6%.[164] Despite these improvements, the device performance of PTs remains noncompetitive, relative to those of other conjugated polymers, such as PPVs, PPPs, and PFs.

5.2.6   Conjugated Polymers with Electron-Deficient Heterocyclic Units

Scheme 5.35 shows representative examples of polymers containing electron-deficient heterocycles. Typical structures are

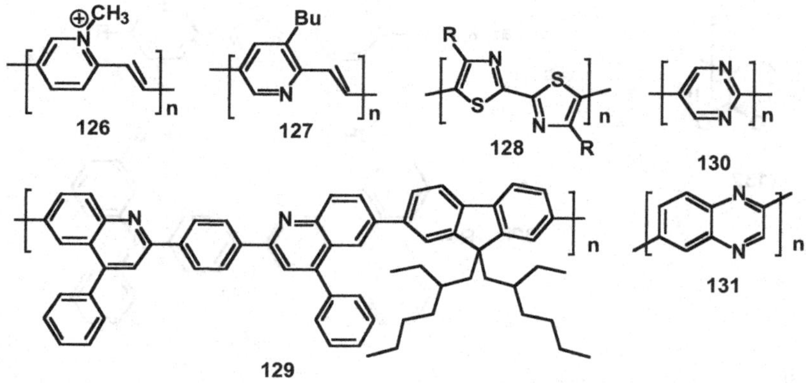

Scheme 5.35 Conjugated polymers with electron deficient hetero-cyclic units.

those containing pyridine (**126** and **127**),[165] pyrimidine **130**,[166] quinoline **129**,[167] polyquinoxaline **131**, or bisthiazole **128** units within the conjugated chain.[168] Among them, it is worth mentioning the use of polyquinoxalines for electron-injecting materials. Similar to regioregular and regioirregular poly-thiophenes, three different regioisomers have also been reported for pyridine-based PPVs, with EL emission maxima of 575 nm (random), 584 nm (head to tail), and 605 nm (head to tail), respectively.[169] In addition, protonated N-H or *N*-alkylpyridinium derivatives such as **126** have been used for device fabrication by self-assembly techniques.[170]

5.2.7 Polymers with Side Chain Chromophores

Well-defined fluorophores have been used as side chains in nonconjugated polymers for PLED fabrication. Polymers with well-defined short arylenevinylene segments in the side chain provide amorphous materials with high glass transition temperatures.[171] Furthermore, any type of functionalizable dye or charge transport material may be used as a side chain, allowing a range of emission wavelengths and charge transport properties.[172]

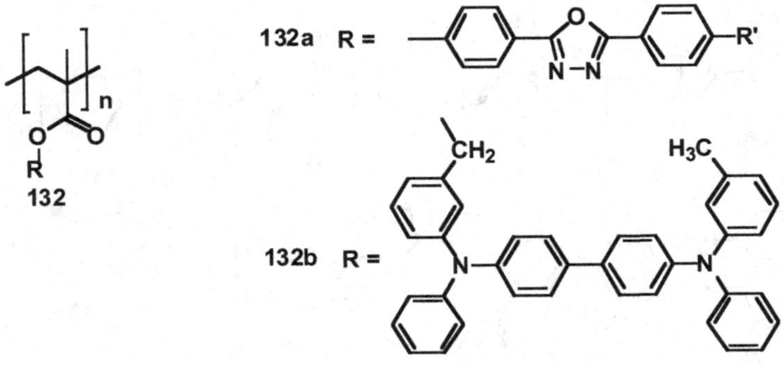

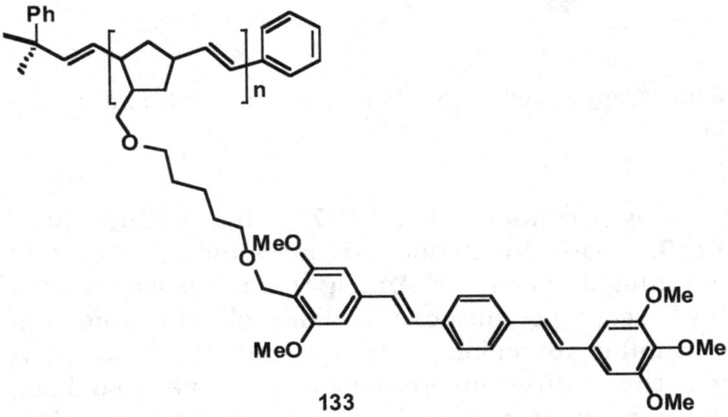

Scheme 5.36 Examples of nonconjugated polymers with side chain chromophores.

It is possible to obtain poly(methacrylate)s with different chromophores laterally fixed to the polymer backbone by using AIBN-initiated radical polymerization protocols. Thus, electron transport polymers containing oxadiazole side chain chromophore **132a** (Scheme 5.36) have been used as the electron transport layer in multilayer devices.[173] Hole transport polymers have also been synthesized by introducing TPD (*N*'-bis(3-methylphenyl)-*N*,*N*'-diphenyl-1,1'-biphenyl-4,4'-diamine) as the side group in **132b**.[172]

The ring-opening metathesis polymerization of a norbornene monomer with a distyrylbenzene chromophore was used to produce polymer **133**, which shows blue emission with a maximum at 475 nm.[174] A single-layer device of ITO/**133**/Ca showed an internal efficiency of 0.3% and a turn-on voltage of 12 V; while a double-layer device with oxadiazole as the electron-transporting layer displayed an efficiency of 0.55%. Other examples of side chain polymers were recently reviewed.[175]

5.3 SURVEY OF THE SYNTHETIC CHEMISTRY FOR ELECTROLUMINESCENT POLYMERS

In this section we examine the most common methodologies for preparing electroluminescent conjugated polymers. The section begins by looking at historically important polymerization methods. We then cover polymerization reactions that operate under the control of transition metal catalysts. Finally, we catalog organic polymerization reactions. Examples from polymers covered in the previous discussion on structure–property relationships are given for all reactions.

5.3.1 Historically Relevant Reactions

5.3.1.1 FeCl₃-Catalyzed Condensation Polymerization (for **3, 113, 114, 115, 116, 120, 121, 122, 123, 125**)

Polymerizations based on oxidative coupling reactions were widely used during the early stages of conjugated polymer research. These reactions are best used with aryl and heterocyclic compounds having electron-rich sites around at their ring systems. For example, polycondensation of 9,9-dialkylfluorene (**134**) or substituted thiophenes (**135**) can be carried out in the presence of $FeCl_3$ as the oxidizing agent in chloroform and yields the corresponding polymers, as shown in Scheme 5.37. This method is convenient because of the simplicity of monomer synthesis; however, the polymer structures are poorly defined. PFs have linkages mainly at 2,7-positions, and PTs at the 2,5-positions.[109,110,142] Typical degrees of polymerization are in the order of 10 to 100. The stereoregularity of the PTs obtained by this method can be improved when the monomers are disubstituted at the 3,4-positions.[176] The molecular weight

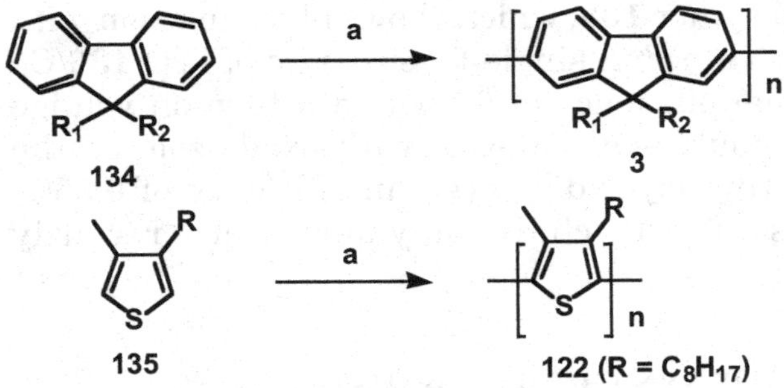

Scheme 5.37 Oxidative polymerization route to PFs and PTs. Reaction conditions: (A) $FeCl_3/CHCl_3$

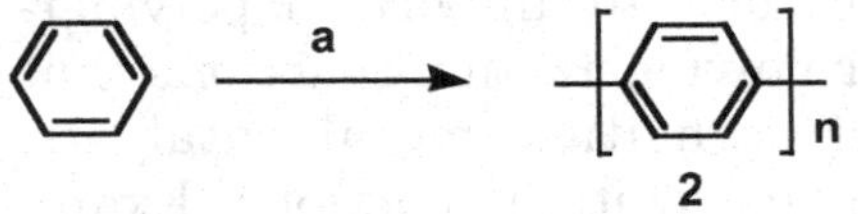

Scheme 5.38 Direct synthesis of PPP via Friedel–Crafts polycondensation. Reaction conditions: (A) $AlCl_3/CuCl_2$, H_2O.

distributions of the polymers synthesized in this method are usually broad.

5.3.1.2 Friedel–Crafts Polymerization (for **2, 4**)

In a manner similar to the direct oxidative polymerization, Lewis acids, such as $AlCl_3$, $AlBr_3$, $FeCl_3$, can promote condensation polymerizations.[177] Both cross-linked and branched polymers can be obtained, depending on the reaction conditions and the nature of the polymerization agent. As shown in Scheme 5.38, only oligomeric materials can be obtained by this method.[178]

5.3.2 Metal-Mediated Polycondensation Reactions

5.3.2.1 Polycondensation of Organomagnesium Monomers (for **4, 137**)

Grignard reagents refer to organomagnesium complexes, which can be prepared by direct reaction of an organic halide

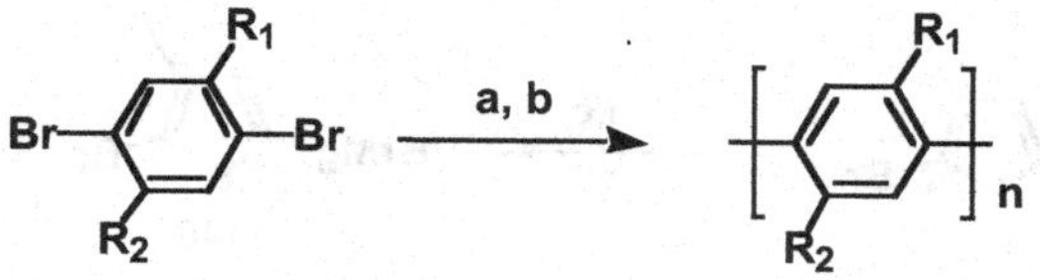

Scheme 5.39 Synthesis of PPP through coupling of Grignard reagents. Reaction conditions: (A) Mg; (B) Ni(dppf)$_2$Cl$_2$, THF.

with magnesium metal.[179] PPPs with long alkyl side chains at every repeat unit (**137**) can be prepared through nickel-assisted coupling of organomagnesium reagents, as shown in Scheme 5.39.[180] The products are characterized by exclusive 1,4-linkages of benzene rings along the polymer main chain. However, an average degree of polymerization of 8 to 20 units is typical for polymers prepared by this method.[178]

5.3.2.2 The McCullough Method for PTs (for **119**)

The McCullough method utilizes bromination at the 2-position of 3-alkylthiophene precursors (**138**) and selective lithiation at the 5-position with lithium diisopropylamine (LDA), followed by quenching with MgBr$_2$.OEt$_2$. The polymerization is catalyzed by a nickel catalyst, such as Ni(dppf)$_2$Cl$_2$ (dichloro[1,1'-bis(diphenylphosphino)ferrocene]nickel(II)). The difunctional monomers (**140**) are stable, but air sensitive, and the overall reaction is controlled by the metal site to give stereoregular structures.[176] The reaction sequence in Scheme 5.40 is one of the most successful methods for generating regioregular PTs.

5.3.2.3 The Rieke Method for PTs (for **119**)

Another important methodology for regioregular PTs is the Rieke method, which relies on organozinc reagents for orienting linkage selectivity.[181] As shown in Scheme 5.41, the higher reactivity of zinc toward iodide, relative to bromide (**141**), leads to selective incorporation at the site opposite alkyl substitution.[181]

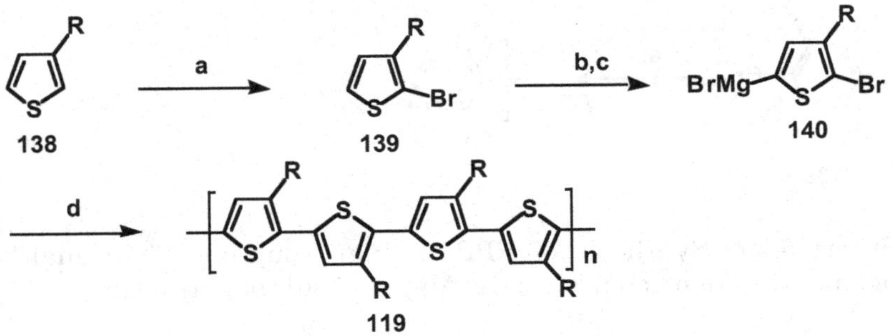

Scheme 5.40 McCullough method for the regioregular synthesis of poly(3-alkylthiophenes). Reaction conditions: (A) Br_2, AcOH, or NBS, AcOH, $CHCl_3$ or NBS, THF; (A) LDA, THF; (C) $MgBr_2$, diethylether; (D) $Ni(dppf)_2Cl_2$.

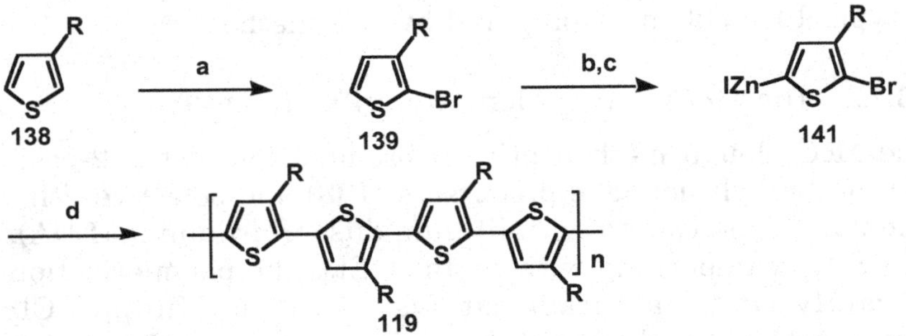

Scheme 5.41 Rieke method for the regioregular synthesis of poly(3-alkylthiophenes). Reaction conditions: (A) Br_2, AcOH, or NBS, AcOH, $CHCl_3$, or NBS, THF; (B) I_2, HNO_3, CH_2Cl_2; (C) Rieke Zn; (D) $Ni(dppe)_2Cl_2$.

The polymerization is carried out using Ni(0) or Pd(0) catalysts, such as $Ni(dppe)_2Cl_2$ (1,2-bis(diphenylphosphino)ethane nickel (II) chloride) and $Pd(PPh_3)_4$.

5.3.2.4 The Heck Reaction (for 6, 32, 59, 62)

The Heck reaction is a palladium-catalyzed vinylation of organic halides, which leads to the formation of carbon–carbon double bonds in mostly *trans* configuration, with very few side reactions.[182] Instead of forming the olefinic linkage in the polymerization step, the functionality can be prebuilt, in the form of styrene-type monomers.[183,184] The reaction has the further advantage of being compatible with a suitable range of chemical functionalities. In addition, the use of aryl dibromides or diiodides as monomers is particularly useful, as it allows the synthesis of PPVs containing alkyl rather than alkoxy side chains.[185]

PPV derivatives, such as poly(2,5-dialkoxy-1,4-phenylenevinylene) and poly(2-phenyl-1,4-phenylenevinylene),[59,183b,186] can be prepared through the coupling of 1,4-dibromobezene derivatives (2-phenylene, 2,5-dialkoxy) with ethylene or other olefins by using palladium acetate, $PdCl_2$, $Pd(dba)_2$ (dibenzylideneacetone palladium), Pd/C, or $Pd(PPh_3)_4$. The number-average molecular weight of the polymers is typically in the range of 5000. Some side reactions, such as reductive dehalogenations and 1,1-diarylations at the olefin moiety, have been detected. The regioselectivity of Heck-type coupling can be controlled by the reaction conditions, by the leaving group, by the substituents on the arylene component, and by the choice of the olefinic monomer.[185]

Scheme 5.42 shows three different methodologies that have been employed to extend the use the Heck reaction to the synthesis of the PPVs: bubbling of ethylene gas through a solution of activated Pd(0) catalyst containing dihalo-aromatic compounds (**136**); coupling of both divinylbenzene (**142**) and dibromobenzene (**136**); or the self-coupling of the difunctional monomer (**143**) containing both halide and vinylene groups.[178]

From a practical perspective the reaction between dihalide and divinylbenzene monomers is useful when the synthesis of bisaldehyde, bissulfonium, or bisphosphonium monomers is difficult. A drawback of the Heck-type reaction is that it is not strictly regioselective.[187] Depending on the substituents, more than 1% of 1,1-diarylation can be observed.

The Heck reaction is of prime importance in the synthesis of oligophenylenevinylene structures. For example, the tetrahedral compound (**59**) was prepared by reaction of

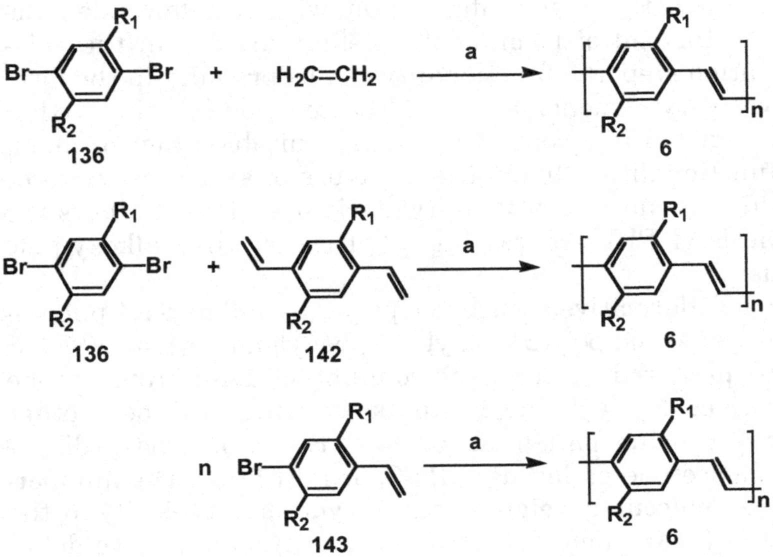

Scheme 5.42 Heck reaction for the synthesis of PPVs. Reaction conditions: (A) Pd (PPh$_3$)$_4$, NEt$_3$, THF.

tetrakis(4-iodo-phenyl)methane (**144**) and an excess of 2,5-dioctyloxy-1-styryl-4-(4'-vinylstyryl)benzene (**145**), using palladium acetate as the catalyst with a yield of 73% (*cis* and *trans* mixture);[82] see Scheme 5.43. The all-*trans* structure can be obtained by addition of a catalytic amount of iodine.[188]

5.3.2.5 The Suzuki Reaction (for 3, 6, 60, 61, 70, 71, 72, 73, 74, 75, 76, 77, 78, 79, 81, 82, 90, 92, 93, 95, 96, 97, 98, 99, 100, 101, 102, 103, 104, 105, 106, 107, 108, 109, 110, 125, 129, 137, 159)

The Suzuki reaction refers to the cross coupling of aryl or vinylic halides (or triflates) with organoboron reagents.[189] The coupling of aryl-boronic acid derivatives and haloaryl structures, using a Pd(0) catalyst has been widely used for the synthesis of substituted PPPs and some PPVs. The most commonly used palladium catalysts are Pd(PPh$_3$)$_4$, Pd(PPh$_3$)$_2$Cl$_2$, and Pd(OAc)$_2$ with

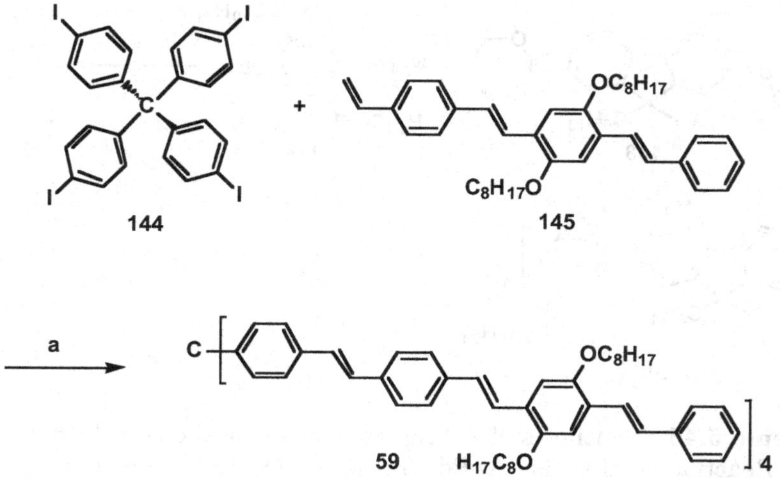

Scheme 5.43 Synthesis of tetrahedral molecules with Heck coupling. Reaction conditions: (A) $Pd(OAc)_2$, DMF.

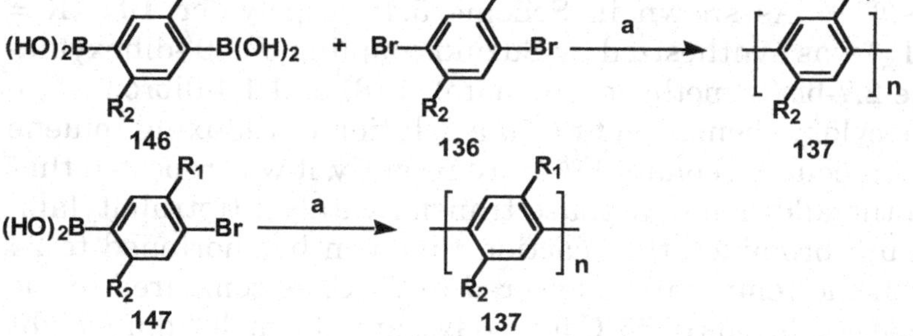

Scheme 5.44 Suzuki coupling for the polymerization. Reaction conditions: (A) $Pd(PPh_3)_4$, toluene/H_2O, K_2CO_3.

triphenylphosphine. Typical reactions are illustrated in Scheme 5.44.[190] The Suzuki reaction has the advantages of being tolerant of a large variety of functional groups and is insensitive to water. Syntheses of alkyl-substituted PPPs are often carried by coupling of alkyl-substituted arylbis(boronic acids) (**146**) with

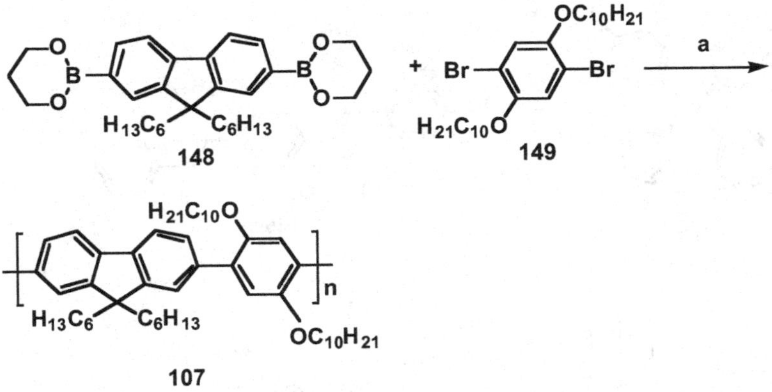

Scheme 5.45 Synthesis of PF copolymers using the Suzuki reaction. Reaction conditions: (A) Pd(PPh$_3$)$_4$, K$_2$CO$_3$, H$_2$O/toluene.

aryl dibromides (**136**)[190] or by use of 2,5-dialkyl-4-bromobenzeneboronic acid (**147**). 2,5-Dialkylbenzene-1,4-bis(trimethyleneborate)s polymerize smoothly to give higher-molecular-weight polymers than those obtained with the parent boronic acids.[191,192] As shown in Scheme 5.45, copolymer **107** (R = C$_{10}$H$_{21}$) was synthesized by Suzuki coupling of 9,9′-dihexylfluorene-2,7-bis(trimethyeneboronate) (**148**) and 1,4-dibromo-2,5-bis(decyloxy) benzene (**149**) in a solution of refluxing toluene and aqueous carbonate.[139h] More recently, it was reported that with the addition of a phase transfer catalyst (tetrabutylammonium bromide), the reaction time can be shortened to 24 h with the temperature lowered to 75°C, as compared to the typical conditions of 85°C for 3 days, and the molecular weight could be improved to over 100,000, provided that the monomers are purified rigorously.[193]

PPP featuring propoxysulfonate groups (**74**), shown in Scheme 5.46, can be obtained by the polymerization of 1,4-dibromo-2,5-bis(3-sulfonatopropoxy) benzene (**150**) with benzene bisboronic acid (**151**) as outlined above.[194] These water-soluble polymers have found use in multilayer device fabrication through layer by layer self-assembly.[98]

In the synthesis of the ladder-type polymers, a precursor polymer (**153**) is prepared through the Suzuki coupling of an

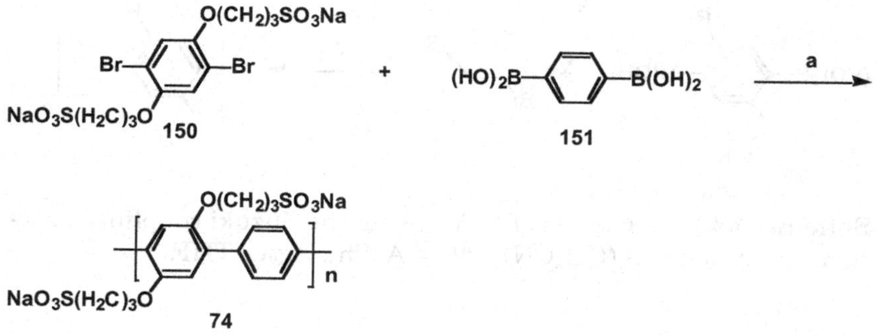

Scheme 5.46 Synthesis of water-soluble PPPs through Suzuki coupling. Reaction conditions: (A) $Pd(PPh_3)_4$, $NaHCO_3$, DMF.

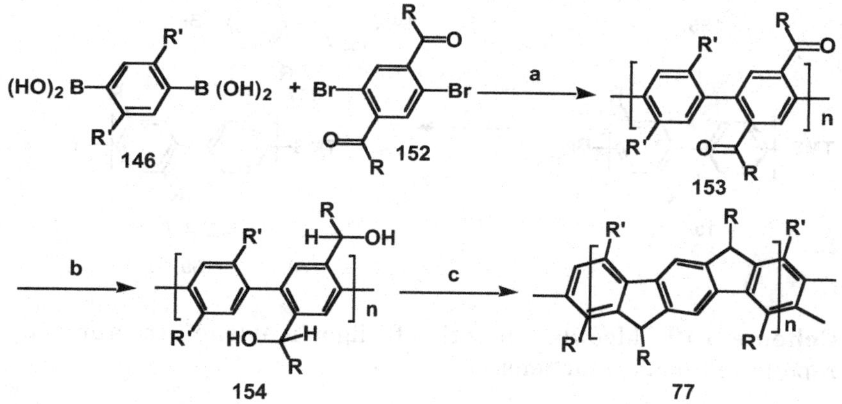

Scheme 5.47 Synthesis of LPPP. Reaction conditions: (A) $Pd(PPh_3)_4$, K_2CO_3, H_2O/toluene; (B) LAH; (C) BF_3.

aromatic diboronic acid (**146**) and an aromatic dibromoketone (**152**). Cyclization to the soluble LPPP takes place in a two-step sequence, consisting of reduction of the keto group, followed by ring closure of the secondary alcohol groups in a Friedel–Crafts alkylation. The overall synthetic sequence for LPPP (**77**) is shown in Scheme 5.47.

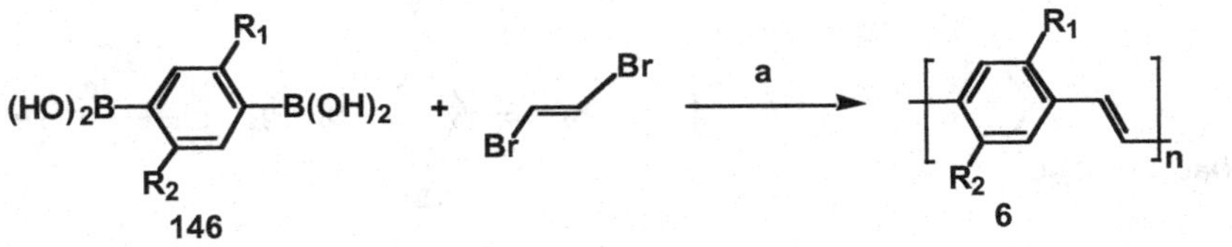

Scheme 5.48 Synthesis of PPVs using the Suzuki coupling. Reaction conditions: (A) $(CH_3CN)_2PdCl_2$, $AsPh_3$, Ag_2O, THF.

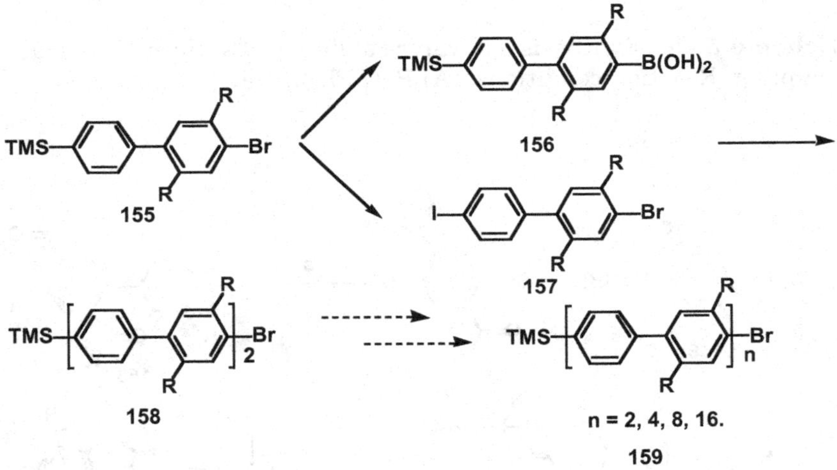

Scheme 5.49 Modular growth of oligophenylene structures by repetitive Suzuki reactions.

When the Suzuki reaction was used to synthesize polymers of general structure **6**,[195] about 3% of biaryl defect structures were observed in the coupling products (M_n up to 12,000), resulting from homo-coupling of boronic acid functions, although no 1,1-diarylated moieties were found; see Scheme 5.48.

A strategy to well-defined oligophenylene structures involves the repetition of protection/coupling/deprotection sequences in a convergent process to minimize the number of reaction steps (Scheme 5.49).[196] With this approach,

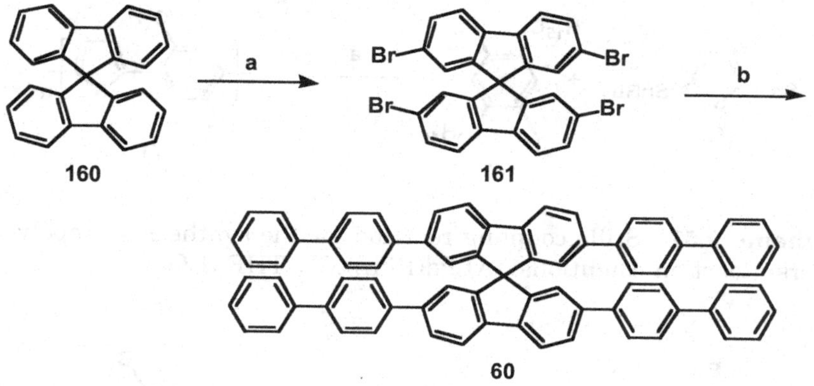

Scheme 5.50 Spirofluorene derivatives prepared by the Suzuki reaction. Reaction conditions: (A) Br_2, $CHCl_3$; (B) biphenyl boronic acid, $Pd(PPh_3)_4$, K_2CO_3, H_2O/toluene.

monodisperse oligo(phenylene) molecules with up to 16 phenylene rings and with well-defined functional end groups have been obtained. The key principle rests upon the significantly faster coupling of the iodoaryl functionality in comparison to the corresponding bromoaryl in the aryl-aryl cross-coupling reaction. Accordingly, building blocks (**157**) containing both iodo and bromo functionalities undergo coupling with an arylboronic acid preferentially at the iodide site. The unreacted bromo site can be used for subsequent chemical derivatization, such as conversion into a boronic acid functionality (**156**).[185]

The Suzuki coupling has also been used in the synthesis of spirofluorene derivatives (**60** in Scheme 5.50).[197] The molecular topology of the molecules reduces the crystallization tendency of the bulk material.

5.3.2.6 The Stille Coupling Reaction (for **31, 71, 126, 127**)

The Stille coupling is the palladium-catalyzed coupling between an organostannane and an electrophile to form a new C–C bond.[198] Advantages of this method are that organostannanes are not oxygen or moisture sensitive and tolerate a

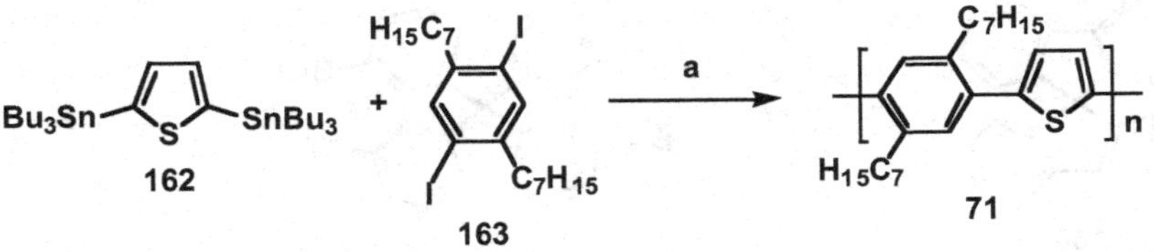

Scheme 5.51 Stille coupling reaction for the synthesis of copolymers. Reaction conditions: (A) $Pd(PPh_3)_2Cl_2$, THF/H_2O.

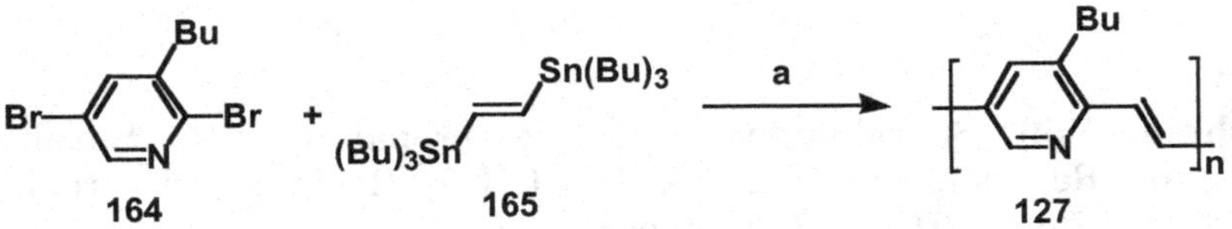

Scheme 5.52 Synthesis of pyridine-based PPVs through Stille coupling. Reaction conditions: (A) $Pd(PPh_3)_2Cl_2$, THF/H_2O.

wide variety of functionalities. Organostannanes are, however, toxic and difficult to purify.

Polymer **71** was prepared by the reaction of 1,4-diiodo-2,5-diheptylbenzene (**163**) and 2,5-bis(tributylstannyl) thiophene (**162**) in the presence of $Pd(PPh)_3Cl_2$ (Scheme 5.51). From GPC analysis, the resulting polymer exhibits a number-average molecular weight near 9500 and a polydispersity of 2.8.[199] As shown in Scheme 5.52, coupling of 3-butyl-2,5-dibromopyridine (**164**) and 1,2-bis(tributylstannyl)ethylene (**165**) gives the substituted polymer (**127**), which is soluble in conventional organic solvents.[170]

5.3.2.7   The McMurry Reaction (for **167**, **170**)

The coupling of aromatic dialdehydes with low-valent titanium reagents to form olefinic units is known as the McMurry coupling reaction.[200] This method was used to synthesize polymer **167** featuring a *cis/trans* ratio of about 0.4, and an average degree of polymerization of 30 (Scheme 5.53).[201]

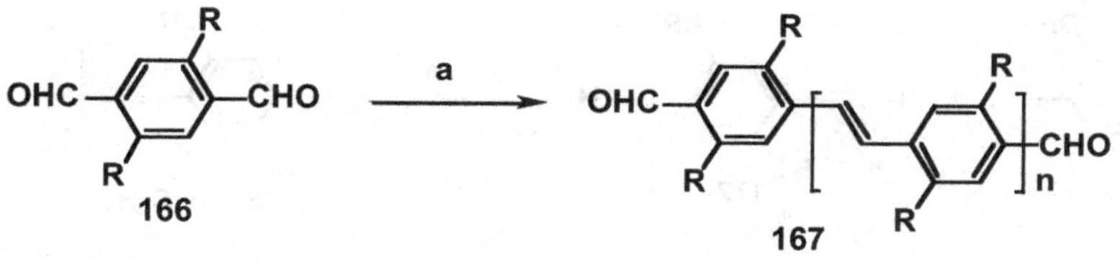

Scheme 5.53 An example of the McMurry polycondensation reaction. Reaction conditions: (A) $TiCl_4$, Zn, THF.

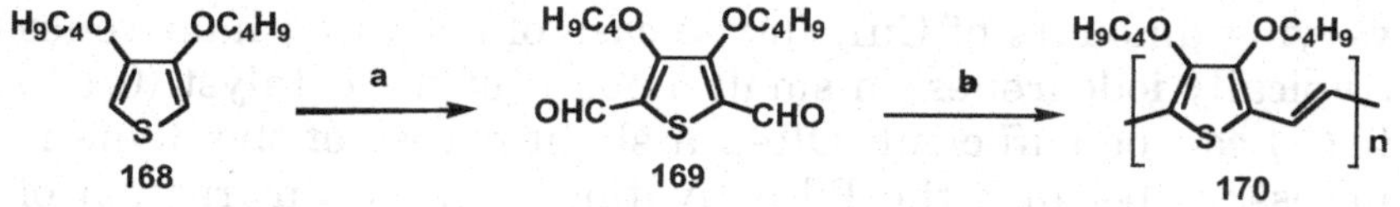

Scheme 5.54 McMurry coupling route to dibutoxy PTV. Reaction conditions; (A) BuLi, TMEDA, DMF; (B) $TiCl_4$, Zn, THF.

To synthesize poly(3,4-dibutoxy-2,5-thienylenevinylene) (**170**), the dialdehyde (**169**) is polymerized in the presence of titanium tetrachloride and zinc dust. This technique can produce polymers with molecular weights of up to 35,000 (Scheme 5.54).[202]

5.3.2.8 The Sonogashira–Hagihara Reaction (for **63**, **65**, **66**, **67**, **68**)

The Pd-catalyzed coupling of terminal alkynes with aromatic bromides or iodides in amine solvents is known as the Sonogashira–Hagihara reaction.[203,204] This reaction is a powerful method to form C-C single bonds between sp- and sp^2-hybridized carbon centers.[204] In most cases, commercially available $Pd(PPh_3)_2Cl_2$ is used as the Pd source. Although both bromo- and iodo-aromatic compounds work in this coupling, the corresponding iodides react considerably faster, in higher yield, and at lower reaction temperatures. The milder reaction conditions with the diiodide reagents yield polymers with less cross-linking and fewer structural defects.[205]

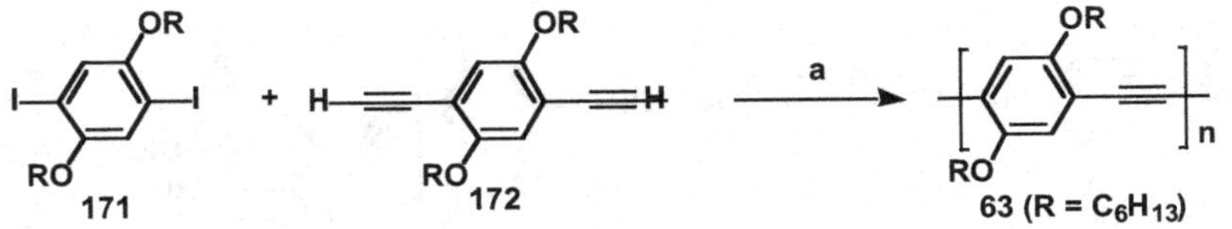

Scheme 5.55 Synthesis of PPE by the Sonogashira–Hagihara route. Reaction conditions: (A) $Pd(PPh_3)_4$, CuI, toluene, diisopropylamine.

Most reactions require 0.1 to 5 mol% $Pd(PPh_3)_2Cl_2$ and varying amounts of CuI. In the case of activated haloarenes (typically iodoarenes), a small amount of the catalyst (0.1 to 0.3%) can be sufficient. Often a slight excess of the diyne is necessary because the Pd activation step uses a fraction of this reagent.[206] CuI is necessary for the conversion of dibro-moarenes into the corresponding alkynylated products,[207] and diisopropylamine, piperidine, pyrrolidine, and morpholine are often suitable solvents.[208] Both the yield and purity of the coupling reaction depend on the choice of amine and co-solvent. Furthermore, electron-withdrawing substituents on the halide improve both the yield and the rate of the coupling reaction. *Ortho-* and *para*-positioned dihalides are more efficient than those placed in the *meta*-position.[205]

Polymers obtained by this coupling have molecular weights varying from thousands to hundred thousands. It is difficult to obtain PPEs of high molecular weight, unless aromatic diiodides are utilized.[205] In addition, the end groups of PPE are often ill defined as a result of dehalogenation and phosphonium salt formation.[209] The Pd-made polymers may contain diyne defects, formed by either reduction of the Pd^{2+} catalyst precursor or the adventitious presence of the oxygen (Scheme 5.55).

5.3.3 Metathesis Polymerization

5.3.3.1 Acyclic Diyne Metathesis Polymerization (for **64, 174, 176, 178**)

The alkyne metathesis reaction corresponds to the permutation of groups across an alkynyl linkage. It can be used to

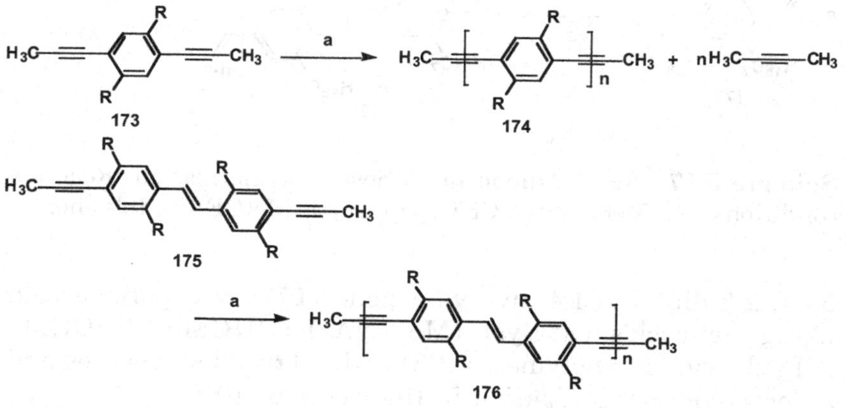

Scheme 5.56 Synthesis of PPEs through ADMET polymerization. Reaction conditions: (A) Mo(CO)$_6$, 4-chlorophenol or 4-(trifluoromethyl)phenol.

form PPEs by taking advantage of the acyclic diyne metathesis polymerization reaction. Although several homogeneous catalysts exist to effect this transformation, such as Schrock's tungsten-alkylidyne complex[210] and the Mortreux system,[211] the reagent developed by Bunz has been used to generate the widest range of polymer structures (Scheme 5.56).[212] The Bunz method uses Mo(CO)$_6$ and 4-chlorophenol with reaction temperatures in the range of 105 to 150°C. During the reaction, it is important to remove the 2-butyne by-product to shift the equilibrium toward polymer formation. Removal is typically achieved by a purge of nitrogen as the reaction proceeds. The degree of polymerization ranges from 300 to 2000 repeat units. Sufficiently bulky side chains are required to ensure that the polymer does not precipitate.[86] In some cases where 4-chlorophenol is not effective as the co-solvent, the more acidic 4-(trifluoromethyl)phenol may be used (Scheme 5.56).[213]

    Soluble PPV derivatives can be prepared by acyclic diene metathesis (ADMET) polymerization through the use of dialkyl divinyl benzene monomers.[214] As shown in Scheme

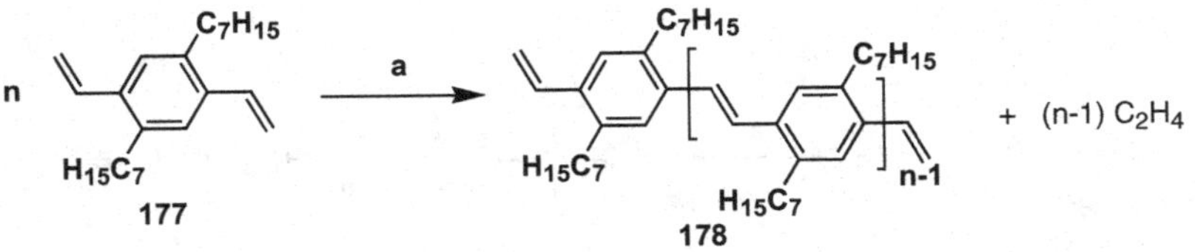

Scheme 5.57 Acyclic diene metathesis polymerization. Reaction conditions: (A) Mo(=NAr)(=CHCMe$_2$Ph)-[OCMe(CF$_3$)$_2$]$_2$, toluene.

5.57, 2,5-diheptyl-1,4-divinylbenzene (**177**) was polymerized using Schrock's catalyst (Mo(=NAr)(=CHCMe$_2$Ph)[OCMe (CF$_3$)$_2$]$_2$) yielding polymers (**178**) with all *trans*-structures and a degree of polymerization in the range of 10.

5.3.3.2 Ring Opening Olefin Metathesis Polymerization (for **1**)

The olefin metathesis reaction allows facile synthesis of PPV derivatives or PPV precursors by a simple single-step process. As shown in Scheme 5.58, starting from bicyclic monomers possessing a bicyclo[2.2.2]octadiene skeleton (**179**), the polymerization with Mo(=NAr)(=CHCMe$_2$Ph)-[OCMe(CF$_3$)$_2$]$_2$ yields well-defined, soluble precursor polymers containing carboxylic ester functions (**180**).[215] The paracyclophane route to PPV was also successfully applied to the synthesis of PPV precursors.[216] Because the polymerization is living, it is possible to prepare block copolymers of PPV and polynorbornene segments. These copolymers are soluble in organic solvents and the average degree of polymerization of the PPV blocks can be controlled and optimized to maximize PL efficiency in solution.[217,218]

5.3.4 Wittig and Wittig–Horner Reactions (for **6, 14, 25, 29, 30, 34, 35, 36, 38, 41, 42, 43, 57, 58, 62, 188, 193**)

The reaction of an alkylidene phosphorane with an aldehyde or ketone to yield alkene and a phosphine oxide is called the Wittig reaction.[219] The Wittig reaction has been applied for

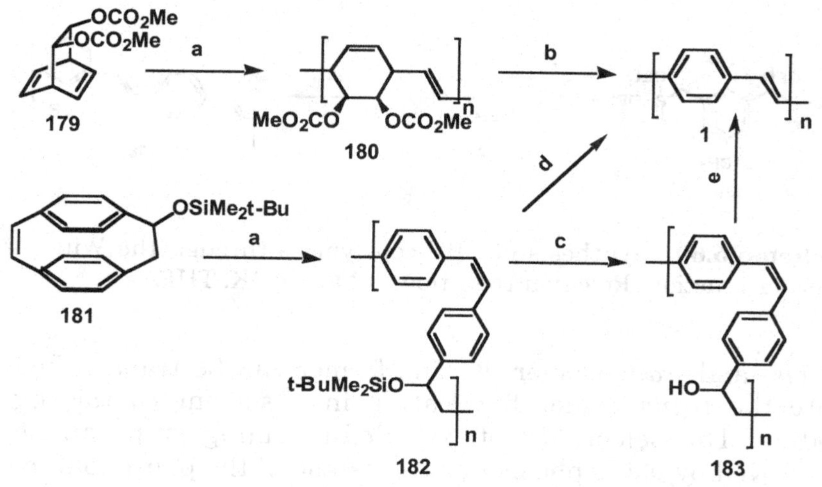

Scheme 5.58 ROMP routes to PPV. Reaction conditions: (A) Mo(=NAr)(=CHCMe$_2$Ph)-[OCMe(CF$_3$)$_2$]$_2$; (B) 280°C; (C) Bu$_4$NF; (D) HCl (g), 190°C; (E) 105°C.

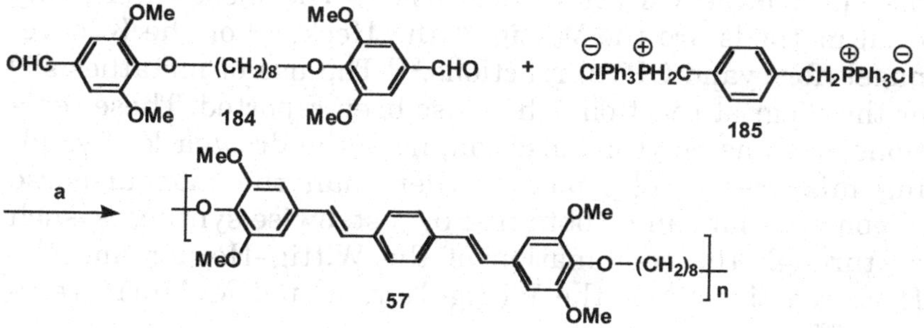

Scheme 5.59 Synthesis of PPV copolymers through the Wittig reaction. Reaction conditions: (A) EtONa, CHCl$_3$, EtOH.

polymer synthesis through the coupling of aldehydes with benzyl triphenyl phosphonium salts to form alkenes (Scheme 5.59). The overall reaction yields achieved are moderate to high, but the molecular weights are not high (10,000 or less). One drawback of this method is the simultaneous production

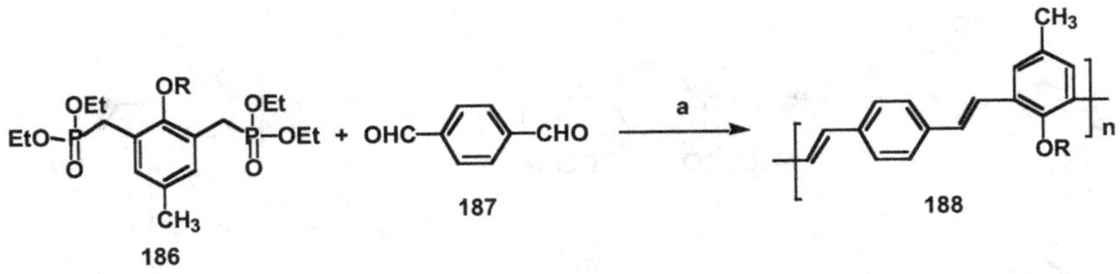

Scheme 5.60 Synthesis of PPV copolymers through the Wittig–Horner reaction. Reaction conditions: (A) t-BuOK, THF.

of *cis*- and *trans*-isomers.[188] The former can be transformed into the *trans*-isomer by heating in a solvent containing iodine. The selectivity of the olefin-forming step can be improved by using phosphonates instead of the phosphonium salts, namely, the Wittig–Horner reaction. This variation produces almost all *trans*-alkenes.[188,221] As shown in Scheme 5.60, polymer **188** obtained through the Wittig–Horner reaction has a molecular weight of 54,000.[222]

Oligo(1,4-phenylene vinylene)s are, in principle, accessible via a large variety of methods.[223] The most extensively used methods are the Wittig,[224] the Heck,[183a] or the Knoevenagel (for cyano OPVs) reactions.[225] The use of metathesis[226] or the Siegrist reaction[227] has also been reported. These reactions, used as polycondensation, have the drawback of yielding mixtures of oligomers rather than the monodisperse oligomers that can be obtained by a stepwise syntheses, such as through the combination of the Wittig–Horner and the Heck reactions[228] or the Wittig–Horner and McMurry reactions.[229]

The combination of the Wittig–Horner reaction and the Heck reaction is shown in Scheme 5.61. Starting from a monofunctional iodostilbene derivative (**189**), two types of bifunctional stilbene monomers [one with a vinyl and an aldehyde functionalities (**190**), and the other with an iodo and a phosphonate functionalities (**192**)] are added alternatively, beginning with **190**. In this way, the iodoarene sites are coupled with the olefin groups, and the phosphonate groups react with the aldehyde functionalities. Similarly, oligomers **58a** and **58b**

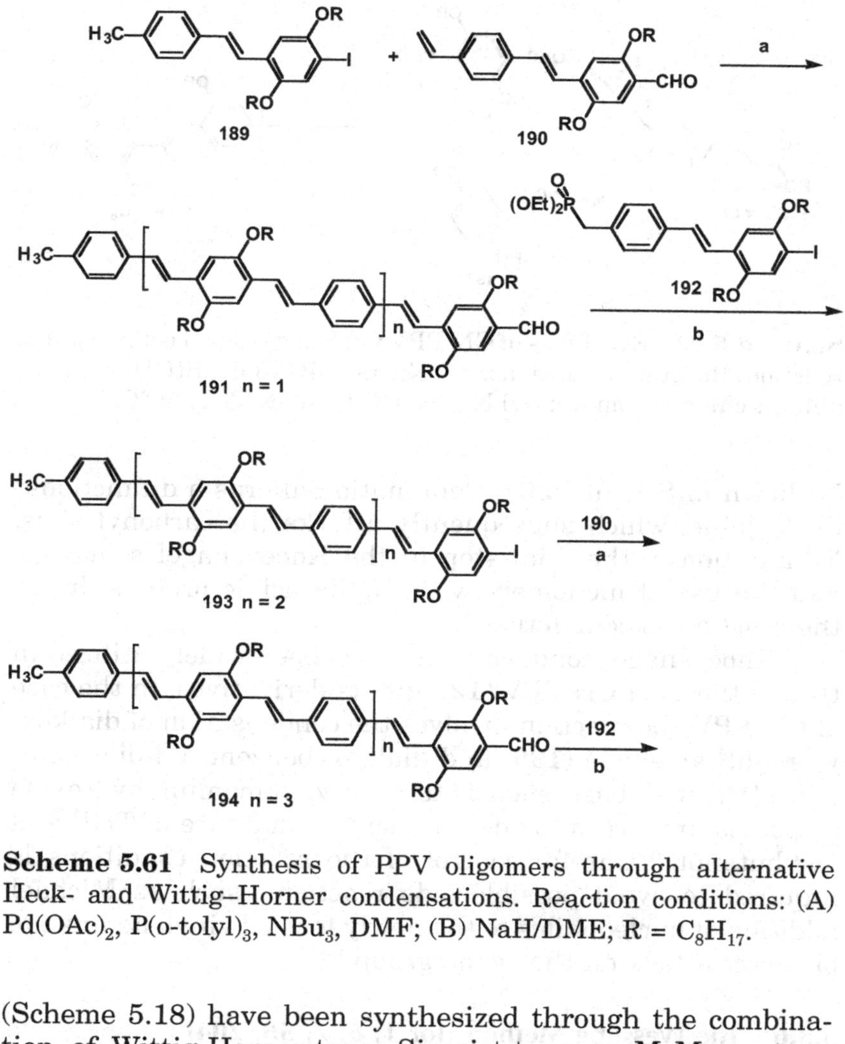

Scheme 5.61 Synthesis of PPV oligomers through alternative Heck- and Wittig–Horner condensations. Reaction conditions: (A) Pd(OAc)$_2$, P(o-tolyl)$_3$, NBu$_3$, DMF; (B) NaH/DME; R = C$_8$H$_{17}$.

(Scheme 5.18) have been synthesized through the combination of Wittig-Horner-type, Siegrist-type or McMurry-type reactions.[77,229,230]

5.3.5 The Knoevenagel Reaction (for 12, 19, 20, 21, 22, 23)

The Knoevenagel reaction is a base-catalyzed condensation between a dialdehyde and an arene possessing two relatively acidic sites, such as benzylic protons.[231] A typical polymerization

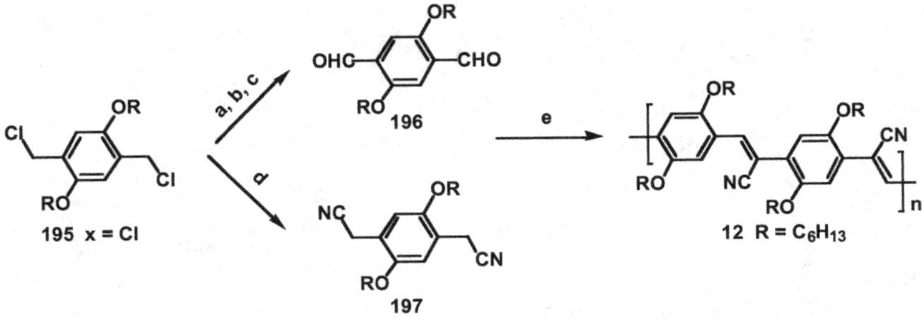

Scheme 5.62 Synthesis of CN-PPV via Knoevenagel condensation reaction. Reaction conditions: (A) NaOAc; (B) KOH, EtOH; (C) pyridinium chlorochromate; (D) NaCN; (E) t-BuOK THF, 50°C.

is shown in Scheme 5.62. Depronation affords a difunctional nucleophile, which subsequently attacks the carbonyl sites. Elimination is the final step in the Knoevenagel sequence, and the use of monomers with highly acidic protons drives the reaction to completion.

Knoevenagel condensation[232] has been widely utilized in the synthesis of CN-PPV (**12**) and its derivatives. In the case of CN-PPV, the reaction involves the condensation of dialkoxyterephthaldehyde (**196**) and dialkoxybenzene-1,4-diacetonitrile (**197**) with base, such as tetrabutylammonium hydroxide or potassium *tert*-butoxide, in a solvent mixture of THF and *tert*-butanol.[20] Careful control of the reaction conditions is required to avoid possible side reactions, such as Michael addition of nucleophiles to the newly formed vinylene linkage or direct attack on the cyano group.[188]

5.3.6   The Wessling Method (for 1, 6, 7, 56, 204)

The Wessling method was first used for the synthesis of PPV and its derivatives through a water-soluble polyelectrolyte precursor. The Wessling procedure is one of the most convenient methods of obtaining high-quality films and fibers of high-molecular-weight PPVs.[233] The sulfonium polyelectrolyte (**5** in Scheme 5.2) is the original Wessling precursor, while the modified Wessling route utilizes organic precursors involving

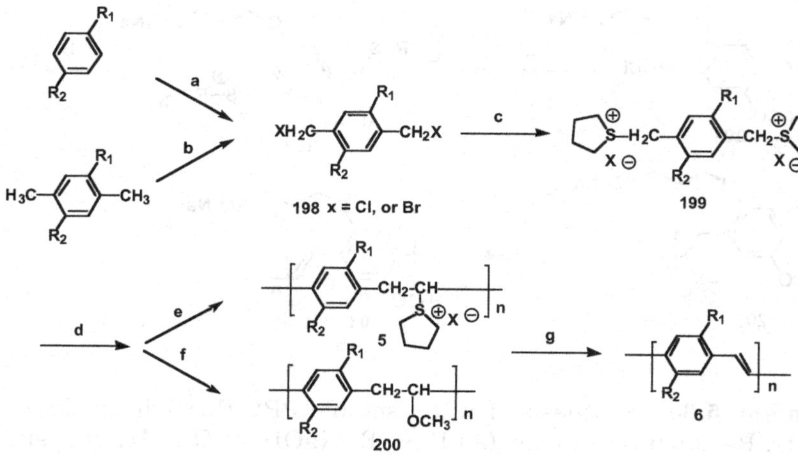

Scheme 5.63 Synthesis of PPV derivatives through the Wessling route. Reaction conditions: (A) HCHO, HX; (B) NCS or NBS, CCl$_4$; (C) tetrahydrothiophene, methanol; (D) base; (E) H$_2$O; (F) MeOH; (G) heat. (R = alkyl).

the alkoxy-substituted non-ionic form (**200**).[234] Other precursor routes, such as using sulfoxide and sulfone functionalized precursor polymers, have also been used.[235] The advantage of using the organic soluble precursor is that the precursor polymer can be purified and fabricated into the desired form prior to conversion to the final molecular structure. Representative synthetic routes for PPVs through the Wessling and modified Wessling routes are shown in Scheme 5.63. Water-soluble propoxysulfonated-PPV has also been prepared through the precursor route, as shown in Scheme 5.64.[236]

There are some disadvantages with synthesizing PPV through the precusor bis(sulfonium salt) methods. The yield is relatively low, which is associated with the difficulties in the purifying of the precursor polymer, although dialysis has been widely used. Polymerization of sulfonium salts often generates alcohol functionalities when performed under basic conditions, and these groups may undergo oxidation to carbonyl groups upon thermal elimination, ultimately decreasing the conjugation length.[237]

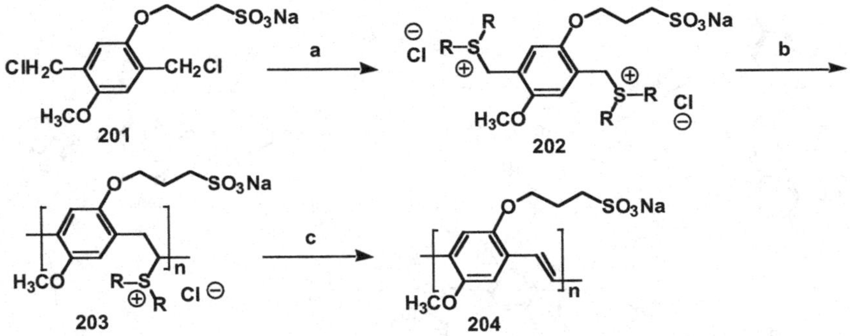

Scheme 5.64 Synthesis of water-soluble PPV through precursor route. Reaction conditions: (A) R_2S; (B) NaOH, H_2O/DMF, dialysis; (C) heat.

5.3.7 Dehydrohalogenation Condensation Polymerization (for 6, 7, 8, 9, 10, 11, 13, 15, 16, 17, 18, 24, 26, 27, 28, 33, 37, 39, 40, 45, 46, 47, 52, 53, 54, 55)

The dehydrohalogenation condensation polymerization route to alkoxy-substituted PPVs is sometimes favored over the sulfonium precursor route.[238] The polymerization, pioneered by Gilch, proceeds via a similar reaction intermediate, but the synthesis of the conjugated polymer requires only two steps and often proceeds with improved yields. The synthesis of MEH-PPV (**7**) is outlined in Scheme 5.65.[239] The concentrations of base and monomer must be carefully controlled to prevent cross-linking and gelation.[239,240]

DMOS-PPV (**16**) and other silicon-containing PPVs were synthesized through analogous methods. Silyl substitution to yield the monomer precursor (**206**) is accomplished by use of magnesium halide reagents. A typical synthetic route is shown in Scheme 5.66.

Other poly(2,5-dialkoxy-1,4-phenylene vinylene)s have been prepared in a similar fashion.[224c] Alternatively, a soluble α-halo precursor polymer (**210**) may be obtained by using less than one equivalent of base (Scheme 5.67). This halo-precursor

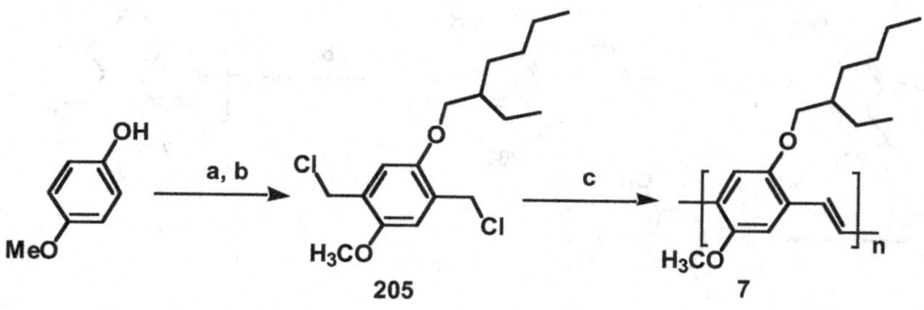

Scheme 5.65 Synthesis of MEH-PPV through the Gilch route. Reaction conditions: (A) 2-ethylhexylbromide, KOH, EtOH; (B) HCHO, HCl (37%), dioxane; (C) t-BuOK, THF.

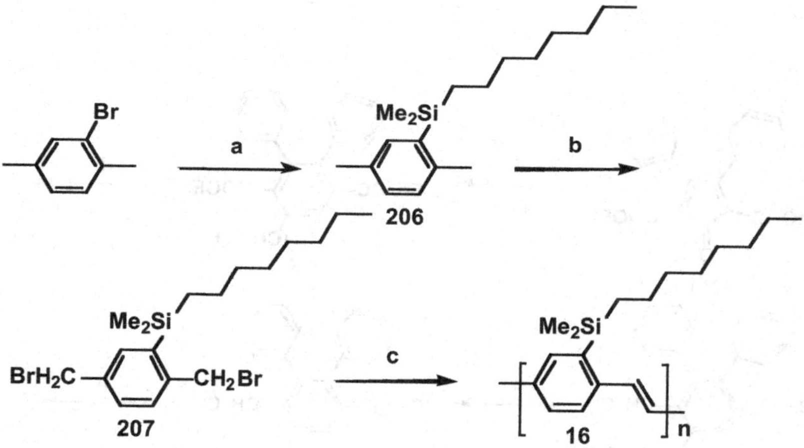

Scheme 5.66 Synthesis of DMOS-PPV (16). Reaction conditions: (A) Mg, THF, $C_8H_{17}SiMe_2Cl$; (B) NBS, CCl_4; (C) t-BuOK, THF.

route may be preferred if the fully conjugated material has limited solubility or if incomplete conversion is desired.

The synthesis of aryl-substituted PPVs involves the synthesis of arylated monomer precursors via the Dies-Alder reaction, followed by the polymerization of the monomer via a modified Gilch route (Scheme 5.68).[241]

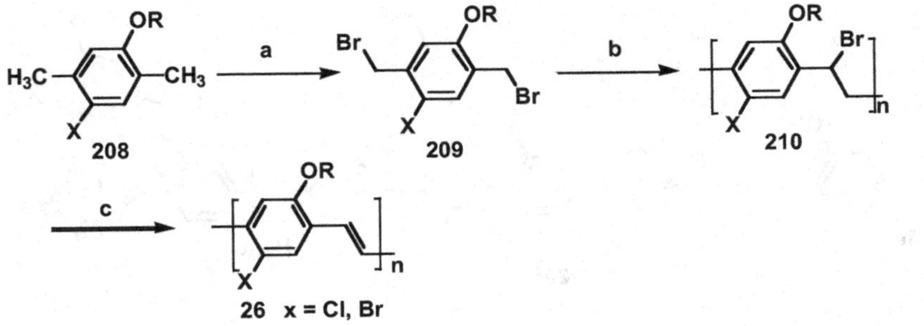

Scheme 5.67 Synthesis of substituted PPV through the halo-precursor route: (A) NBS, CCl$_4$; (B) t-BuOK, THF; (C) 160 to 220°C, vacuum.

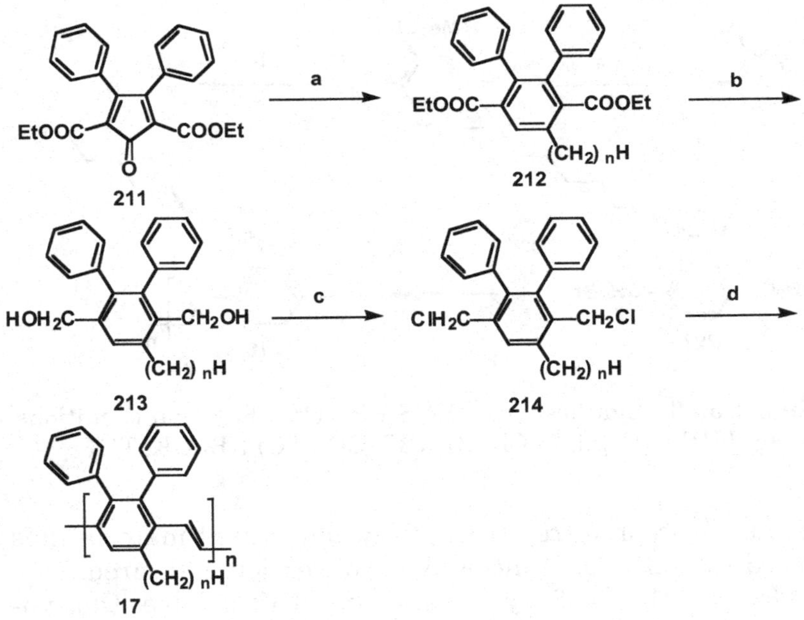

Scheme 5.68 Preparation of diphenyl-PPV derivatives. Reaction conditions: (A) 1-octyne, or 1-decyne or 1-dodecyne; (B) LiAlH$_4$; (C) SOCl$_2$; (D) t-BuOK, THF.

5.3.8   Reductive Polymerization

5.3.8.1   Yamamoto-Type Aryl–Aryl Coupling (for **69, 80, 81, 83, 84, 85, 86, 87, 88, 89, 91, 94, 125, 128, 130, 131**)

Yamamoto coupling refers to the nickel-promoted C–C coupling through the condensation of dihaloaromatic compounds.[111] The most widely used reagents employed for condensation include $NiBr_2$ or $NiCl_2$ in the presence of Zn, PPh_3, and bipyridine, or the nickel(0)/1,5-cyclooctadiene complex $Ni(COD)_2$ with the co-reagents (2,2′-bipyridine and 1,5-cyclooctadiene (COD)) in dimethylacetamide or dimethylformamide as solvent.[242]

Yamamoto methodologies can be used for the synthesis of random copolymers and homopolymers. A major advantage of this method is the straightforward preparation of the dihalogenated reactants. As compared to the $FeCl_3$-catalyzed oxidative polymerization, Yamamoto coupling improves specificity and minimizes branching. In addition, one can obtain very high molecular weights of up to 200,000 (M_n).[243] A disadvantage of this method is that catalyst residues may be difficult to remove from the product. In addition, these nickel-catalyzed coupling reactions are not suitable for the synthesis of alternating copolymers.

As shown in Scheme 5.69, the reductive polymerization of 2,7-dibromo-9,9′-bis(3,6-dioxaheptyl)fluorene (**215**) in DMF using zinc as the reductant in the presence of $NiCl_2$ catalyst for 24 to 72 h produces **80** as a light yellow powder, with a molecular weight of 94,000.[118a] An application of Yamamoto coupling for the synthesis of random copolymers (**85**) is shown in Scheme 5.70.[139d]

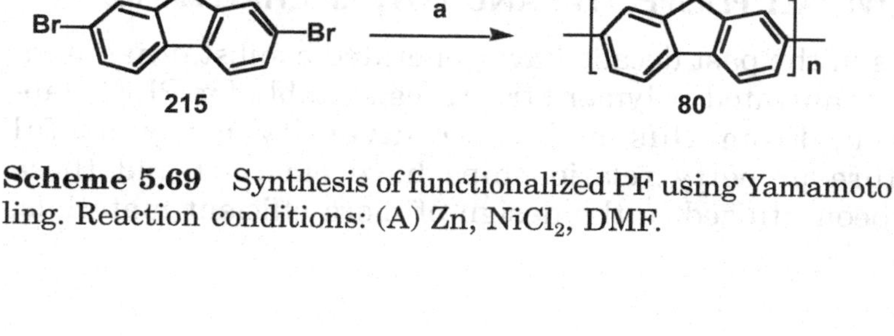

Scheme 5.69 Synthesis of functionalized PF using Yamamoto coupling. Reaction conditions: (A) Zn, $NiCl_2$, DMF.

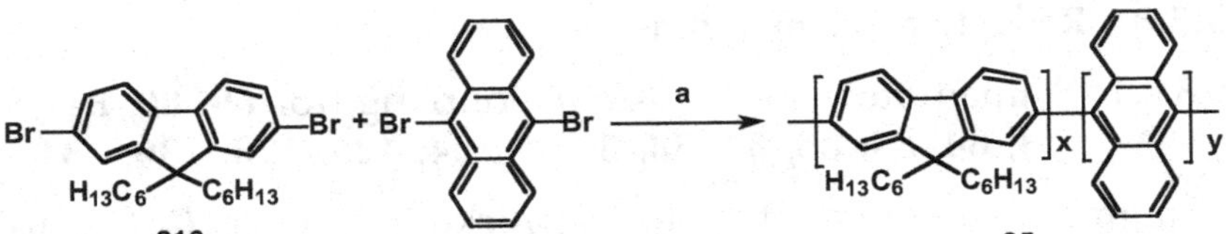

Scheme 5.70 Synthesis of random copolymers using Yamamoto coupling. Reaction conditions: (A) Zn, Ni(COD)$_2$, 2,2′-bipyridine, COD.

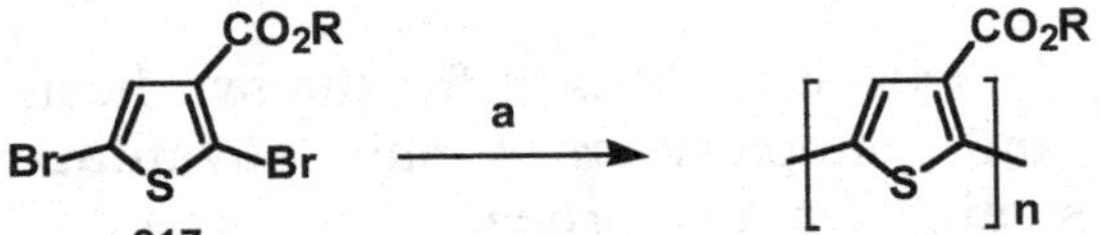

Scheme 5.71 An example of Ullmann coupling polymerization. Reaction conditions: (A) Cu, DMF.

5.3.8.2 Ullmann Coupling Reaction (for 218)

The copper-promoted C–C coupling of aromatic dihalides is known as the Ullmann coupling.[244] This methodology is an attractive alternative for the polymerization of heterocycles bearing strong electron-withdrawing groups, as they generally enhance the reaction.[245] The molecular weight of the polymers is usually low by this method. Scheme 5.71 shows the polymerization of alkyl 2,5-dibromothiophene-3-carboxylate (**217**) through Ullmann coupling. The molecular weight of **218** is about 3000.[246]

5.4 OVERALL PERSPECTIVE AND FUTURE CHALLENGES

Efforts in the past decade have generated a substantial number of conjugated polymer structures suitable for PLED fabrication. From this molecular diversity many useful structure–property relationships have emerged and these have been utilized in the design of more efficient materials.

The means for chemical synthesis have enabled the generation of specific materials with predetermined molecular structures. Indeed, the ability to correlate structure with properties, together with the ability to obtain specific molecular structures, have been of paramount importance in developing PLEDs suitable for commercial applications.

Several challenges remain. An obvious target is materials with longer operational lifetimes, in particular among blue emitters. It would be also desirable to have blue, green, and red materials with similar aging rates so that multicolor displays do not lose their appearance over time. Also of interest are polymers with substantially different solubility properties that would enable the fabrication of devices with multilayer structures by casting from solution. It is also anticipated that fabrication by ink-jet techniques will require polymers with unique solution properties. Finally, we note that organometallic complexes that are efficient triplet emitters have been used in the fabrication of the most efficient devices reported thus far.[247] Introduction of similar phosphorescent functionalities within novel polymer structures should increase substantially the efficiency of the PLED devices.

REFERENCES

1. (a) A. Bernanose, M. Comte, and P. Vouaux, *J. Chim. Phys.*. 50, 64, 1953. (b) A. Bernanose and P. Vouaux, *J. Chim. Phys.*, 50, 261, 1953. (c) A. Bernanose, *J. Chim. Phys.*, 52, 396, 1955. (d) A. Bernanose and P. Vouaux, *J. Chim. Phys.*, 52, 509, 1955.

2. C. W. Tang and S.A. VanSlyke, *Appl. Phys. Lett.*, 51, 913, 1987.

3. J.H. Burroughes, D.D.C. Bradley, A.R. Brown, R.N. Marks, K. Mackey, R.H. Friend, P.L. Burn, and A. B. Holmes, *Nature*, 347, 539, 1990.

4. A.B. Holmes, D.D.C. Bradley, A.R. Brown, P.L. Burn, J.H. Burroughes, R.H. Friend, N.C. Greenham, R.W. Gymer, D.A. Halliday, R.W. Jackson, A. Kraft, J.H.F. Martens, K. Pichler, and I.D.W. Samuel, *Synth. Met.*, 57, 4031, 1993.

5. R.A. Wessling, *J. Polym. Sci. Polym. Symp.*, 72, 55, 1985.

6. P.L Burn, D.D.C. Bradley, R.H. Friend, D.A. Halliday, A.B. Holmes, R.W. Jackson, and A. Kraft, *J. Chem. Soc. Perkin Trans. I*, 3225, 1992.

7. (a) A. Beerden, D. Vanderzande, and J. Gelan, *Synth. Met.*, 52, 387, 1992. (b) R.O. Garay, U. Baier, C. Bubeck, and K. Mullen, *Adv. Mater.*, 5, 561, 1993.

8. A. Kraft, A.C. Grimsdale, and A.B. Holmes, *Angew. Chem. Int. Ed.*, 37, 402, 1998.

9. (a) B.R. Hsieh, H. Antoniadis, M.A. Abkowitz, and M. Stolka, *Polym. Prepr.*, 33, 414, 1992. (b) F. Papadimitrakopoulos, K. Konstandinidis, T.M. Miller, R. Opila, E.A. Chandross, and M.E. Galvin, *Chem. Mater.*, 6, 1563, 1994.

10. M. Herold, J. Gmeiner, and M. Schwoerer, *Acta Polym.*, 47, 436, 1996.

11. D. Braun and A. Heeger, *Appl. Phys. Lett.*, 58, 1982, 1991.

12. G. Yu, *Synth. Met.*, 80, 143, 1996.

13. (a) F. Wudl, S. Höger, C. Zhang, K. Pakbaz, and A.J. Heeger, *Polym. Prepr.*, 34, 197, 1993. (b) M.R. Andersson, G. Yu, and A.J. Heeger, *Synth. Met.*, 85, 1275, 1997.

14. S. Doi, M. Kuwabara, T. Noguchi, and T. Ohnishi, *Synth. Met.*, 57, 4174, 1993.

15. J. Salbeck and Ber. Bunsenges, *J. Phys. Chem.*, 100, 1666, 1996.

16. N. Greenham, F. Cacialli, D. Bradley, R. Friend, S. Moratti, and A. Holmes, *Mater. Res. Soc. Symp. Proc.*, 328, 351, 1994.

17. (a) H. Meng, W.L. Yu, and W. Huang, *Macromolecules*, 32, 8841, 1999. (b) J. Kido, G. Harada, and K. Nagai, *Chem. Lett.*, 161, 1996.

18. (a) V. Boucard, T. Benazzi, D. Ades, G. Sauvet, and A. Siove, *Polymer*, 38, 3697, 1997. (b) J. Lu, A.R. Hlil, Y. Sun, and A.S. Hay, *Chem. Mater.*, 11, 2501, 1999. (c) S. Pfeiffer, H. Rost, and H.H. Hörhold, *Macromol. Chem. Phys.*, 200, 2471, 1999. (d) X.C. Li, Y. Liu, M.S. Liu, and A.K.Y. Jen, *Chem. Mater.*, 11, 1568, 1999.

19. S. Höger, J. McNamara, S. Schricker, and F. Wudl, *Chem. Mater.*, 6, 171, 1994.

20. N.C. Greenham, S.C. Moratti, D.D.C. Bradley, R.H. Friend, and A.B. Holmes, *Nature,* 365, 628, 1993.

21. X.C. Li, A. Kraft, R. Cervini, G.C.W. Spencer, F. Cacialli, R.H. Friend, J. Grüner, A.B. Holmes, J.C. Demello, and S.C. Moratti, *Mat. Res. Soc. Symp. Proc.,* 413, 13, 1996.

22. S.C. Moratti, D.D.C. Bradley, R.H. Friend, N.C. Greenham, and A.B. Holmes, *Mat. Res. Soc. Symp. Proc.,* 328, 371, 1994.

23. (a) D.R. Baigent, N.C. Greenham, J. Grüner, R.N. Mark, R.H. Friend, S.C. Moratti, and A.B. Holmes, *Synth. Met.,* 67, 3, 1994. (b) J.J.M. Halls, D.R. Baigent, F. Cacialli, N.C. Greenham, R.H. Friend, S.C. Moratti, and A.B. Holmes, *Thin Solid Films,* 276, 13, 1996.

24. D.R. Baigent, P.J. Hamer, R.H. Friend, N.C. Greenham, and A.B. Holmes, *Synth. Met.,* 71, 2175, 1995.

25. Y. Xiao, W.L. Yu, S.J. Chua, and W. Huang, *Chem. Eur. J.,* 6, 1318, 2000.

26. A. Lux, A.B. Holmes, R. Cervini, J.E. Davies, S.C. Moratti, J. Gruner, F. Cacialli, and R.H. Friend, *Synth. Met.,* 84, 293, 1997.

27. S.E. Döttinger, M. Hohloch, J. L. Segura, E. Steinhuber, M. Hanack, A. Tompert, and D. Oelkrug, *Adv. Mater.,* 9, 233, 1997.

28. S.E. Döttinger, M. Hohloch, D. Hohnholz, J.L. Segura, E. Steinhuber, and M. Hanack, *Synth. Met.,* 84, 293, 1997.

29. P.M. Lahti, A. Sarker, R.O. Garay, R.W. Lenz, and F.E. Karasz, *Polymer,* 35, 1312, 1994.

30. (a) R.K. McCoy, F.E. Karasz, A. Sarker, and P.M. Lahti, *Chem. Mater.,* 3, 941, 1991. (b) J.L. Jin, J.C. Kim, and H.K. Shim, *Macromolecules,* 25, 5519, 1992. (c) J.I. Lee, J.Y. Han, H.K. Shim, S.C. Jeonng, and D. Kim, *Synth. Met.,* 84, 261, 1997. (d) I. Benjamin, E.Z. Faraggi, Y. Avny, D. Davidov, and R. Neumann, *Chem. Mater.,* 8, 352, 1996.

31. C.B. Yoon and H.K. Shim, *Macromol. Chem. Phys.,* 197, 3689, 1996.

32. A.C. Grimsdale, F. Cacialli, J. Gruner, X.C. Li, A.B. Holmes, S.C. Moratti, and R.H. Friend, *Synth. Met.,* 76, 165, 1996.

33. (a) M. Strukelj, F. Papadimitrakopoulos, T.M. Miller, and L.J. Rothberg, *Science*, 267, 1969, 1995. (b) Y. Yang and Q. Pei, *J. Appl. Phys.*, 77, 4807, 1995.

34. Z. Peng, Z. Bao, and M.E. Galvin, *Adv. Mater.*, 10, 680, 1998.

35. Z.K. Chen, H. Meng, Y.H. Lai, and W. Huang, *Macromolecules*, 32, 4351, 1999.

36. (a) Z. Peng and J. Zhang, *Chem. Mater.*, 11, 1138, 1999. (b) Z.H. Peng and J.H. Zhang. *Synth. Met.*, 105, 73, 1999.

37. S.Y. Song, M.S. Jang, H.K. Shim, D.H. Hwang, and T. Zyung. *Macromolecules*, 32, 1482, 1999.

38. Y. Lee, X. Chen, S. Chen, P. Wei, and W. Fann, *J. Am. Chem. Soc.*, 123, 2296, 2001.

39. D.W. Lee, K.Y. Kwon, J.I. Jin, Y. Park, Y.R. Kim, and I.W. Hwang, *Chem. Mater.*, 13, 565, 2001.

40. S. Jin, J. Jung, I. Yeom, S. Moon, K. Koh, S. Kim, and Y. Gal, *Eur. Polym. J.*, 35, 227, 1999.

41. PEDOT: poly(3,4-ethylenedioxythiophene); PSS: poly(styrene sulfonic acid).

42. Y. Pu, M. Soma, J. Kido, and H. Nishide, *Chem. Mater.*, 13, 3817, 2001.

43. K. Kim, Y. Hong, S.W. Lee, J. Jin, Y. Park, B.H. Sohn, W. Kim, and J.K. Park, *J. Mater. Chem.*, 11, 3023, 2001.

44. C. Zhang, S. Hoger, K. Pakbaz, F. Wudl, and A.J. Heeger, *J. Electron. Mater.*, 23, 453, 1994.

45. S.T. Kim, D.H. Hwang, X.C. Li, J. Gruner, R.H. Friend, A.B. Holmes, and H.K. Shim, *Adv. Mater.*, 8, 979, 1996.

46. H. Hwang, S.T. Kim, H.K. Shim, A.B. Holmes, S.C. Moratti, and R.H. Friend, *Chem. Commun.*, 2241, 1996.

47. K.D. Kim, J.S. Park, H.K. Kim, T.B. Lee, and K.T. No, *Macromolecules*, 31, 7267, 1998.

48. (a) H. Li and R. West. *Macromolecules*, 31, 2866, 1998. (b) R.D. Miller and J. Michl, *Chem. Rev.*, 90, 1359, 1989. (c) J. Ohshita and A. Kunai, *Acta Polym.*, 49, 379, 1998.

49. (a) J.K. Herrema, P.F. Hutten, R.E. Gill, J. Wildeman, R.H. Wieringa, and G. Hadziioannou, *Macromolecules,* 28, 8102, 1995. (b) F. Garten, A. Hilberer, F. Cacialli, E. Esselink, Y. Dam, B. Schlatmann, R.H. Friend, T.M. Klapwijk, and G. Hadziioannou, *Adv. Mater.,* 9, 127, 1997. (c) H.K. Kim, M.K. Ryu, K.D. Kim, S.M. Lee, S.W. Cho, and J.W. Park, *Macromolecules,* 31, 1114, 1998.

50. F. Garten, A. Hilberer, F. Cacialli, F.J. Esselink, Y. Dam, A.R. Schlatmann, R.H. Friend, T.M. Klapwijk, and G. Hadziioannou, *Synth. Met.,* 85, 1253, 1997.

51. A. Pohl and J.L. Bredas, *Int. J. Quantum Chem.,* 63, 437, 1997.

52. B.S. Chuah, D.H. Hwang, S.T. Kim, S.C. Moratti, A.B. Holmes, J.C. Demello, and R.H. Friend, *Synth. Met.,* 91, 279, 1997.

53. Z. K. Chen, W. Huang, L.H. Wang, E.T. Kang, B.J. Chen, C.S. Lee, and S.T. Lee, *Macromolecules,* 2000, 33, 9015.

54. T. Ahn, M.S. Jang, H.K. Shim, D.H. Hwang, and T. Zyung, *Macromolecules,* 32, 3279, 1999.

55. H.N. Cho, D.Y. Kim, Y.C. Kim, J.Y. Lee, and C.Y. Kim, *Adv. Mater.,* 9, 326, 1997.

56. (a) L. Liao, Y. Pang, L. Ding, and F. Karase, *Macromolecules,* 34, 7300, 2001. (b) Y. Pang, J. Li, B. Hu, and F. Karase, *Macromolecules,* 32, 3946, 1999.

57. S. Son, A. Dodabalapur, A.J. Lovinger, and M.E. Galvin, *Science,* 269, 376, 1995.

58. (a) H.H. Hörhold, R. Bergmann, J. Gottschaldt, and G. Drefahl, *Acta Chim. Acad. Sci. Hung.,* 81, 239, 1974. (b) A. Greiner and W. Heitz, *Makromol. Chem., Rapid Commun.,* 9, 581, 1988.

59. A. Greiner, and W. Heitz, *Makromol. Chem., Rapid Commun.,* 9, 581, 1988.

60. B.R. Hsieh, Y. Yu, E.W. Forsythe, G.M. Schaaf, and W.A. Feld, *J. Am. Chem. Soc.,* 120, 231, 1998.

61. M.R. Andersson, G. Yu, and A.J. Heeger, *Synth. Met.,* 85, 1383, 1997.

62. D.M. Johansson, G. Srdanov, G. Yu, M. Theander, O. Inganäs, and M.R. Andersson, *Macromolecules,* 33, 2525, 2000.

63. H. Becker, H. Spreitzer, K. Ibrom, and W. Kreuder, *Macromolecules,* 32, 4925, 1999.

64. D.M. Johansson, X.J. Wang, T. Johansson, O. Inganäs, G. Yu, G. Srdanov, and M.R. Andersson, *Macromolecules,* 35, 4997, 2002.

65. H. Spreitzer, H. Becker, E. Kluge, W. Kreuder, H. Schenk, R. Demandt, and H. Schoo, *Adv. Mater.,* 10, 1340, 1998.

66. H. Becker, H. Spreitzer, W. Kreuder, E. Kluge, H. Schenk, I. Parker, and Y. Cao, *Adv. Mater.,* 12, 42, 2000.

67. (a) M.R. Andersson, D.M. Johansson, M. Theander, and O. Inganäs, *Synth. Met.,* 119, 63, 2001. (b) D.M. Johansson, M. Theander, G. Srdanov, G. Yu, O. Inganäs, and M.R. Andersson, *Macromolecules,* 34, 3716, 2001.

68. S. Lee, B. Jang, and T. Tsutsui, *Macromolecules,* 35, 1356, 2002.

69. B. Sohn, K. Kim, D. Choi, Y. Kim, S. Jeoung, and J. Jin. *Macromolecules,* 35, 2876, 2002.

70. (a) S.J. Chung, I.J. Jin, C.H. Lee, and C.I. Lee, *Adv. Mater.,* 10, 684, 1998. (b) I.J. Jin, S.J. Chung, and S.H. Yu, *Macromol. Symp.,* 128, 79, 1998.

71. L.J. Rothberg and A.J. Lovinger, *J. Mater. Res.,* 11, 3174, 1996.

72. P.L. Burn, A.B. Holmes, A. Kraft, D.D.C. Bradley, A.R. Brown, R.H. Friend, and R.W. Gymer, *Nature,* 356, 47, 1992.

73. A. Kraft, P.L. Burn, A.B. Holmes, D.D.C. Bradley, A.B. Brown, R.H. Friend, and R.W. Gymer, *Synth. Met.,* 55, 936, 1993.

74. G. Padmanaban and S. Ramakrishnan, *J. Am. Chem. Soc.,* 122, 2244, 2000.

75. M.M. de Kok, A.J.J. van Breeman, P.J. Adriaensens, van Dixhoorn, J.M. Gelan, and D.J. Vanderzande, *Acta Polym.,* 49, 510, 1998.

76. (a) I. Sokolik, Z. Yang, F.E. Karasz, and D.C. Morton, *J. Appl. Phys.,* 74, 3584, 1993. (b) Z. Yang, I. Sokolik, and F.E. Karasz, *Macromolecules,* 26, 1188, 1993. (c) Z. Yang, F.E. Karasz, and H.J. Geise, *Macromolecules,* 26, 6570, 1993. (d) T. Zyung, D.H. Hwang, I.N. Kang, H.K. Shim, W.Y. Hwang, and J.J. Kim, *Chem. Mater.,* 16, 135, 1995.

77. R. Schenk, H. Gregorius, K. Meerholz, J. Heinze, and K. Müllen, *J. Am. Chem. Soc.*, 113, 2635, 1991.

78. V. Gebhardt, A. Bacher, M. Thelakkat, U. Stalmach, H. Meier, H. Schmidt, and D. Haarer, *Adv. Mater.*, 11, 119, 1999.

79. E. Han, L. Do, Y. Niidome, and M. Fujihura, *Chem. Lett.*, 969, 1994.

80. M.D. Joswick, I.H. Cambell, N.N. Barashkov, and J.P. Ferraris, *J. Appl. Phys.*, 80, 2883, 1996.

81. S. Tokito, H. Tanaka, K. Noda, A. Okada, and T. Taga, *Appl. Phys. Lett.*, 70, 1929, 1997.

82. (a) S. Wang, W. Oldham, R. Hudack, and G.C. Bazan. *J. Am. Chem. Soc.*, 122, 5695, 2000, and the references therein. (b) M. Robinson, S. Wang, A. J. Heeger, and G.C. Bazan, *Adv. Func. Mater.*, 11, 413, 2001.

83. C. Weder, M.J. Wagner, and M.S. Wrighton, *Mater. Res. Soc. Symp. Proc.*, 413, 77, 1996.

84. A. Montali, P. Smith, and C. Weder, *Synth. Met.*, 97, 123, 1998.

85. L. S. Swanson, J. Shinar, Y. Ding, and T. Barton, *Synth. Met.*, 55, 1, 1993.

86. U.H.F. Bunz, *Acc. Chem. Res.*, 34, 998, 2001.

87. C. Schmitz, P. Pösch, M. Thelakkat, H. Schmidt, A. Montali, K. Feldman, P. Smith, and C. Weder, *Adv. Func. Mater.*, 11, 41, 2001.

88. (a) J. Li and Y. Pang, *Macromolecules*, 31, 5740, 1998. (b) K. Tada, M. Onoda, M. Hirohata, T. Kawai, and K. Yoshino, *Jpn. J. Appl. Phys.*, 35, L251, 1996. (c) G. Jégou and S. Jenekhe, *Macromolecules*, 34, 7926, 2001.

89. K. Miyashita and M. Kaneko, *Synth. Met.*, 68, 161, 1995.

90. D.L. Gin and V.P. Conticello, *Trends Polym. Sci.*, 4, 217, 1996.

91. M. Hamaguchi, H. Sawada, J. Kyokane, and K. Yoshino, *Chem. Lett.*, 527, 1996.

92. J.L. Brédas, Springer Series in Solid State Science, Vol. 63, Springer, Berlin, 1985, 166.

93. G. Grem and G. Leising, *Synth. Met.*, 57, 4105, 1993.

94. Y. Yang, Q. Pei, and A.J. Heeger, *J. Appl. Phys.*, 79, 934, 1996.

95. (a) J. Huang, H. Zhang, W. Tian, J. Hon, Y. Ma, J. Shen, and S. Liu, *Synth. Met.*, 87, 105, 1997. (b) J. Birgersson, K. Kaeriyama, P. Barta, P. Bröms, M. Fahlman, T. Granlund, and W.R. Salaneck, *Adv. Mater.*, 8, 982, 1996.

96. M. Remmers, D. Neher, J. Gruner, R.H. Friend, G.H. Gelinck, J.M. Warman, C. Quattrocchi, D.A. dos Santos, and J.L. Brédas, *Macromolecules*, 29, 7432, 1996.

97. (a) M.R. Pinto and K.S. Schanze, *Synthesis*, 9, 1293, 2002. (b) T.I. Wallow and B.M. Novak, *J. Am. Chem. Soc.*, 113, 7411, 1991. (c) P.B. Balanda, M.B. Ramey, and J.R. Reynolds, *Macromolecules*, 32, 3970, 1999.

98. (a) J.W. Baur, S. Kim, P.B. Balanda, J.R. Reynolds, and M.F. Rubner, *Adv. Mater.*, 10, 1452, 1998. (b) R. Helgeson, F. Wudl, M.B. Ramey, and J.R. Reynolds, *Appl. Phys. Lett.*, 73, 2561, 1998.

99. (a) U. Scherf and K. Müllen, *Markromol. Chem. Rapid Commun.*, 12, 489, 1991. (b) U. Scherf and K. Müllen, *Macromolecules*, 25, 3546, 1992.

100. G. Grem, C. Paar, J. Stampfl, G. Leising, J. Huber, and U. Scherf, *Chem. Mater.*, 7, 2, 1995.

101. Y. Geerts, U. Keller, U. Scherf, M. Schneider, and K. Müllen, *Polym. Prepr.*, 38, 315, 1997.

102. J. Grüner, H.F. Wittmann, P.J. Hamer, R.H. Friend, J. Huber, U. Scherf, K. Müllen, S.C. Moratti, and A.B. Holmes, *Synth. Met.*, 67, 181, 1994.

103. J. Gruner, P.J. Hamer, R.H. Friend, H.J. Huber, U. Scherf, and A.B. Holmes, *Adv. Mater.*, 6, 748, 1994.

104. G. Leising, S. Tasch, F. Meghdadi, L. Athouel, G. Froyer, and U. Scherf, *Synth. Met.*, 81, 185, 1996.

105. W. Graupner, S. Eder, S. Tasch, G, Leising, G. Lanzani, M. Nisoli, S. de Silvestri, and U. Scherf, *J. Fluoresc.*, 7, 195s, 1995.

106. S. Tasch, A. Niko, G. Leising, and U. Scherf, *Appl. Phys. Lett.*, 68, 1090, 1996.

107. S. Tasch, A. Niko, G. Leising, and U. Scherf, *Mat. Res. Soc. Symp. Proc.*, 413, 71, 1996.

108. G. Leising, F. Meghdadi, S. Tasch, C. Brandstätter, W. Graupner, and G. Kranzelbinder, *Synth. Met.*, 85, 1209, 1997.

109. M. Fukuda, K. Sawaka, and K. Yoshino, *Jpn. J. Appl. Phys.*, 28, L1433, 1989.

110. M. Fukuda, K. Sawaka, and K. Yoshino, *J. Polym. Sci. A Polym. Chem.*, 31, 2465, 1993.

111. T. Yamamoto, *Prog. Polym. Sci.*, 17, 1153, 1992.

112. N. Miyaura, T. Yanagi, and A. Suzuki, *Synth. Commun.*, 11, 513, 1981.

113. (a) M. Bernius, M. Inbasekaran, E. Woo, W.S. Wu, and L. Wujkowski, *J Mater. Sci. Mater. Electron.*, 11, 111, 2000. (b) M. Inbasekaran, E. Woo, W.S. Wu, M. Bernius, and L. Wujkowski, *Synth. Met.*, 111, 397, 2000. (c) D.D.C. Bradley, M. Greel, A. Grice, A. R. Tajbakhsh, D.F. O'Brien, and A. Bleyer, *Opt. Mater.*, 9,1, 1998.

114. (a) B. Liu, W.L. Yu, Y. Lai, and W. Huang, *Chem. Mater.*, 13, 1984, 2001. (b) A.W. Grice, D.D.C. Bradley, M.T. Bernius, M. Inbasekaran, and W.W. Wu, *Appl. Phys. Lett.*, 73, 629, 1998.

115. M. Grell, D.D.C. Bredley, M. Inbasekaran, and E.P. Woo, *Adv. Mater.*, 9, 798, 1997.

116. M. Grell, W. Knoll, D. Luo, A. Meisel, T. Miteva, D. Neher, H.G. Nothofer, U. Sherf, and A. Yasuda, *Adv. Mater.*, 11, 671, 1999.

117. Y. Ohmori, M. Uchida, K. Muro, and K. Yoshino, *Jpn. J. Appl. Phys.*, 30, 1941, 1991.

118. (a) Q. Pei and Y. Yang, *J. Am. Chem. Soc.*, 118, 7416, 1996. (b) Y. Yang and Q. Pei, *J. Appl. Phys.*, 81, 3294, 1997.

119. T. Nguyen, I.B. Martini, J. Liu, and B.J. Schwartz, *J. Phys. Chem. B*, 104, 237, 2000.

120. J.I. Lee, G. Klærner, and R.D. Miller, *Synth. Met.*, 101, 126, 1999.

121. W.L. Yu, J. Pei, W. Huang, and A.J. Heeger, *Adv. Mater.* 2000, 12, 828.

122. S. Setayesh, A.C. Grimmsdale, T. Wil, V. Enkelmann, K. Müllen, F. Meghdadi, E.J.W. List, and G. Leising, *J. Am. Chem. Soc.*, 123, 946, 2001.

123. (a) D. Marsitzky, R. Vestberg, P. Blainey, B.T. Tang, C.J. Hawker, and K.R. Carter, *J. Am. Chem. Soc.*, 123, 6965, 2001. (b) H.Z. Tang, M. Fujiki, Z.B. Zhang, K. Torimitsu, and M. Motonaga, *Chem. Commun.*, 2426, 2001.

124. D. Neher, *Macromol. Rapid Commun.*, 22, 1365, 2001.

125. J.L. Lee, G. Klaerner, M.H. Davey, and R.D. Miller, *Synth. Met.*, 102, 1087, 1999.

126. (a) D. Marsitzky, J.C. Scott, J.P. Chen, V.C. Lee, R.D. Miller, S. Setayesh, and K. Müllen, *Adv. Mater.*, 13, 1096, 2001. (b) G. Klarner, M.H. Davey; W.D. Chen, J.C. Scott, and R.D. Miller, *Adv. Mater.*, 10, 993, 1998.

127. J.I. Lee, G. Klærner, and R.D. Miller, *Chem. Mater.*, 11, 1083, 1999.

128. K.H. Weinfurtner, H. Fujikawa, S. Tokito, and Y. Taga, *Appl. Phys. Lett.*, 76, 2502, 2000.

129. E.J.W. List, R. Guentner, P.S. Freitas, and U. Scherf, *Adv. Mater.*, 14, 374, 2002.

130. V.N. Bliznyuk, S.A. Carter, J.C. Scott, G. Klarner, R.D. Miller, and D.C. Miller, *Macromolecules*, 32, 361, 1999.

131. G. Klaerner, J.I. Lee, V.Y. Lee, E. Chan, J.P. Chen, A. Nelson, D. Markiewicz, R. Siemens, J.C. Scott, and R.D. Miller, *Chem. Mater.*, 11, 1800, 1999.

132. J.P. Chen, G. Klaerner, J.I. Lee, D. Markiewicz, V.Y. Lee, R.D. Miller, and J.C. Scott, *Synth. Met.*, 107, 129, 1999.

133. D. Sainova, T. Miteva, H.G. Nothofer, U. Scherf, I. Glowacki, J. Ulanski, H. Fujikawa, and D. Neher, *Appl. Phys. Lett.*, 76, 1810, 2000.

134. M. Redecker, D.D.C. Bradley, M. Inbasekaran, W.W. Wu, E.P. Woo, *Adv. Mater.*, 11, 241, 1999.

135. B. Ye, M. Trudeau, and D.M. Antonelli, *Adv. Mater.*, 13, 29, 2001.

136. T. Miteva, A. Meisel, W. Knoll, H.G. Nothofer, U. Scherf, D.C. Muller, K. Meerholz, A. Yasuda, and D. Neher, *Adv. Mater.*, 13, 565, 2001.

137. C. Ego, A.C. Grimsdale, F. Uckert, G. Yu, G. Srdanov, and K. Müllen, *Adv. Mater.*, 14, 809, 2002.

138. M. Leclerc, *J. Polym. Sci. Polym. Chem.*, 39, 2867, 2001.

139. (a) M. Ranger and M. Leclerc, *J. Chem. Soc. Chem. Commun.*, 1597, 1997. (b) M. Ranger, D. Rondeau, and M. Leclerc, *Macromolecules*, 30, 7686, 1997. (c) M. Inbasekaran, W. Wu, and E.P. Woo, U S. Patent 5,777,070, 1998. (d) M. Kreyenschmidt, G. Klaerner, T. Fuhrer, J. Ashenhurst, S. Karg, W.D. Chen, V.Y. Lee, J.C. Scott, and R.D. Miller, *Macromolecules*, 31, 1099, 1998. (e) M. Ranger and M. Leclerc, *Can. J. Chem.*, 76, 1571, 1998. (f) M. Grell, M. Redecker, K.S. Whitehead, D.D.C. Bradley, M. Inbasekaran, E.P. Woo, and W. Wu, *Liq. Cryst.* 26, 1403, 1999. (g) H.N. Cho, J.K. Kim, D.Y. Kim, C.Y. Kim, N.W. Song, and D. Kim, *Macromolecules*, 32, 1476, 1999. (h) W.L. Yu, J. Pei, Y. Cao, W. Huang, and A.J. Heeger, *Chem. Commun.*, 1837, 1999. (i) B. Tsuie, J.L. Reddinger, G.A. Sotzing, J. Soloducho, A.R. Katritzky, and J.R. Reynolds, *J. Mater. Chem.*, 9, 2189, 1999. (j) T. Virgili, T.D.G. Lidzey, and D.D.C. Bradley, *Adv. Mater.*, 12, 58, 2000. (k) B. Liu, W. L. Yu, Y. H. Lai, and W. Huang, *Macromolecules*, 33, 8945, 2000. (l) S. Beaupre, M. Ranger, and M. Leclerc, *Macromol. Rapid Commun.*, 15, 1013, 2000. (m) X. Jiang, S. Liu, H. Ma, and A.K.Y. Jen, *Appl. Phys. Lett.*, 76, 1813, 2000. (n) S. Liu, X. Jiang, P. Herguth, and A.K.Y. Jen, *Chem. Mater.*, 13, 3820, 2001. (o) S. Liu, X. Jiang, H. Ma, M.S. Liu, and A.K.Y. Jen, *Macromolecules*, 33, 3514, 2000. (p) M. Imbasekaran, E.P. Woo, W. Wu, and M.T. Bernius, *PCT WO* 00/46321, 2000. (q) J.F. Ding, M. Day, G. Robertson, and J. Roovers, *Macromolecules*, 35, 3474, 2002. (r) S. Beaupré and M. Leclerc. *Adv. Funct. Mater.*, 12, 192, 2002.

140. M. Redecker, D. Bradley, M. Inbasekaran, and E. Woo, *Appl. Phys. Lett.*, 74, 1400, 1999.

141. M.T. Bernius, M. Inbasekaran, J. Brien, and W. Wu, *Adv. Mater.*, 12, 1737, 2000.

142. M. Leclerc, F.M. Diaz, and G. Wegner, *Macromol. Chem.*, 190, 3105, 1989.

143. H. Mao and S. Holdcroft, *Macromolecules*, 25, 554, 1992.

144. H. Mao, B. Xu, and S. Holdcroft, *Macromolecules*, 26, 1163, 1993.

145. (a) Y. Ohmori, M. Uchida, K. Muro, and K. Yoshino, *Jpn. J. Appl. Phys.*, 30, L1938, 1991. (b) Y. Ohmori, M. Uchida, K. Muro, and K. Yoshino, *Solid State Commun.*, 80, 605, 1991. (c) D. Braun, G. Gustafsson, D. McBranch, and A.J. Heeger, *J. Appl. Phys.*, 72, 564, 1992. (d) N.C. Greenham, A.R. Brown, D.D.C. Bradley, and R.H. Friend, *Synth. Met.*, 57, 4134, 1993.

146. (a) M. Berggren, G. Gustafsson, O. Inganäs, M.R. Andersson, O. Wennerström, and T. Hjertberg, *Adv. Mater.*, 6, 488, 1994. (b) M. Berggren, O. Inganäs, G. Gustafsson, J. Rasmusson, M. R. Andersson, T. Hjertberg, and O. Wennerström, *Nature*, 372, 444, 1994. (c) O. Inganäs, M. R. Andersson, G. Gustafsson, T. Hjertberg, O. Wennerström, P. Dyreklev, and M. Granström, *Synth. Met.*, 71, 2121, 1995.

147. M.R. Andersson, M. Berggren, O. Inganäs, G. Gustafsson, J.C. Gustafsson-Carlberg, D. Selse, T. Hjertberg, and O. Wennerström, *Macromolecules*, 28, 7525, 1995.

148. R.E. Gill, G.G. Malliaras, J. Wildeman, and G. Hadziioannou, *Adv. Mater.*, 6, 132, 1994.

149. M.R. Andersson, D. Selse, M. Berggren, H. Järvinen, T. Hjertberg, O. Inganas, O. Wennerström, and J.E. Õsterholm, *Macromolecules*, 27, 6503, 1994.

150. (a) R.D. McCullough and R.D. Lowe, *J. Chem. Soc. Chem. Commun.*, 70, 1992. (b) R.D. McCullough, R.D. Lowe, M. Jayaraman, and D.L. Anderson, *J. Org. Chem.*, 58, 904, 1993.

151. R.D. McCullough, S.P. Williams, S. Tristram-Nagle, M. Jayaraman, P.C. Ewbank, and L. Miller, *Synth. Met.*, 69, 279, 1995.

152. (a) T.A. Chen and R.D. Rieke, *J. Am. Chem. Soc.*, 114, 10087, 1992. (b) T.A. Chen, X. Wu, and R. D. Rieke, *J. Am. Chem. Soc.*, 117, 233, 1995.

153. B. Xu and S. Holdcroft, *Macromolecules*, 26, 4457, 1993.

154. C.A. Sandstedt, R.D. Rieke, and C.J. Eckhardt, *Chem. Mater.*, 7, 1057, 1995.

155. (a) F. Chen, P.G. Mehta, L. Takiff, and R.D. McCullough, *J. Mater. Chem.*, 6, 1763, 1996. (b) P. Barta, F. Cacialli, R.H. Friend, and M. Zugorska, *J. Appl. Phys.*, 84, 6279, 1998.

156. M. R. Andersson, O. Thomas, W. Mammo, M. Svensson, M. Theander, and O. Inganäs, *J. Mater. Chem.*, 9, 1933, 1999.

157. N.C. Greenham, I.D.W. Samuel, G.R. Hayes, R.T. Phillips, Y.A.R.R. Kessener, S.C. Moratti, A. B. Holmes, and R.H. Friend, *Chem. Phys. Lett.*, 241, 89, 1995.

158. D. Moses, J. Wang, A. Dogariu, D. Fichou, and C. Videlot, *Phys. Rev. B*, 59, 7715, 1999.

159. M. Muccini, E. Lunedei, A. Bree, G. Horowitz, F. Garnier, and C. Taliani, *J. Chem. Phys.*, 108, 7327, 1998.

160. M.R. Andersson, M. Berggren, T. Olinga, T. Hjertberg, O. Inganäs, and O. Wennerström, *Synth. Met.*, 85, 1383, 1997.

161. U. Lemmer, R.F. Mahrt, Y. Wada, A. Greiner, H. Bässer, and E.O. Göbler, *Appl. Phys. Lett.*, 62, 2827, 1993.

162. A. Fujii, H. Kawahara, M. Yoshida, Y. Ohmori, and K. Yoshino, *J. Phys. D. Appl. Phys.*, 28, 2135, 1995.

163. G. Gigli, M. Ani, G. Barbarella, L. Favaretto, F. Cacialli, and R. Cingolani, *Physica E*7, 612, 2000.

164. (a) J. Pei, W.L. Yu, J. Ni, Y.H. Lai, W. Huang, and H.J. Heeger, *Macromolecules*, 34, 7241, 2001. (b) J.M. Xu, S.C. Ng, and H.S.O. Chan, *Macromolecules*, 34, 4314. 2001. (c) J. Pei, W.L. Yu, W. Huang, and A.J. Heeger, *Macromolecules*, 33, 2462, 2000.

165. (a) D.D. Gebler, Y.Z. Wang, J.W. Blatchford, S.W. Jessen, H.L. Wang, and A.J. Epstein, *J. Appl. Phys.*, 78, 4264, 1995. (b) H.L. Wang, A.G. Macdiarmid, Y.Z. Wang, D.D.C. Gebler, and A.J. Epstein, *Synth. Met.*, 78, 33, 1996. (d) S.W. Jessen, J.W. Blatchford, L.B. Lin, T.L. Gustafson, J. Partee, J. Shinar, D.K. Fu, M.J. Marsella, T.M. Swager, A.G. Macdiarmid, and A.J. Epstein, *Synth. Met.*, 84, 501, 1997. (e) D.D. Gebler, Y.Z. Wang, S.W. Jessen, A.G. Macdiarmid, T.M. Swager, D.K. Fu, and A.J. Epstein, *Synth. Met.*, 85, 1205, 1997. (f) A.J. Epstein, Y.Z. Wang, S.W. Jesson, J.W. Blatchford, T.M. Swager, and A.J. MacDiarmid, *Macromol. Symp.*, 116, 27, 1997. (g) M.Y. Hwang, M.Y. Hua, and S.A. Chen, *Polymer*, 40, 396, 1999. (h) R.S. Montani, A.S. Diez, and R.O. Garay, *Macromolecules*, 5, 396, 2000.

166. T. Kanbara, T. Kushida, N. Saito, I. Kuwajima, K. Kubota, and T. Yamamoto, *Chem. Lett.*, 583, 1992.

167. (a) N. Satio, T. Kanbara, Y. Nakamura, T. Yamamoto, and K. Kubota, *Macromolecules*, 27, 756, 1994. (b) Y.Q. Liu, H. Ma,

and A.K.Y. Jen, *J. Mater. Chem.*, 11, 1800, 2001. (c) X.W. Zhan, Y.Q. Liu, X. Wu, S.A. Wang, and D.B. Zhu, *Macromolecules*, 35, 2529, 2002.

168. (a) T. Yamamoto, K. Sugiyama, T. Kushida, T. Inoue, and T. Kanbara, *J. Am. Chem. Soc.,* 9, 1217, 1997. (b) T. Fukuda, T. Kanbara, T. Yamamoto, K. Ishikawa, H. Takezoe, and A. Fukuda, *Synth. Met.,* 85, 1195, 1997.

169. H.L. Wang, M.J. Marsella, D.K. Fu, T.M. Swager, A.G. Mac-Diarmid, and A.J. Epstain, *Polym. Mater. Sci. Eng.,* 73, 473, 1995.

170. (a) J. Tian, C.C. Wu, M.E. Thompson, J.C. Sturm, R.A. Register, M.J. Marsella, and T.M. Swager, *Adv. Mater.,* 7, 395, 1995. (b) H. Hong, D. Davidov, M. Tarabia, H. Chayet, I. Benjamin, E.Z. Faraggi, Y. Avny, and R. Neumann, *Synth. Met,* 1997, 85, 1265.

171. P. Heseman, H. Vestweber, J. Pommerehne, R.F. Mahrt, and A. Greiner, *Adv. Mater.,* 7, 388, 1995.

172. C.M. Bouché, P. Berdagué, H. Facoetti, P. Robin, P.L. Barny, and M. Schott, *Synth. Met.,* 81, 191, 1996.

173. (a) X.C. Li, F. Cacialli, M. Giles, J. Grüner, R.H. Friend, A.B. Holmes, S.C. Moratti, and T.M. Yong, *Adv. Mater.,* 7, 898, 1995. (b) X.C. Li, T.M. Yong, J. Grüner, A.B. Holmes, S.C. Moratti, F. Cacialli, and R.H. Friend, *Synth. Met.,* 84, 437, 1997.

174. J.K. Lee, R.R. Schrock, D.R. Baigent, and R.H. Friend, *Macromolecules,* 28, 1966, 1995.

175. J.L. Segura, *Acta Polym.,* 49, 319, 1998.

176. R.D. McCullough, *Adv. Mater.,* 10, 93, 1998.

177. P. Kocacic and A. Kyriakis, *J. Am. Chem. Soc.,* 85, 454, 1963.

178. J.L. Reddinger and J.R. Reynolds, *Adv. Polym. Sci.,* 145, 58, 1999.

179. V. Grignard, *C.R. Acad. Sci.,* 130, 1322, 1900.

180. M. Rehahn, A.D. Schlüter, G. Wegner, and W.J. Feast, *Polymer,* 30, 1054, 1989.

181. (a) T.A. Chen, X.M. Wu, and R.D. Rieke, *J. Am. Chem. Soc.,* 117, 233, 1995. (b) X.M. Wu and T.A. Chen, *Macromolecules,* 28, 2101, 1995.

182. (a) R.F. Heck, *J. Am. Chem. Soc.*, 90, 5518, 1968. (b) R.F. Heck, in *Organic Reactions*, Vol. 27, W. G. Dauben et al., Eds., John Wiley & Sons, New York, 1982, 345.

183. (a) S. Klingelhöfer, C. Schellenberg, J. Pommerehne, H. Bässler, A. Greiner, and W. Heitz, *Macromol. Chem. Phys.*, 198, 1511, 1997. (b) R.F. Heck, in *Organic Reactions*, Vol. 27, W.G. Dauben et al., Eds., John Wiley & Sons, New York, 1982, chap. 2.

184. A. Greiner and W. Heitz, *Polym. Prepr.*, 32, 333, 1991.

185. U. Scherf, *Top. Curr. Chem.*, 201, 164, 1999.

186. M. Brenda, A. Greiner, and W. Heitz, *Makromol. Chem.*, 191, 1083, 1990.

187. H. Martelock, A. Greiner, and W. Heitz, *Makromol. Chem.*, 192, 967, 1991.

188. G. Hadziioannou and P.F. van Hutten, *Semiconducting Polymers*, Wiley, Germany, 2000, chap. 16.

189. A. Suzuki, *Pure Appl. Chem.*, 63, 419, 1991.

190. (a) M. Rehahn, A.D. Schlüter, G. Wegner, and J. Feast, *Polymer*, 30, 1060, 1989. (b) R. Rulkens, M. Schulze, and G. Wegner, *Macromol. Rapid Commun.*, 15, 669, 1994. (c) U. Lauter, W.H. Meyer, and G. Wegner, *Macromolecules*, 30, 2092, 1997.

191. (a) N. Tanigaki, H. Masuda, and K. Kaeriyama, *Polymer*, 38, 1221, 1997. (b) M. Remers, M. Schulze, and G. Wegner, *Macromol. Rapid Commun.*, 17, 239, 1996.

192. M. Remers, M. Schulze, and G. Wegner, *Macromol. Rapid Commun.*, 17, 239, 1996.

193. S. Janietz, D. Bradley, M. Grell, M. Inbasekaran, and E.P. Woo, *Appl. Phys. Lett.*, 73, 2453, 1998.

194. A.D. Child and J.R. Reynold, *Macromolecules*, 27, 1975, 1994.

195. F. Koch and W. Heitz, *Macromol. Chem. Phys.*, 198, 1531, 1997.

196. P. Liess, V. Hensel, and A.D. Schlüter, *Liebigs Ann. Chem.*, 614, 1996.

197. J. Saalbeck, *Ber. Bunsenges Phys. Chem.*, 100, 1667, 1996.

198. J.K. Stille, *Angew. Chem. Int. Ed.*, 25, 508, 1986.

199. H. Saadeh III, T. Goodson, and L. Yu, *Macromolecules*, 30, 4608, 1997.

200. (a) J.E. McMurry and M.P. Fleming, *J. Am. Chem. Soc.*, 96, 4708, 1974. (b) J.E. McMurry, *Chem. Rev.*, 89, 1513, 1989.

201. M. Rehahn and A.D. Schluter, *Makromol. Chem. Rapid Commun.*, 11, 375, 1990.

202. S. Iwatsuki, M. Kubo, and Y. Itoh. *Chem. Lett.*, 1085, 1993.

203. (a) H.A. Dieck and R.F. Heck, *J. Organomet. Chem.*, 93, 259, 1975. (b) J.I. Cassar, *J. Organomet. Chem.*, 93, 253, 1975.

204. K. Sonogashira, Y. Tohda, and N. Hagihara, *Tetrahedron Lett.*, 16, 4467, 1975.

205. U.H.F. Bunz, *Chem. Rev.*, 100, 1605, 2000.

206. Q. Zhou and T.M. Swager, *J. Am. Chem. Soc.*, 117, 12593, 1995.

207. K. Osakada, R. Sakata, and T. Yamamoto, *Organometallics*, 16, 5354, 1997.

208. (a) R. Giesa, *J.M.S. Rev. Macromol. Chem. Phys.*, 36, 631, 1996. (b) M. Alami, F. Ferri, and G. Linstrumelle, *Tetrahedron Lett.*, 34, 6403, 1993.

209. F.E. Goodson, T.I. Wallow, and B.M. Novak, *J. Am. Chem. Soc.*, 119, 12441, 1997.

210. R.R. Schrock, D.N. Clark, J. Sancho, J.H. Wengrovius, and S.F. Pederson, *Organometallics*, 1, 1645, 1982.

211. (a) U.H.F. Bunz and L. Kloppenburg, *Angew. Chem.*, 38, 478, 1999. (b) N. Kaneta, K. Hikichi, S. Asaka, M. Uemura, and M. Mori, *Chem. Lett.*, 1055, 1995.

212. L. Kloppenburg, D. Song, and U.H.F. Bunz, *J. Am. Chem. Soc.*, 120, 7973, 1998.

213. W. Steffen and U.H.F. Bunz, *Macromolecules*, 33, 9518, 2000.

214. E. Csányi and P. Kraxner, *Macromol. Rapid Commun.*, 16, 147, 1995.

215. V.P. Conticello, D.L. Gin, and R.H. Grubbs, *J. Am. Chem. Soc.*, 114, 9708, 1992.

216. Y.J. Miao and G.C. Bazan, *Macromolecules*, 27, 1063, 1994.

217. G.C. Bazan, Y.J. Miao, M.L. Renak, and B.J. Sun, *J. Am. Chem. Soc.,* 11, 2618, 1996.

218. M.L. Renak, G.C. Bazan, and D. Roitman, *Adv. Mater.,* 9, 392, 1997.

219. T. Laue and A. Plagens, *Named Organic Reactions,* 2nd Ed., John Wiley & Sons, New York, 1999, 271.

220. H.E. Katz, S.F. Bent, W.L. Wilson, M.L. Schilling, and S.B. Ungashe, *J. Am. Chem. Soc.,* 116, 6631, 1994.

221. N.J. Lawrence, *The Wittig Reactions and Related Methods in Preparation of Alkenes: A Practical Approach,* J.M.J. Williams, Ed., Oxford University Press, New York, 1996.

222. L. Liao, Y. Pang, L.M. Ding, and F.E. Karasz, *Macromolecules,* 35, 3819, 2002.

223. K.B. Becker, *Synthesis,* 341, 1983.

224. (a) G. Drefahl, R. Kuhmstedt, H. Oswald, and H.H. Horhold, *Makromol. Chem.,* 131, 89, 1970. (b) R. Schenk, H. Gregorius, K. Meerholz, J. Heinze, and K. Müllen, *J. Am. Chem. Soc.,* 113, 2634, 1991. (c) N.N. Barashkov, D.J. Guerrero, H.J. Olivos, and J.P. Ferraris, *Synth. Met.,* 75, 153. 1995.

225. (a) R.E. Gill, P.F. van Hutten, A. Meetsma, and G. Hadziioan-nou, *Chem. Mater.,* 8, 1341, 1996. (b) S.E. Dottinnger, M. Hohloch, J.L. Segura, E. Steinhuber, M. Hanack, A. Tompert, and D. Oelkrug, *Adv. Mater.,* 9, 233, 1997.

226. E. Thorn-Csanyi and K.P. Pflug, *J. Mol. Catal.,* 90, 69, 1994.

227. H. Kretzschmann and H. Meier, *J. Prakt. Chem.,* 336, 247, 1994.

228. T. Maddux, W. Li, and L. Yu, *J. Am. Chem. Soc.,* 119, 844, 1997.

229. U. Stalmach, H. Kolshorn, I. Brehm, and H. Meier, *Liebigs Ann.,* 1449, 1996.

230. M.H. Meier, U. Stalmach, and H. Kolshorn, *Acta Polym.,* 48, 379, 1997.

231. T. Laue and A. Plagens, *Named Organic Reactions,* 2nd Ed., John Wiley & Sons, New York, 1999, 164.

232. H.H. Hörhold and M. Helbig, *Makromol. Chem. Macromol. Symp.*, 12, 229, 1987.

233. J.I. Jin, C.K. Park, and H.K. Shim, *Macromolecules*, 26, 1779, 1993. (b) D.R. Gagnon, J.D. Capistran, F.E. Karasz, and R.W. Lenz, *Polymer*, 28, 567, 1987.

234. (a) G. Gustafasson, Y. Cao, G.M. Treacy, F. Klauetter, N. Colaneri, and A.J. Heeger, *Nature*, 357, 477, 1992. (b) I.N. Kang, H.K. Shim, and T. Zyung, *Chem. Mater.*, 9, 746, 1997.

235. F. Louwet, D. Vanderzande, J. Gelan, and J. Mullens, *Macromolecules*, 28, 1330, 1995.

236. S. Shi and F. Wudl, *Macromolecules*, 23, 2119, 1990.

237. I. Murase, T. Ohnishi, T. Noguchi, and M. Hirooka, *Polym. Commun.*, 25, 327, 1984.

238. H.G. Gilch and W.L. Wheelwright, *J. Polym. Sci. A-1*, 4, 1337, 1966.

239. (a) A.J. Heeger and D. Braun, *Chem. Abstr.*, 118, 15740j, 1993. (b) F. Wudl, P.M. Allemand, G. Srdanov, Z. Ni, and D. Bobranch, *ACS Symp. Ser.*, 455, 683, 1991.

240. G.J. Sarnecki, P.L. Burn, A. Kraft, R.H. Friend, and A.B. Holmes, *Synth. Met.*, 55, 914, 1993.

241. B.R. Hsieh, Y. Yu, H.K. Lee, and A. C. Vanlaeken, *Macromolecules*, 30, 8094, 1997.

242. U. Scherf and E.W.J. List, *Adv. Mater.*, 14, 477, 2002.

243. H.G. Nothofer, A. Meisel, T. Miteva, D. Neher, M. Forster, M. Oda, G. Lieser, D. Sainova, A. Yasuda, D. Lupo, W. Knoll, and U. Scherf, *Macromol. Symp.*, 154, 139, 2000.

244. P.E. Fanta, *Chem. Rev.*, 64, 613, 1964.

245. S. Claesson, R. Gehm, and W. Kern, *Makromol. Chem.*, 7, 46, 1949.

246. M. Pomerantz, H. Yang, and Y. Cheng, *Macromolecules*, 28, 5706, 1995.

247. (a) M.A. Baldo, M.E. Thompson, and S.R. Forrest, *Nature*, 403, 750, 2000. (b) X. Gong, M.R. Robinson, J.C. Ostrowski, D. Moses, G.C. Bazan, and A.J. Heeger, *Adv. Mater.*, 14, 585, 2002.

6

Organic Electrophosphorescence

MARC A. BALDO, STEPHEN R. FORREST,
AND MARK E. THOMPSON

CONTENTS

6.1 INTRODUCTION

To be widely adopted, electroluminescent devices must not only match the color purity and long-term stability of competing technologies, but they must also provide a significant advantage in efficiency, especially in low-power, portable applications. In this chapter, we examine the fundamental limits to the performance of organic light-emitting diodes (OLEDs), and concentrate on methods to improve their efficiencies significantly. We first explain the various factors that comprise the power efficiency of an OLED. Then electrophosphorescence, a method for significantly improving the efficiency of organic devices, is described and characterized. Finally, the status of electrophosphorescent technology is discussed as of writing in 2002; we concentrate on three device criteria: power efficiency, emission color purity, and stability.

6.2 POWER EFFICIENCY

The power efficiency of an OLED is the ratio of the luminous power out, as detected by the human eye, to the electrical power in. Because the responsivity of the eye varies significantly across the visible spectrum, a specific unit, the lumen (lm), is defined to quantify perceived luminous intensity. The standardized response of the eye to bright light, also known as the photopic response,[1] is shown in Figure 6.1. By definition, a

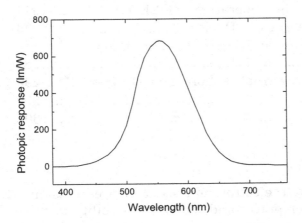

Figure 6.1 The photopic response is the standardized efficiency of the cone photoreceptor cells in the eye. By definition, it peaks at 683 lm per incident optical watt at a wavelength of $\lambda = 555$ nm.

typical human eye has a maximum luminous efficacy of 683 lm/W at a wavelength of $\lambda = 555$ nm. Response in the blue and red is much weaker.[2] The responsivity, Φ, of the eye to incident light with an arbitrary spectrum $\phi(\lambda)$ can be calculated using Equation 6.1,

$$\Phi = \int \phi(\lambda) P(\lambda) d\lambda \Big/ \int \phi(\lambda) d\lambda \, , \qquad (6.1)$$

where $P(\lambda)$ is the photopic response. The related luminous intensity is defined as the candela (cd), which is the luminous flux per solid angle in steradians, i.e., 1 cd = 1 lm/sr.* The sensation of color is standardized and quantified similarly. The spectral responses of three different color receptive cells in the eye are shown in Figure 6.2. Chromaticity coordinates, also known as Commission Internationale d'Éclairage (CIE) coordinates, $(\bar{x}, \bar{y})$ are defined by calculating the "tristimulus" response of the color receptive cells, i.e.:

---

* Note that the luminous intensity is dependent on viewing angle. For a Lambertian emitter within an organic material sandwiched between a perfectly reflective metal and air, the angular distribution of emitted light is[3] $F(\theta) = (1/2\pi)(n_2{}^2/n_1{}^2)\cos\theta \Big/ \sqrt{1 - (n_2{}^2/n_1{}^2)\sin^2\theta}$, where n_2 and n_1 are the refractive indices of the air and organic, respectively, and θ is the angle to the normal. The total fractional intensity emitted into air is obtained by integrating $F(\theta)$ over the forward hemisphere. The resulting conversion factor between intensity (in cd) and flux (in lm) is $F(\theta) \Big/ \left(1 - \sqrt{1 - n_2^2/n_1^2}\right)$. If $n_1 = 1.7$ and $n_2 = 1$, perpendicular to the device 1 cd $\approx$ 1 lm/π. Note that this calculation is an approximate solution that ignores the microcavity formed by the OLED metal contact.

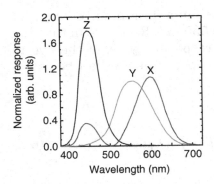

Figure 6.2 (Color figure follows p. 274.) The standardized response of the three-color-sensitive photoreceptors in the eye. Color coordinates can be calculated by overlapping each of these responses with the emission spectrum of the OLED.

$$x = \int \phi(\lambda) X(\lambda) d\lambda$$

$$y = \int \phi(\lambda) Y(\lambda) d\lambda$$

$$z = \int \phi(\lambda) Z(\lambda) d\lambda \; , \qquad (6.2)$$

$$\bar{x} = \frac{x}{x + y + z}$$

$$\bar{y} = \frac{y}{x + y + z}$$

where $X(\lambda)$, $Y(\lambda)$, and $Z(\lambda)$ are the responsivities of the three receptors. When normalized the $(\bar{x}, \bar{y})$ coordinates can be plotted on a color chart such as Figure 6.3.

To determine the power efficiency of an OLED, we first define the external quantum efficiency, η_Q:

$$\eta_Q = \frac{\text{photons out}}{\text{electrons in}} \; . \qquad (6.3)$$

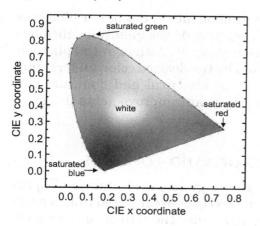

Figure 6.3 (Color figure follows p. 274.) The CIE chart used to quantify the sensation of color. A full-color display should possess red, green, and blue pixels. Positioning these on the chart above forms a triangle that circumscribes all the colors available to the display as a combination of the red, green, and blue elements. Thus, the individual colored pixels should be as saturated as possible (i.e., their emission should contain the minimum amount of white) and they should be located on the perimeter of the chart.

From the definitions of η_Q and Φ in Equations 6.1 and 6.3, the luminous flux, L, measured in lumens, is

$$L = \Phi \frac{hc}{\lambda} \eta_Q \frac{I}{q}, \qquad (6.4)$$

where Φ is the photopic response defined by Equation 6.1, h is Planck's constant, c is the speed of light, I is the current, and q is the electron charge. Dividing both sides of Equation 6.4 by the electrical power, VI, yields:

$$\eta_P = \frac{L}{VI} = \Phi \eta_Q \frac{V_\lambda}{V}, \qquad (6.5)$$

where η_P is the power efficiency and $qV_\lambda = hc/\lambda$ is the emissive photon energy.

   Thus, there are three components to the power efficiency
of an OLED: the photopic response, Φ; the quantum efficiency,
η_Q; and the electrical efficiency, V_λ/V. Although the photopic
response, Φ, is typically fixed by the desired color, understanding the physics underlying the electrical and quantum efficiency factors can lead to significant improvements in device
performance.

6.3   THE STRUCTURE AND OPERATION OF OLEDS

An OLED directs electrical energy to the excitation of light-
emitting organic molecules; electrons are injected from a cathode, holes from an anode, and under the influence of an electric field the carriers hop toward one another. If both charges
arrive on a single molecule, a molecular excited state may be
formed. The binding energy of the state may approach
~1 eV in low-molecular-weight materials due to Coulomb
interactions within confined molecular orbitals.[4] Consequently, a molecular excited state is not readily dissociated
and its properties are conserved as it diffuses between molecules, allowing it to be treated as a particle. These states are
known as "excitons." Not all excitons can efficiently decay and
emit light. The ground state of most molecules has a total
spin, $S = 0$, and because the emission of a photon conserves
spin, typically only $S = 0$ excited states can emit light in a
fast and efficient process known as fluorescence. The probability of luminescence from the remaining $S = 1$ excited states
is generally so low that almost all their energy is lost to nonradiative processes. Because the spin multiplicity of $S = 0$ and
$S = 1$ states is 1 and 3, respectively, they are known as singlet
and triplet excitons. We define the fraction of emissive excitons
as χ; in a fluorescent OLED the restriction to singlet excited
states gives $\chi = 25\%$. The singlet-to-triplet ratio has been
directly measured for Alq_3-based electrophosphorescent devices
and is very close to the expected 1:3 ratio.[5] Several measurements of the singlet-to-triplet ratio in polymer-based devices
have also been performed.[6-8] While those experiments suggest
that the singlet fraction in polymers is greater than 25%, the
remaining excitons are still triplets, highlighting the need to

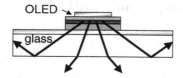

Figure 6.4 (Color figure follows p. 274.) The fraction of photons emitted in the forward, or viewing, direction is reduced by absorption losses and waveguiding within the device and its substrate.

efficiently harvest triplet excitons for high-efficiency OLEDs, regardless of the choice of molecular or polymeric materials.

The other major limitations to the quantum efficiency of OLEDs are the light output coupling fraction η_X, the charge balance fraction η_C, and the photoluminescent efficiency of the emissive material, ϕ_{PL}. In high-efficiency devices, heterostructures may be used to maximize η_C by preventing the leakage of charge carriers from the luminescent region,[9,10] thereby ensuring that each injected charge combines with its opposite and forms an emissive state. The photoluminescent efficiency of a molecule under optical excitation is defined as the number of emitted photons per absorbed photon. Since the triplet absorption is typically very weak, ϕ_{PL} is generally a measurement of the efficiency of the radiative relaxation of photoexcited singlet excitons. It can be optimized by chemical design of the emissive material, and is frequently close to 100%. For example, the addition of non-π-conjugated spacer groups to luminescent molecules can reduce intermolecular overlap, restraining excimer formation and the transport of excitons to defect sites. The output coupling fraction is limited by absorption losses and guiding of electroluminescence within the device and its substrate; see Figure 6.4. A variety of techniques have been employed to increase the output coupling fraction, notably: roughening the substrate to increase scattering, and reduce guiding;[11] etching trenches in the substrate to reflect guided light in the viewing direction;[3] use of a low refractive index substrate;[12] and the use of transparent contacts to reduce absorption losses.[13]

In summary, the quantum efficiency of an OLED may be expressed as

$$\eta_Q = \chi \eta_X \eta_C \phi_{PL} , \qquad (6.6)$$

where χ is the fraction of emissive excitons, η_X is the output coupling fraction, η_C is the charge balance fraction, and ϕ_{PL} is the photoluminescent efficiency of the emissive molecule.

6.4 ELECTROPHOSPHORESCENT DEVICES

Although radiation from triplet states is rare, the process can be quite efficient in some materials. For example, the decay of the triplet state is partially allowed if the excited singlet and triplet states are mixed such that the triplet gains some singlet character. The emission of light in this case is still significantly slower than fluorescence, but if singlet–triplet mixing yields a radiative decay rate that is faster than the nonradiative rate, then the luminescence can be efficient. This emission of light from a "disallowed" transition is known as phosphorescence, and if the molecule is excited electrically, it is known as "electrophosphorescence."

Triplet excitons possess a spatial wave function that is antisymmetric under exchange of the two constituent electrons. It follows from this symmetry that the electron–electron repulsion in a triplet exciton is lower than for a singlet exciton. Thus, the energy of a molecular triplet exciton is always lower than the corresponding singlet; and given mixing between singlet and triplet states, singlet to triplet energy transfer, known as intersystem crossing (ISC), will occur. In an efficient phosphor this ensures that all singlet excitons are rapidly converted to the luminescent triplet state.

To exploit an efficiently phosphorescent material in an OLED, we require the transfer of both singlet and triplet excitons from the charge transport layer (henceforth called the host) to the phosphorescent guest. Several common host materials are shown in Figure 6.5. In the absence of a phosphorescent host, nonradiative dipole–dipole coupling (Förster transfer)[14,15] is restricted to spin-allowed transitions and the

Electron Transporting
Hosts

Hole Transporting
Hosts

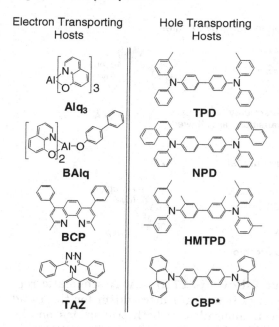

Figure 6.5 A selection of the host materials used in conjunction with phosphorescent dyes. CBP is highlighted because it may also transport electrons.

transfer of singlet excitons. However, energy transfer may occur by the parallel combination of Förster transfer to the singlet state, along with electron exchange (hopping, or Marcus transfer); see Figure 6.6. The latter is the physical transfer of an exciton from a donor to an acceptor site at a rate proportional to the frontier orbital overlap of the donor and acceptor molecules.[16] Consequently, it is a short-range process, attenuating exponentially with distance. But it allows for the transfer of triplet excitons because only the total spin of the donor–acceptor complex is conserved.

The first demonstration of electrophosphorescence used benzophenone mixed into a poly(methylmethacrylate) host.[17] The singlet–triplet mixing in benzophenone is weak, and nonradiative decay surpasses phosphorescence at room temperature. Although phosphorescence from benzophenone can be

A Dipole-dipole coupled (Förster) energy transfer

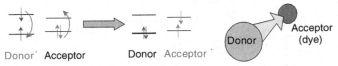

Donor   Acceptor   Donor   Acceptor

Exciton nonradiatively transferred by   up to ~100 Å
dipole - dipole coupling if transitions are allowed.

B Electron exchange energy transfer

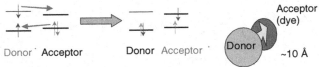

Donor   Acceptor   Donor   Acceptor

Exciton hops from donor to acceptor.

Figure 6.6 (Color figure follows p. 274.) (A) A schematic representation of Förster energy transfer, a mechanism for the rapid transfer of energy between molecules. If both transitions on the donor and acceptor are allowed, then the range of transfer may extend to 100 Å. This is the dominant method for the transfer of singlet excitons. (B) If one of the transitions is disallowed, then Förster energy transfer is not possible; for example, we may not excite the triplet state of the acceptor. However, it is possible to transfer triplets by exciton hopping from one molecule to the next. The rate constant of this process is determined by the rate of Marcus electron and hole transfer; see Closs et al.[16]

observed at $T \sim 100$ K, no triplet transfer from the host was observed in this device, and benzophenone was found to be ill suited for OLED applications.

Another relatively well studied red-emitting phosphorescent dye is 2,3,7,8,12,13,17,18-octaethyl-21H,23H-porphine platinum(II) (PtOEP); see Figure 6.7. Porphine complexes are known to possess long-lived triplet states useful in oxygen detection.[18] The addition of platinum to the porphyrin ring reduces the phosphorescence lifetime by increasing spin-orbit coupling; the triplet states gain additional singlet character, and vice versa. As described above, this also enhances the

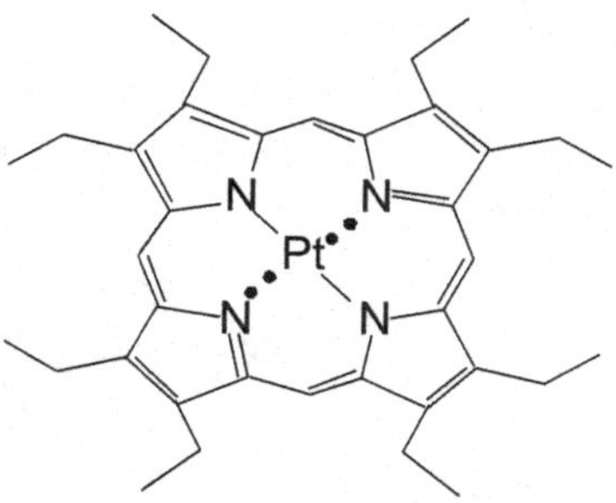

PtOEP

Figure 6.7 The chemical structure of platinum octaethylporphyrin.

efficiency of intersystem crossing from the first singlet excited state to the triplet excited state. Transient absorption spectrometry gives a singlet lifetime in PtOEP of ~1 ps, and the fluorescence efficiency is extremely weak.[19] In contrast, the room temperature phosphorescence efficiency of PtOEP in a polystyrene matrix is 0.5 with an observed lifetime of 91 μs.[20]

PtOEP shows strong singlet absorption at 530 nm,[21] corresponding to the peak emission of the electron transport material, tris(8-hydroxyquinoline) aluminum (Alq_3) (see Figure 6.5); and in the absence of direct evidence of the triplet energy of Alq_3 it may be assumed that PtOEP is a suitable dopant for Alq_3-based OLEDs. Alq_3 possesses no measurable phosphorescence even at low temperature; therefore the triplet energy must be estimated from that of other, phosphorescent, metal–quinoline complexes. Based on phosphorescence from heavy metal analogs of Alq_3, such as Irq_3 and Ptq_2, the Alq_3 triplet energy is estimated to lie between 1.9 and 2.1 eV.[22] The device structure used to demonstrate PtOEP electrophosphorescence is shown in Figure 6.8.

The electroluminescence spectra of the devices with three different concentrations of PtOEP are shown in Figure 6.9. Emission from the previously identified singlet state (expected at approximately 580 nm)[19] is not observed, but strong emission is observed from the triplet excited state at 650 nm, with weaker emission seen at the vibronic harmonic

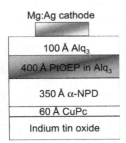

Figure 6.8 (Color figure follows p. 274.) The device structure used to demonstrate PtOEP electrophosphorescence.

overtones at 623, 687, and 720 nm. At high drive currents, the emission at 650 nm saturates, and increasing emission is seen from other features, especially the broad Alq₃ peak centered at 530 nm. This results in a reversible shift in the color from deep red to orange, corresponding to a shift in the device chromaticity coordinates shown in the inset of Figure 6.9.

The transient and steady-state responses of PtOEP are compared in Figure 6.10. Spectral and time-resolved emission measurements were performed with a streak camera on OLEDs excited with current pulses. Slow decay of between 10 and 50 μs was observed for the triplet state, confirming the presence of phosphorescence. The luminescent lifetime is found to decrease with increasing current density, as shown in Figure 6.10A. From the spectral data, the fraction of output power *in the red* (subtracting the Alq₃ emission) can be determined as a function of current to provide the external quantum efficiency shown in Figure 6.10B. The external quantum efficiency reaches a maximum value of 4% at a PtOEP concentration of 6%.

Because PtOEP molecules possess a long exciton lifetime, it is possible that saturation is responsible for the decreased quantum efficiency at high currents and low dopant concentrations. This hypothesis can be examined by comparing the total number of PtOEP sites to the maximum number of photons extracted. For a molar concentration of 1%, the number of PtOEP sites in the ~50-Å-thick recombination zone[23]

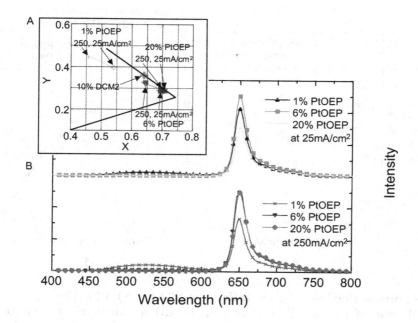

Figure 6.9 (Color figure follows p. 274.) Spectra of the OLEDs with different molar concentrations of PtOEP at different current densities. (A) 1, 6, and 20% PtOEP in Alq_3 OLEDs at 25 mA/cm² (B) 1, 6, and 20% PtOEP in Alq_3 OLEDs at 250 mA/cm². Note the increased Alq_3 emission at 530 nm in the 1% PtOEP OLED. Inset: CIE chromaticity coordinates for the devices in Figure 6.8 at the specified current densities. Only the red corner of the CIE diagram is shown. Note the trend from saturated red to orange with increasing current. The 6% PtOEP in the Alq_3 device at 25 mA/cm² has a luminance of 100 cd/m². The DCM2 result is from Bulović et al.[15]

near the α–NPD/PtOEP:Alq_3 interface is ~8 × 10¹² cm⁻². After adjusting for the output coupling efficiency,[3] the internal emitted photon flux saturates at ~3 × 10¹⁶ s⁻¹cm⁻², yielding a radiative lifetime of ~300 μs, consistent with previous PtOEP phosphorescence efficiency and lifetime measurements.[20] If sites within the recombination zone saturate, then Alq_3 excitons are less likely to transfer to PtOEP, corresponding to an

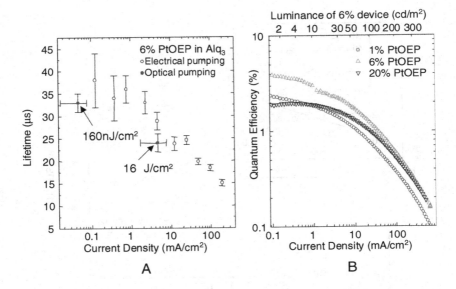

Figure 6.10 (Color figure follows p. 274.) (A) Phosphorescent lifetime of the 6% PtOEP in Alq_3 OLED as function of current density. Both electrical and optical pumping show a decrease in PtOEP lifetime with increasing excitation strength, indicative of a bimolecular quenching process such as triplet–triplet annihilation. (B) Quantum efficiency of PtOEP emission as a function of doping concentration and current density. The top axis shows the luminance of the 6% PtOEP in Alq_3 device.

increase in the probability for radiative recombination in Alq_3. This is supported by Figure 6.9, which shows increased Alq_3 emission for devices with low concentrations (~1%) of PtOEP.

Saturation of emissive sites is alleviated by increasing the concentration of PtOEP. Yet from Figure 6.10B, it is evident that at molar concentrations >6%, the quantum efficiency still decreases with current density. Because high densities of porphyrin complexes often exhibit bimolecular quenching,[24] it is likely that the PtOEP/Alq_3 system is also affected by this process.[25] To explore the influence of bimolecular interactions such as triplet-triplet annihilation (shown in Figure 6.11), PtOEP quenching was investigated using photoluminescence; see Figure 6.9A. This experiment was

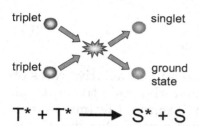

$$T^* + T^* \longrightarrow S^* + S$$

Figure 6.11 (Color figure follows p. 274.) Total spin is conserved if two excited triplets combine and form an excited singlet and ground-state singlet exciton. The process destroys at least one triplet exciton and is known as triplet–triplet annihilation.

performed by pumping a 500-Å-thick film of 6% PtOEP in Alq_3 with 500-ps pulses from a nitrogen laser at a wavelength of 337 nm. At a pulse energy density of 16 μJ/cm² the measured PtOEP lifetime was (24 ± 2) μs. To check if the optical and electrical decay times have a similar origin, it is noted that the laser output is absorbed principally by Alq_3 with an absorption coefficient[26] of ~3 × 10⁴ cm⁻¹. Hence, ~26% of the nitrogen laser pulse is absorbed, giving an exciton density of 1.4 × 10¹⁸ cm⁻³. Assuming 100% transfer of injected carriers to PtOEP excitons, the current density corresponding to this pumping level is ~5 mA/cm². When the pulse energy is decreased to ~160 nJ/cm² the lifetime increases to (33 ± 2) μs. Thus, the trend in the lifetime of optically pumped samples, which has been previously attributed[24] to bimolecular quenching, is consistent with the trend of the electroluminescent lifetime data and the quantum efficiency results of Figure 6.10B, suggesting a common physical origin. In addition, the phosphorescent decay becomes biexponential as the current increases, supporting the possibility of a bimolecular quenching process such as triplet–triplet annihilation.

Accounting for photons lost via emission through the sides of the device and total internal reflection within the substrate,[3] a maximum external efficiency of 4% corresponds to an internal efficiency of 23%. The photoluminescent efficiency is related to the lifetime by

$$\phi_{PL} = \frac{k_R}{k_R + k_{NR}}, \qquad (6.7)$$

where k_R and k_{NR} are the radiative and nonradiative rates of PtOEP, respectively. Assuming that the radiative rate of PtOEP is constant, a maximum exciton lifetime in the PtOEP/Alq$_3$ devices of 45 μs implies a peak phosphorescent yield of ≈25%. This implies ≈90% energy transfer from Alq$_3$ to PtOEP, confirming that *both* triplet and singlet excitons must participate in energy transfer.

PtOEP doped in polystyrene has a lifetime of 91 μs and a photoluminescent efficiency of 50%,[20] or twice that of PtOEP in Alq$_3$. It is thought that the efficiency of PtOEP is related to how tightly PtOEP triplets are confined by the surrounding host material. Thus, superior electrophosphorescent OLED architectures should employ the phosphorescent guest in a host material with significantly larger triplet energy to confine the triplet states. One such host material is 4,4'-*N,N'*-dicarbazole-biphenyl (CBP; see Figure 6.5), which has a triplet energy of (2.6 ± 0.1) eV, as compared to the PtOEP triplet energy of (1.9 ± 0.1) eV.[22] The phosphorescent lifetime of PtOEP in CBP is ~100 μs, or approximately twice that of PtOEP in Alq$_3$, indicating that the phosphorescent yield of PtOEP in CBP will also be approximately twice that of PtOEP:Alq$_3$.

The lifetime of CBP triplets is ~1 s, enabling them to diffuse significant distances.[22] In addition to confining triplets once guest molecules capture them, it is necessary to minimize the loss of host triplets into surrounding materials. This can be achieved by increasing the rate of triplet capture by guest molecules, for example, by adding the guest molecules at a high concentration (~10%) into the host; or the device can be designed with organic layers that block the diffusion of triplets.

There are two possible architectures for triplet blocking layers. If the host material is principally hole transporting (e.g., CBP), then the exciton blocking layer must also be an electron transport layer (ETL) and block holes. However, the common ETL Alq$_3$ cannot be used as an exciton blocking material because its triplet energy is too low ((2.0 ± 0.1) eV).[22]

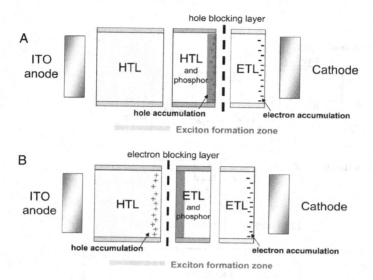

Figure 6.12 (Color figure follows p. 274.) (A) The electrophosphorescent device architecture used for a predominantly hole-transporting host material. A high density of holes are accumulated at the HTL/ETL interface because transport is typically limited by electron-injection. Thus, it is essential to include a hole blocking layer. This layer should also retard the transport of excitons into the ETL. (B) Devices limited by electron-injection are better suited to emit from an ETL layer. In this case, an electron blocking layer is not as important since the majority of electron charge is located near the cathode. Less charge in the exciton formation zone should also reduce losses from exciton-charge quenching.

An alternative ETL blocking layer is 2,9-dimethyl-4,7-diphenyl-1,10-phenanthroline (bathocuproine, or BCP; see Figure 6.5).[27] Close relatives of bathocuproine have good electron-transporting properties,[28] and BCP itself has a large ionization potential of 6.4 eV,[29] making it suitable for use as a hole-blocking layer in OLEDs, and a triplet energy of (2.5 ± 0.1) eV.[22] In the device architecture shown in Figure 6.12A, the BCP allows the transport of electrons into the luminescent CBP layer while preventing holes from leaking into the Alq_3.[30] It is also intended to retard the diffusion of triplets into the low-energy Alq_3 triplet sink, where they are effectively lost to luminescence.

A

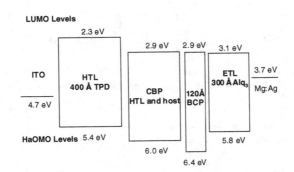

B

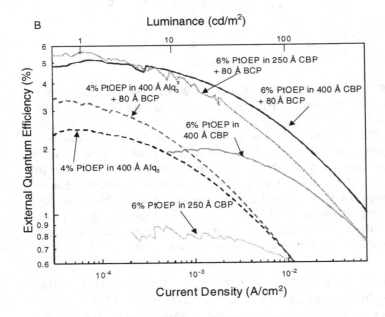

Figure 6.13 (See caption on opposite page.)

The alternative device structure is shown in Figure 6.12B.[31] Here, the host is used as an ETL, and it is required that the hole transport layer (HTL) possess a triplet energy larger than that of the phosphorescent guest, and act as an exciton blocking layer. Transport in OLEDs is typically limited by electron injection.[32] Consequently, there is a large density

Figure 6.13 (Opposite)
(A) The energy levels of the PtOEP:CBP electrophosphorescent device. Electron and hole transport layers are labeled ETL and HTL, respectively. The highest occupied molecular orbital (HOMO) obtained for each material corresponds to its ionization potential (IP). As calculated, the lowest unoccupied molecular orbital (LUMO) is equal to the IP plus the optical energy gap determined from absorption spectra. The vacuum level is assumed to be constant and relative alignments of the energies in the fully assembled devices will differ from those shown if there is charge transfer at the interfaces. (Based on I.G. Hill and A. Kahn, *J. Appl. Phys.*, 86, 4515–4519, 1999.) (B) External quantum efficiencies of PtOEP:CBP and PtOEP:Alq$_3$ devices as a function of current density with and without a BCP blocking layer. The top axis shows the luminance of a device with a 80-Å-thick BCP layer and a 400-Å-thick CBP luminescent layer containing 6% PtOEP.

of electrons in the vicinity of the cathode. These electrons are balanced by an equal density of holes in the HTL at the organic HTL/ETL interface. Because of the possibility of triplet-polaron quenching,[25,32] it is preferable for these holes not to be located in the luminescent region. Thus, given the constraint of electron-injection limited transport, the structure in Figure 6.12B is preferable for high-efficiency OLEDs, although high fields across the ETL may increase exciton dissociation in this structure also.

The architecture of Figure 6.12A can be demonstrated by examining the electrophosphorescent efficiency of PtOEP in CBP. The OLED structure employed was[30] an ITO anode, a 450-Å-thick HTL of α-NPD, a luminescent HTL containing PtOEP in CBP, an 80-Å-thick blocking layer of BCP, and a further 200-Å-thick cap layer of Alq$_3$ to prevent nonradiative quenching of PtOEP excitons by the cathode. A shadow mask with 1-mm-diameter openings was used to define the cathodes consisting of a 1000-Å-thick layer of 25:1 Mg:Ag, with a further 500-Å-thick Ag cap.

As shown in Figure 6.13, devices using CBP as the host without the BCP layer possess poor external quantum efficiencies due to the leakage of holes and excitons into the Alq$_3$

layer. The BCP barrier layer substantially improves the quantum efficiency in PtOEP:CBP devices to a peak of $\eta_Q = (5.6 \pm 0.1)\%$, almost twice that of the PtOEP:Alq$_3$ devices. This was expected from the comparison of triplet lifetimes for PtOEP in Alq$_3$ and CBP, respectively.

Although PtOEP demonstrates the efficiency improvements made possible by the participation of triplet excitons in phosphorescence, the long lifetime of PtOEP (>10 μs) causes saturation of emission at low dopant concentrations and exacerbates triplet–triplet annihilation, which increases quadratically with triplet lifetime. But the high efficiency of PtOEP demonstrates the principal advantage of phosphors, encouraging further work on these materials.

6.5 IRIDIUM COMPLEXES: THE SECOND GENERATION OF PHOSPHORS

The presence of a metal atom in complexes such as PtOEP increases singlet–triplet mixing and adds excited states to the electronic spectrum. In addition to states localized on the organic groups (ligands), the metal atom can participate in an excited state formed by the transfer of an electron from the metal to a ligand. These are known as metal–ligand charge transfer (MLCT) states. Relative to the ligand excited states, MLCT states have a larger overlap with the heavy metal atom; consequently, the spin orbit coupling is higher, and there is greater mixing between singlets and triplets.

Often phosphorescence occurs from a mixture of MLCT and ligand states. The relative importance of the MLCT and ligand components depends on their energetic separation. For example, the MLCT component of PtOEP phosphorescence is very small because the porphyrin ligand has a significantly lower triplet energy.[24]* This affects device performance because, localized on the ligand, a PtOEP triplet experiences

---

* Nevertheless, Pt reduces the triplet lifetime of the porphyrin from ~20 ms to ~100 μs.[24]

less spin orbit mixing due to the central Pt atom, and consequently, its lifetime may be as long as 100 μs. Triplet–triplet annihilation increases with the square of the triplet lifetime,[25] (see Figure 6.14) and PtOEP-based OLEDs experience strong

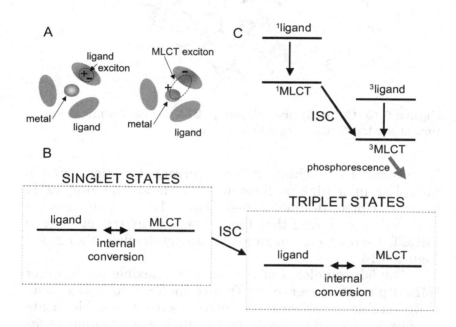

Figure 6.14 (Color figure follows p. 274.) (A) In metal–organic complexes the emissive state is generally a mixture of a ligand-centered exciton and a metal–ligand charge transfer (MLCT) exciton. The MLCT state has superior singlet-triplet mixing because it overlaps with the heavy-metal atom. The overall energy level diagram for a metal-organic phosphor is shown in B. For strong spin orbit coupling, the intersystem crossing and internal conversion rates are extremely fast. Thus, emission occurs predominantly from the lowest energy triplet state. In a high performance phosphor, such as that shown in C, the ³MLCT state is often the lowest state, thereby minimizing the triplet lifetime and triplet–triplet annihilation. Here we show a system where the ³MLCT is the lowest energy exciton. Alternately, the ligand-centered exciton may be lower in energy, in which case it will dominate emission.

fac tris(2-phenylpyridine) iridium

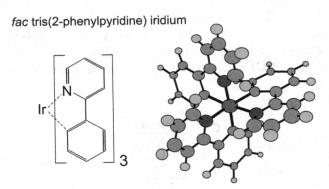

Figure 6.15 (Color figure follows p. 274.) The chemical structure of *fac* tris(2-phenylpyridine) iridium.

triplet-triplet quenching at large current densities. Thus, a phosphor in a high-performance OLED should preferably have a triplet lifetime of less than ~10 μs. This may be achieved by ensuring that the lowest energy triplet state is ^{3}MLCT, thereby maximizing the singlet–triplet mixing; see Figure 6.14.

The 5d^6 complexes of Ir^{3+} provide a flexible platform for ^{3}MLCT phosphorescence.[33,34] These complexes possess a chemically stable octahedral symmetry, with three bidentate ligands complexed to the central Ir atom. For example, in *fac* tris-(2-phenylpyridine) iridium (*fac*-Irppy$_3$) (see Figure 6.15), the lowest triplet state of the ligand emits at λ = 460 nm, and ^{3}MLCT phosphorescence is at λ = 510 nm. Thus, the lowest energy state is the ^{3}MLCT, and Irppy$_3$ exhibits efficient green phosphorescence with a transient lifetime of ~1 μs.[35–37] Indeed, the spin orbit coupling in Irppy$_3$ is sufficient to give a "forbidden" ^{3}MLCT absorption cross section that is almost identical to that of the "allowed" ^{1}MLCT transition.[33,34] A slightly higher luminescent efficiency and a simplified synthesis[33] are possible for the Irppy$_3$ variant ppy$_2$Ir(acac). Its luminescence spectrum is shown in Figure 6.16A. It is also possible to tune the color of these molecules; see Figure 6.16B to C and Figure 6.17. Ligands with lower triplet energies can be used to shift phosphorescence into the red.[34] For example,

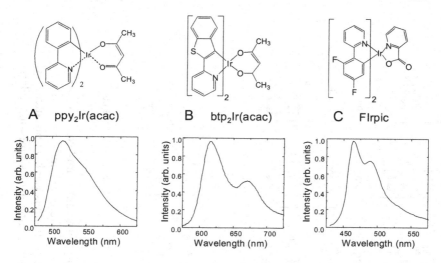

Figure 6.16 The chemical structures and room-temperature phosphorescent spectra of (A) green, (B) red, and (C) blue, iridium-based metal-organic phosphors. FIrpic = (Fppy)$_2$Ir(pic).

btp$_2$Ir(acac) emits from the btp ligand at $\lambda = 610$ nm, albeit with a longer triplet lifetime of ~10 μs because the lowest energy state is no longer the ^{3}MLCT. Alternatively, if the ligand is made strongly electron accepting, then ^{3}MLCT may be blue shifted. An example of a blue phosphor is (Fppy)$_2$Ir(pic) (abbreviated FIrpic), shown in Figure 6.16C.[38] By using this approach it has been possible to prepare phosphors that span the entire visible spectrum, from blue to red. A representative sample of highly emissive Ir complexes is shown in Figure 6.17.

Figure 6.18 shows the chromaticity coordinates of OLEDs prepared with several Ir-based phosphorescent dopants. The Ir phosphors possess broad (>60 nm) emission spectra. This does not adversely influence the purity of red and blue colors, as the spectral peaks of these phosphors can be shifted toward the nonvisible spectrum. For example, the coordinates of btp$_2$Ir(acac) are in the deep red[39] despite its spectral width of >75 nm, because much of the emission is in the near infrared at $\lambda > 675$ nm. However, obtaining pure green colors using

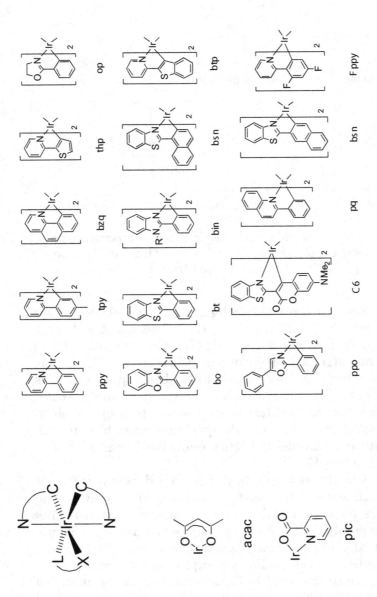

Figure 6.17 A widely tunable class of octahedral Ir phosphorescent complexes has been prepared; the complexes have the general structure shown in the upper left-hand corner; see Lamansky et al.[34] A representation set of organometallic ligands (right) and ancillary ligands (acac and pic) are shown. By using different ligand combinations emission colors ranging from blue to red have been achieved.

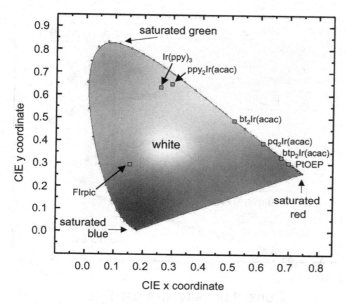

Figure 6.18 (Color figure follows p. 274.) The chromaticity coordinates of six Ir-based phosphors together with the coordinates of PtOEP.

organic materials will require narrower spectra than that exhibited by ppy$_2$Ir(acac).

In Figure 6.19 the external quantum efficiencies are shown for several Ir(ppy)$_3$-based OLEDs.[40] A CBP HTL was used as the host for Ir(ppy)$_3$ in the architecture of Figure 6.12A, requiring the use of a BCP hole-blocking layer. The efficiency of each Ir(ppy)$_3$:CBP/BCP device decreases slowly with increasing current, but the triplet lifetime of Ir(ppy)$_3$ in CBP is 0.5 µs and consequently triplet–triplet annihilation is much less severe than in PtOEP. Similar to the results for the Alq$_3$:PtOEP system, the Ir(ppy)$_3$:CBP devices achieve a maximum efficiency ($\eta_Q \sim 8\%$) for mass ratios of approximately 6 to 8%. The energy transfer mechanism from host to phosphor is different in these two devices: PtOEP in Alq$_3$ emits after singlet and triplet energy transfer; but Ir(ppy)$_3$ acts as a charge trap in CBP, and excitons are formed on it directly.[22,41] It is probable that high concentrations of the

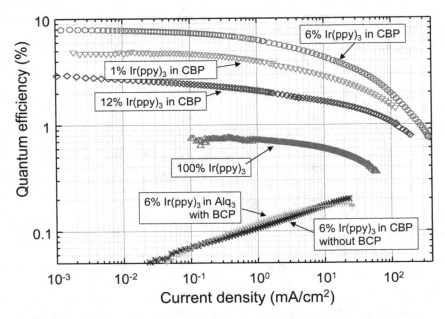

Figure 6.19 (Color figure follows p. 274.) The external quantum efficiency of OLEDs using Ir(ppy)$_3$:CBP luminescent layers. Peak efficiencies are observed for a mass ratio of 6% Ir(ppy)$_3$:CBP. The efficiency of a Ir(ppy)$_3$:CBP device grown without a blocking layer is also shown.

phosphor are required in both systems because the triplet density in the host must be minimized to prevent triplet–triplet annihilation.[25]

In addition to the Ir(ppy)$_3$:CBP/BCP devices, heterostructures were fabricated using a homogeneous film of Ir(ppy)$_3$ as the luminescent layer. The reduction in efficiency (to ~0.8%) is caused by "concentration quenching"* and is reflected in the transient decay, which has a lifetime of only ~100 ns, and deviates significantly from mono-exponential behavior. A 6% Ir(ppy)$_3$:CBP device without a BCP barrier layer is also shown

---

* "Concentration quenching" is a qualitative description of the efficiency lowering that often accompanies increases in the concentration of a luminescent dye.

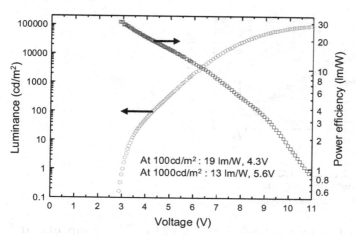

Figure 6.20 (Color figure follows p. 274.) The power efficiency and luminance of the 6% Ir(ppy)$_3$:CBP device. At 100 cd/m^2, the device requires 4.3 V and its power efficiency is 19 lm/W.

together with a 6% Ir(ppy)$_3$:Alq$_3$ device with a BCP barrier layer. Here, triplets are lost to the low energy Alq$_3$ sink. The quantum efficiencies are observed to increase with current as the sink is saturated.

In Figure 6.20, the luminance and power efficiency are plotted as functions of voltage. The peak power efficiency is 31 lm/W with a quantum efficiency of 8% (28 cd/A). At 100 cd/m^2, a power efficiency of 19 lm/W with a quantum efficiency of 7.5% (26 cd/A) is obtained at a voltage of 4.3 V. As observed in PtOEP:Alq$_3$, slow triplet relaxation can form a bottleneck in electrophosphorescence. This is loosened substantially here, resulting in a maximum luminance of ~100,000 cd/m^2. Similar results have been obtained by mixing Ir(ppy)$_3$ into the hole-transporting polymer poly(9-vinylcarbazole), demonstrating that molecular phosphors may also be used in polymer-based devices.[42,43]

An alternative architecture for the green phosphor ppy$_2$Ir(acac) employs the electron transport material 3-phenyl-4-(1′-naphthyl)-5-phenyl-1,2,4-triazole (triazole or TAZ; see Figure 6.5) as a host.[31,44] As shown in Figure 6.21, this device

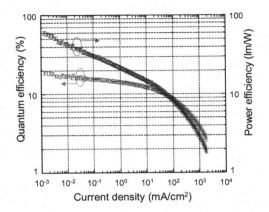

Figure 6.21 (Color figure follows p. 274.) The quantum and power efficiency of 12% ppy$_2$Ir(acac) in TAZ. Inset: The chemical structure of the electron-transporting host material TAZ.

exhibits a much higher quantum efficiency of $(19.0 \pm 1.0)\%$. The maximum external quantum efficiency can be calculated by considering the radiative modes in the OLED optical microcavity[8,45] employing the theory of Chance et al.[46] Dyadic Green's functions are used to compute the radiative decay rates, allowing us to consider arbitrarily complex structures. This treatment also accounts for nonradiative losses due to dipole coupling with surface plasmon modes at the metal–organic cathode interface. For a calculated output coupling efficiency of $\eta_p = (22 \pm 2)\%$, the internal efficiency is $\eta_{int} = (87 \pm 7)\%$.[44] The power efficiency peaks at (60 ± 5) lm/W; this corresponds to a wall plug efficiency (i.e., the ratio of optical power out to electrical power in) of over 10%. In contrast, the peak quantum efficiency of ppy$_2$Ir(acac) in an HTL such as CBP is 13%.[31] As discussed above, using an ETL as a host material should lower polaron quenching in electron injection-limited devices, and this may partially account for the high efficiency of ppy$_2$Ir(acac):TAZ vs. ppy$_2$Ir(acac):CBP.

Similar results are obtained for the red phosphor btp$_2$Ir(acac) in CBP, and as shown in Figure 6.22A its external quantum efficiency peaks at $\eta_Q = 7\%$.[39] Because its chromaticity coordinates are located in the deep red, the photopic

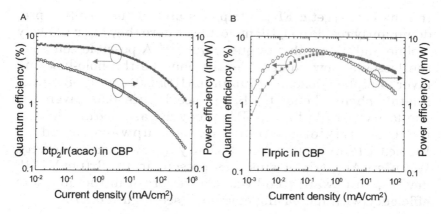

Figure 6.22 (Color figure follows p. 274.) The quantum and power efficiency of 6% btp₂Ir(acac) in CBP.

response of the eye to btp₂Ir(acac) phosphorescence is lower than for the green phosphors; thus the peak power efficiency of btp₂Ir(acac) is 4 lm/W. OLEDs have also been fabricated with bt₂Ir(acac) and pq₂Ir(acac), which emit yellow and orange, respectively. Optimized devices with these two dopants emit with external efficiencies >11%.[39]

Finally, in Figure 6.22B, the quantum and power efficiencies are shown for the blue phosphor FIrpic in CBP. The trend of the quantum efficiency is clearly different from that of the red and green phosphors shown in Figure 6.21 and Figure 6.22A. Instead of a monotonic decrease in efficiency, the quantum efficiency of FIrpic at first increases at low current density. The origin of this effect may be related to the unusual energy transfer required in this particular device. Unlike the other phosphor and host combinations discussed here, the triplet energy of FIrpic[38] (2.62 ± 0.1) eV is *greater* than that of CBP (2.56 ± 0.1) eV.[22] Endothermic energy transfer is characterized by thermally activated emission from FIrpic and the appearance of a long-lived (~10 ms) species in the FIrpic phosphorescent transient; the triplet lifetime of FIrpic is typically only ~1 μs.[38] Because of weak singlet–triplet mixing, the triplet transition of CBP is strongly disallowed and consequently the most efficient decay path for CBP triplets is via the more energetic FIrpic triplet state.* Thus, triplets predominantly reside on CBP molecules where they are relatively mobile and vulnerable to quenching.[22,38] A possible source of the losses at low current density may be the hole-blocking layer 4-biphenyloxolato-aluminum(III)bis(2-methyl-8-quinolinato)4-phenylphenolate (BAlq); see Figure 6.5. Given the similarity of BAlq to Alq_3, BAlq may act as a saturable sink for CBP triplets, explaining the upward trend in FIrpic:CBP/BAlq quantum efficiency at low current density. When an Alq_3 triplet sink was present in the $Ir(ppy)_3$:CBP devices, an upward trend was also observed in the quantum efficiency with increasing current[40] (see Figure 6.19).

6.6 RELIABILITY OF ELECTROPHOSPHORESCENT DEVICES

Achieving long operational and storage lifetimes is crucial if organic materials are to find application in electroluminescent devices. A benchmark lifetime (defined as the time for the luminance to decay to 50 cd/m² from an initial value of 100 cd/m²) of approximately 10,000 h has been realized in fluorescent OLEDs.[47] It is likely that electrophosphorescent devices will have superior reliability than comparable fluorescent devices. There are two reasons for this.[48,49] First, due to spin statistics, phosphorescence has an intrinsic efficiency approximately four times that of fluorescence; this lowers the operating current density required to achieve a given brightness. Second, triplets may live for as long as several seconds in fluorescent devices, but singlet–triplet mixing in phosphorescent devices reduces the lifetime of triplets significantly, preventing the formation of high densities of potentially reactive excited states on the host molecules.

Reliability studies of PtOEP in CBP or Alq_3 in the device architectures of Figure 6.12 determined that the hole-blocking BCP layer was unstable,[50] although the exact mechanism of its degradation remains unclear. Stable devices were achieved by omitting the BCP layer but the quantum efficiency of the

---

* Endothermic energy transfer has also been observed in $Ir(ppy)_3$:TPD.[22]

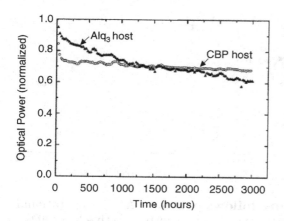

Figure 6.23 Optical power as a function of time for PtOEP-based devices. The initial brightness was 45 cd/m^2 for the CBP host and 35 cd/m^2 for Alq$_3$. (After Burrows et al.[49])

devices using CBP as a host material was reduced; see Figure 6.13B. The device structure employed was ITO, 200 Å of CuPc, 400 Å of α-NPD, 300 Å of either CBP or Alq$_3$ mixed with 6% PtOEP, 200 Å of Alq$_3$, and an Mg:Ag cathode. The operating lifetime for PtOEP in CBP and PtOEP in Alq$_3$ is shown in Figure 6.23. For the CBP-based devices, the light output decreases by 25% in the first 50 h, followed by a further loss of ~5% over the subsequent 3000 h.

Unlike the case of PtOEP, a hole-blocking layer is necessary to prevent almost total loss of luminescent efficiency in iridium-based phosphors. The first report of the lifetime of Ir(ppy)$_3$:CBP devices using an unstable BCP hole-blocking layer showed a lifetime of ~1500 h at an initial luminance at 100 cd/m^2.[50] Subsequently,[51,52] it was found that BAlq, an Alq$_3$ variant with blue rather than green fluorescence, was a stable alternative to BCP. For example, in Figure 6.24 the operational stability of Ir(ppy)$_3$ is shown in CBP. The structure is ITO, 100 Å of CuPc, 300 Å of α-NPD, 300 Å of Ir(ppy)$_3$ in CBP, 100 Å of BAlq, and 400 Å of Alq$_3$. The cathode comprises 10 Å of LiF followed by 1000 Å of Al. The projected lifetime at an operating brightness of ~500 cd/m^2 is 10,000 h. However,

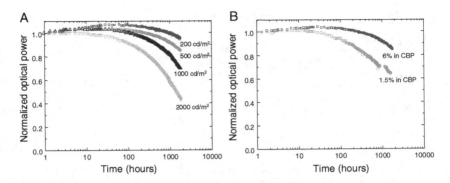

Figure 6.24 (Color figure follows p. 274.) (A) The operational stability of electrophosphorescent devices with an active layer of 6% Ir(ppy)$_3$ in CBP and a BAlq hole-blocking layer. The projected life to 50% of an initial brightness of 500 cd/m^2 is 10,000 h. (B) A comparison of the operational stability of devices with 6 and 1.5% Ir(ppy)$_3$ in CBP at an initial brightness of 500 cd/m^2. The difference in lifetime reflects the quantum efficiency of the two devices. (After Ray Kwong et al.[52])

it is unlikely that BAlq is as efficient at blocking triplet excitons as BCP; thus, devices employing BAlq exhibit lower quantum efficiencies. The 6% Ir(ppy)$_3$ device in Figure 6.24A has a peak quantum efficiency of 5.6%, as compared to 8% when BCP is used. The development of stable hole-blocking materials with large triplet energies will allow phosphorescent devices to be operated at lower current densities, further improving operational stability.

6.7 CONCLUSIONS

The power efficiencies of four phosphors are presented in Table 6.1 as a function of current density; also listed are the three components of power efficiency in Equation 6.5: photopic efficiency Φ, quantum efficiency η_Q, and electrical efficiency V_λ/V. Both the internal and external quantum efficiencies are tabulated assuming a light output coupling fraction of 0.22.[44] The table shows the success of phosphorescence in

Table 6.1 Power Efficiency and Its Components for Four Electrophosphorescent OLEDs

| phosphor | host | Φ (lm/W) | $J = 1 \mu A/cm^2$ | | | | $J = 1 mA/cm^2$ | | | |
			η_P (lm/W)	$\eta_{Q\,ext}$	$\eta_{Q\,int}$	V_λ/V	η_P (lm/W)	$\eta_{Q\,ext}$	$\eta_{Q\,int}$	V_λ/V
ppy$_2$Ir(acac)	TAZ	530	60	0.19	0.87	0.60	20	0.15	0.68	0.25
btpIr(acac)	CBP	170	4	0.07	0.32	0.34	2.2	0.06	0.27	0.22
FIrpic	CBP	260	1.3	0.006	0.027	0.83	5.0	0.057	0.23	0.34
PtOEP	CBP	60	0.3	0.056	0.23	0.09	0.2	0.042	0.19	0.08

Note: Operating at low current densities minimizes losses, and the data for $J \sim 1 \mu A/cm^2$ typically reflects the maximum possible performance of each electrophosphorescent device. Practical biases, however, are likely to be closer to $J \sim 1 mA/cm^2$. By harnessing triplet and singlet excitons the internal efficiency may be increased to nearly 100%; for example, see ppy$_2$Ir(acac) in TAZ, where the "wall plug" efficiency, i.e., the ratio of optical power out to electrical power in, is 11% at $J = 1 mA/cm^2$. Also shown are the electrical efficiencies, V_λ/V, for each device. All devices are excited via a blue host material such as CBP or TAZ; thus, the red devices have particularly low electrical efficiencies due to the energy loss during energy transfer from host to guest. This is an obvious target for future improvements in OLED power efficiencies.

increasing the internal efficiency of OLEDs, but it also high-lights the remaining opportunities for OLED designers, namely, increasing the output coupling fraction and the elec-trical efficiency, which is approximately 0.3 at typical operat-ing current densities of $J \sim 1$ mA/cm^2. Lower voltages can therefore lead to power efficiency improvements of up to a factor of three, and optimizing output coupling could yield power efficiency improvements of up to five.

The remaining considerations for display applications of organic phosphors are reliability and color purity. In fluores-cent devices, triplets may exist for several seconds, and large densities of these potentially reactive excited states can be developed. Phosphorescent molecules should increase the operating life of OLEDs because their strong singlet–triplet mixing and triplet-gathering properties minimize the lifetime of triplet excitons. The increase in efficiency due to phospho-rescence also reduces the necessary drive current, reducing the stress on materials and contacts. Indeed, studies of Ir(ppy)$_3$ and PtOEP have confirmed that at initial brightness of 100 cd/m^2 electrophosphorescent devices can possess oper-ating lifetimes in excess of 10,000 h.

Finally, with suitable chemical adjustments to molecular structure, it is likely that arbitrarily pure blue and red colors will be obtained by shifting phosphorescent spectra toward nonvisible wavelengths, albeit at the cost of power efficiency. However, given the spectral width of most organic phosphors, achieving higher-purity green colors remains a significant challenge.

REFERENCES

1. D.A. Roberts, Radiometry and photometry: lab notes on units, *Photonics Spectra*, 4, 59–63, 1987.

2. E. Hecht, *Optics,* Addison-Wesley, Reading, MA, 1987.

3. G. Gu, D.Z. Garbuzov, P.E. Burrows, S. Venkatesh, S.R. Forrest, and M.E. Thompson, High-external-quantum-efficiency organic light-emitting devices, *Optics Lett.*, 22, 396–398, 1997.

4. M. Pope and C. Swenberg, *Electronic Processes in Organic Crystals*, Oxford University Press, Oxford, 1982.

5. M.A. Baldo, D.F. O'Brien, M.E. Thompson, and S.R. Forrest, The excitonic singlet–triplet ratio in a semiconducting organic thin film, *Phys. Rev. B*, 60, 14422–14428, 1999.

6. Y. Cao, I. Parker, G. Yu, C. Zhang, and A. Heeger, Improved quantum efficiency for electroluminescence in semiconducting polymers, *Nature*, 397, 414–417, 1999.

7. M. Wohlgenannt, K. Tandon, S. Mazumdar, S. Ramasesha, and Z.V. Vardeny, Formation cross-sections of singlet and triplet excitons in pi-conjugated polymers, *Nature*, 409, 494–497, 2001.

8. J.-S. Kim, P.K.H. Ho, N.C. Greenham, and R.H. Friend, Electroluminescence emission pattern of organic light-emitting diodes: implications for device efficiency calculations, *J. Appl. Phys.*, 88, 1073–1081, 2000.

9. C.W. Tang and S.A. VanSlyke, Organic electroluminescent diodes, *Appl. Phys. Lett.*, 11, 913–915, 1987.

10. M.A. Baldo, R.J. Holmes, and S.R. Forrest, The prospects for electrically-pumped organic lasers, *Phys. Rev. B*, 66, 35321, 2002.

11. T. Yamasaki, K. Sumioka, and T. Tsutsui, Organic light-emitting device with an ordered monolayer of silica microspheres as a scattering medium, *Appl. Phys. Lett.*, 76, 1243–1245, 2000.

12. T. Tsutsui, M. Yahiro, H. Yokogawa, and K. Hawano, Doubling coupling-out efficiency in organic light-emitting devices using a thin silica aerogel layer, *Adv. Mater.*, 13, 1149–1152, 2001.

13. G. Parthasarathy, P.E. Burrows, V. Khalfin, V.G. Kozlov, and S.R. Forrest, A metal-free cathode for organic semiconductor devices, *Appl. Phys. Lett.*, 72, 2138–2140, 1998.

14. T. Förster, Transfer mechanisms of electronic excitation, *Disc. Faraday Soc.*, 27, 7–17, 1959.

15. V. Bulovic, R. Deshpande, M.E. Thompson, and S.R. Forrest, Tuning the color emission of thin film molecular organic light emitting devices by the solid state solvation effect, *Chem. Phys. Lett.*, 308, 317–322, 1999.

16. G.L. Closs, M.D. Johnson, J.R. Miller, and P. Piotrowiak, A connection between intramolecular long-range electron, hole, and triplet energy transfers, *J. Am. Chem. Soc.*, 111, 3751–3753, 1989.

17. S. Hoshino and H. Suzuki, Electroluminescence from triplet excited states of benzophenone, *Appl. Phys. Lett.*, 69, 224–226, 1996.

18. A. Mills and A. Lepre, Controlling the response characteristics of luminescent porphyrin plastic film sensors for oxygen, *Anal. Chem.*, 69, 4653–4659, 1997.

19. G. Ponterini, N. Serpone, M.A. Bergkamp, and T.L. Netzel, Comparison of Radiationless Decay Processes in Osmium and Platinum Porphyrins, *J. Am. Chem. Soc.*, 105, 4639–4645, 1983.

20. D.B. Papkovski, New oxygen sensors and their applications to biosensing, *Sensors Actuators B*, 29, 213–218, 1995.

21. H.-Y. Liu, S.C. Switalski, B.K. Coltrain, and P.B. Merkel, Oxygen permeability of sol-gel coatings, *Appl. Spectrosc.*, 46, 1266–1272, 1992.

22. M.A. Baldo and S.R. Forrest, Transient analysis of organic electrophosphorescence. I. Transient analysis of triplet energy transfer, *Phys. Rev. B*, 62, 10958–10966, 2000.

23. C.W. Tang, S.A. VanSlyke, and C.H. Chen, Electroluminescence of doped organic thin films, *J. Appl. Phys.*, 65, 3610–3616, 1989.

24. J. Rodriguez, L. McDowell, and D. Holten, Elucidation of the role of metal to ring charge-transfer states in the deactivation of photo-excited ruthenium porphyrin carbonyl complexes, *Chem. Phys. Lett.*, 147, 235–240, 1988.

25. M.A. Baldo, C. Adachi, and S.R. Forrest, Transient analysis of organic electrophosphorescence. II. Transient analysis of triplet-triplet annihilation, *Phys. Rev. B*, 62, 10967–10977, 2000.

26. D.Z. Garbuzov, V. Bulovic, P.E. Burrows, and S.R. Forrest, Photoluminescence efficiency and absorption of aluminum-tris-quinolate (Alq_3) thin films, *Chem. Phys. Lett.*, 249, 433–437, 1996.

27. H. Nakada, S. Kawami, K. Nagayama, Y. Yonemoto, R. Murayama, J. Funaki, T. Wakimoto, and K. Imai, Blue electroluminescent devices using phenanthroline derivatives as an electron transport layer, *Polym. Prep. Jpn.*, 43, 2450–2451, 1994.

28. S. Naka, H. Okada, H. Onnagawa, and T. Tsutsui, High electron mobility in bathophenanthroline, *Appl. Phys. Lett.*, 76, 197–199, 2000.

29. I.G. Hill and A. Kahn, Organic semiconductor heterointerfaces containing bathocuproine, *J. Appl. Phys.*, 86, 4515–4519, 1999.

30. D.F. O'Brien, M.A. Baldo, M.E. Thompson, and S.R. Forrest, *Appl. Phys. Lett.*, 74, 442–444, 1999.

31. C. Adachi, M.A. Baldo, S.R. Forrest, and M.E. Thompson, High-efficiency organic electrophosphorescent devices with tris(2-phenylpyridine) iridium doped into electron transporting materials, *Appl. Phys. Lett.*, 77, 904–906, 2000.

32. M.A. Baldo and S.R. Forrest, Interface limited injection in amorphous organic semiconductors, *Phys. Rev. B*, 64, 85201, 2001.

33. S. Lamansky, P. Djurovich, D. Murphy, F. Abdel-Razzaq, R. Kwong, I. Tsyba, M. Bortz, B. Mui, R. Bau, and M.E. Thompson, Synthesis and characterization of phosphorescent cyclometalated iridium complexes, *Inorgan. Chem.*, 40, 1704–1711, 2001.

34. S. Lamansky, P. Djurovich, D. Murphy, F. Abdel-Rezzaq, H.-E. Lee, C. Adachi, P.E. Burrows, S.R. Forrest, and M.E. Thompson, Highly phosphorescent bis-cyclometalated iridium complexes: synthesis, photophysical characterization, and use in organic light emitting diodes, *J. Am. Chem. Soc.*, 123, 4304–4312, 2001.

35. K.A. King, P.J. Spellane, and R.J. Watts, Excited-state properties of a triply ortho-metalated iridium(III) complex, *J. Am. Chem. Soc.*, 107, 1431–1432, 1985.

36. K. Dedeian, P.I. Djurovich, F.O. Garces, G. Carlson, and R.J. Watts, A new synthetic route to the preparation of a series of strong photoreducing agents: *fac* tris-ortho-metaled complexes of iridium(III) with substituted 2-phenylpyridines, *Inorgan. Chem.*, 30, 1685–1687, 1991.

37. E. Vander Donckt, B. Camerman, F. Hendrick, R. Herne, and R. Vandeloise, Polystyrene immobilized Ir(III) complex as a new material for oxygen sensing, *Bull. Soc. Chim. Belg.*,103, 207–211, 1994.

38. C. Adachi, R.C. Kwong, P. Djurovich, V. Adamovich, M.A. Baldo, M.E. Thompson, and S.R. Forrest, Endothermic energy transfer: a mechanism for generating very efficient high-energy phosphorescent emission in organic materials, *Appl. Phys. Lett.*, 79, 2082–2084, 2001.

39. C. Adachi, M.A. Baldo, S.R. Forrest, S. Lamansky, M.E. Thompson, and R.C. Kwong, High-efficiency red electrophosphorescence devices, *Appl. Phys. Lett.*, 78, 1622–1624, 2001.

40. M.A. Baldo, S. Lamansky, P.E. Burrows, M.E. Thompson, and S.R. Forrest, Very high-efficiency green organic light-emitting devices based on electrophosphorescence. *Appl. Phys. Lett.*, 75, 4–6, 1999.

41. A.J. Mäkinen, I.G. Hill, and Z.H. Kafafi, Vacuum level alignment in organic guest-host systems, *J. Appl. Phys.*, 92, 1598–1603, 2002.

42. M.-J. Yang and T. Tsutsui, Use of poly(9-vinylcarbazole) as host material for iridium complexes in high-efficiency organic light-emitting devices, *Jpn. J. Appl. Phys.*, Part 2, 39, L828–L829, 2000.

43. S. Lamansky, R.C. Kwong, M. Nugent, P.I. Djurovich, and M.E. Thompson, Molecularly doped polymer light emitting diodes utilizing phosphorescent Pt(II) and Ir(III) dopants, *Organic Electronics*, 2, 53–62, 2001.

44. C. Adachi, M.A. Baldo, M.E. Thompson, and S.R. Forrest, Nearly 100% internal phosphorescence efficiency in an organic light emitting device, *J. Appl. Phys.*, 90, 5048–5051, 2001.

45. V. Bulovic, V.B. Khalfin, G. Gu, P.E. Burrows, D.Z. Garbuzov, and S.R. Forrest, Weak microcavity effects in organic light emitting devices, *Phys. Rev. B*, 58, 3730, 1998.

46. R.R. Chance, A. Prock, and R. Sibley, Molecular fluorescence and energy transfer near metal interfaces, *Adv. Chem. Phys.*, 37, 1–65, 1978.

47. J. Shi and C.W. Tang, Doped organic electroluminescent devices with improved stability, *Appl. Phys. Lett.*, 70, 1665–1667, 1997.

48. M.A. Baldo, M.E. Thompson, and S.R. Forrest, Phosphorescent materials for application to organic light emitting devices, *Pure Appl. Chem.*, 71, 2095–2106, 1999.

49. P.E. Burrows, S.R. Forrest, T.X. Zhou, and L. Michalski, Operating lifetime of phosphorescent organic light emitting devices, *Appl. Phys. Lett.*, 76, 2493–2945, 2000.

50. T. Tsutsui, M.-J. Yang, M. Yahiro, K. Nakamura, T. Watanabe, T. Tsuji, Y. Fukuda, T. Wakimoto, and S. Miyaguchi, High quantum efficiency in organic light-emitting devices with iridium-complex as a triplet emissive center, *Jpn. J. Appl. Phys.*, Part 2, 38, 1502–1504, 1999.

51. T. Watanabe, K. Nakamura, S. Kawami, Y. Fukuda, T. Tsuji, T. Wakimoto, and S. Miyahuchi, *Proc. SPIE*, 4105, 175, 2000.

52. R.C. Kwong, M.R. Nugent, L. Michalski, T. Ngo, K. Rajan, Y.-J. Tung, M.S. Weaver, T.X. Zhou, M. Hack, M.E. Thompson, S.R. Forrest, and J.J. Brown, High operational stability of electrophosphorescent devices, *Appl. Phys. Lett.*, 81, 162–164, 2002.

7

Patterning of OLED Device Materials

MARTIN B. WOLK

CONTENTS

7.1 INTRODUCTION

Organic light-emitting diodes (OLED) technology seems to be mired in the transition from research and development (R&D) prototyping to large-scale manufacturing. Simple monochrome and multicolor OLEDs have been on the market since Pioneer introduced a car audio player with an OLED display in 1997. Multicolor OLEDs are just now entering into a phase of significant incorporation in other small-format consumer electronic products such as mobile phones and portable audio players. In contrast, only a few full-color AMOLED displays have reached significant production levels. SK Display, a joint venture of Sanyo and Kodak, was the first to produce a full-color active-matrix OLED (AMOLED) panel in 2003 and incorporate it into a consumer product (Kodak LS633 digital camera). In the fall of 2004, Sony introduced the Clie PEG-VZ90 Personal Entertainment Organizer, a multimedia personal digital assistant (PDA) with a 3.8-diagonal (D) "Super Top Emission" AMOLED. Product introduction schedules lag somewhat behind the forecasts, so OLED sales projections have been adjusted downward. For example, Display Search, a leading industry analyst, revised their estimates for OLED sales in 2007 from $3.1 billion (2002 estimate) to $2.6 billion (2003 estimate), indicating a ratcheting down of the growth projections.[1,2]

Many factors have contributed to delays in AMOLED manufacturing including those related to backplane manufacture, infrastructure readiness, and device performance. One issue that still concerns the industry is the limited number of technologies for patterning OLED materials in the fabrication of full-color, large-format displays. Scaling of these methods to cover large areas with high-resolution features is particularly problematic.[3]

The scope of this chapter includes a discussion of these difficulties. AMOLED patterning requirements and patterning issues for manufacture are reviewed. In addition, each of the major patterning technologies is discussed. Finally, a comparison of the patterning options is presented.

7.2 ALTERNATIVES TO PATTERNING OLED MATERIALS

There are at least two methods used to achieve full color in AMOLEDs without having to pattern the additive primary emitters themselves. "Color-by-White" AMOLEDs are based on a white-emitting AMOLED aligned to a photolithographically defined RGB color filter.[4] The mode of operation is similar to that of a TFT-LCD — spatially modulated white light is filtered at the subpixel level to produce a range of full colors in the display. Another related method, "Color Conversion," utilizes a blue-emitting AMOLED aligned to a photolithographically defined array of fluorescent subpixels that can convert the blue light into the remaining additive primary colors red and green.[5] Both methods offer considerable advantages for manufacturing by avoiding the OLED patterning issue and enabling the production of larger-format, high-resolution AMOLEDs without relying on further technological developments. However, there are compromises involved as well. Power efficiency, color saturation, or other advantageous OLED display attributes may be adversely affected, making the comparison between AMOLEDs and thin film transistor liquid crystal displays (TFT-LCDs) less favorable. While these display constructions are an important aspect of OLED commercialization, they are not addressed further in this chapter.

7.3 PATTERNING ISSUES

It is worthwhile to consider how display panels are manufactured and how certain panel parameters influence the choice of patterning methods. The discussion illustrates how eight important factors — material type, device design, pixel array pattern, display format, substrate size, placement accuracy, process takt-time, and defect density — can determine the success or failure of an OLED patterning scheme in a manufacturing setting.

7.3.1 Material Type

As described elsewhere in this book and in several excellent reviews, materials development for OLED applications has been bifurcated since birth into two groups according to the method of materials deposition: vacuum evaporation and solution coating.[6,7] Materials for vacuum evaporation are molecular cousins of charge transport materials and visible dyes developed earlier for electrophotography and organic photovoltaic applications. Devices based on these materials tend to have multiple discrete layers dedicated to unique electronic functions such as charge transport, blocking, and injection. Solution-coated materials are most commonly conjugated polymers, but may be small molecular glasses,[8] low-molecular-weight oligomers,[9] dendrimers,[10] or nonconjugated polymers with active pendant groups.[11] The solution-coated devices tend to have fewer discrete layers due to the difficulties in coating from solvents that are mutually orthogonal — a requirement so that coated layers are not detrimentally affected by subsequent coating steps.

More recently, another bifurcation of materials has developed, spawning from the development of phosphorescent organic emitters by Baldo et al.[12,13] These materials now augment a large array of fluorescent materials, both small molecule and polymer. As reported by the group at UDC, substantial boosts in efficiency and device lifetime have been realized in devices based on phosphorescent emitters.[15] Work on soluble phosphorescent materials continues at CDT/Sumitomo Chemical (phosphorescent dendrimers),[15] and elsewhere.

Table 7.1 OLED Panel Makers and Material Type

Solution Coating	Vacuum Evaporation	
CDT	AUO	Picvue
Delta Opto	Chi Mei	RiTDisplay
DuPont Displays	eMagin	ROHM
Fujitsu	Hitachi	Samsung SDI
MED	Hyundai	Sanyo
OSRAM	LGE	SK Display
Philips	Lighttronik	Sony
Samsung SEC	Lite Array	Stanley Electric
Seiko Epson	Luxell	TDK
	Nippon Seiki	TECO
	Optotech (Taiwan)	Tohoku Pioneer

Needless to say, the OLED materials patterning method is highly dependent on the material type. Many panel makers have made their choices and pursue the fabrication of devices based on one of the major materials types (Table 7.1).[2] Only a few of the patterning methods are capable of patterning more than one type of material.

7.3.2 Device Design

OLED devices may be divided into two major classes: passive-matrix OLEDs (PMOLEDs) and active-matrix OLEDs (AMOLEDs). The device construction and overall performance of the two have significant differences that affect the type of product into which they can be incorporated. In general, commercial PMOLEDs tend to be small format (<5-in. D), low information content, monochrome or multicolor panels oriented toward information display (e.g., mobile phone subdisplays). The size and resolution of AMOLEDs are not limited in the same way. Theoretically, panel size could range from a microdisplay format of less than 1-in. D to very large, high-resolution, full-color panels of 80-in. D or more. In many ways, the ideal applications for AMOLEDs are similar to those of its chief competitor, the TFT LCD. Both exceed at data and image display. Because of their rapid response time, good color saturation, and wide viewing angle, full-color AMOLEDs are thought to be superior for use in high-definition video rather than data display or monitor applications.

AMOLEDs can be designed in two basic ways. Bottom emitting displays are configured so that the backplane is located between the emitting layer and the viewer. Top emitting displays are configured so that the emitting layer is located between the backplane and the viewer.[16]

Top emitting configurations become more attractive as brightness level and brightness uniformity specifications advance. Both specifications are intimately connected with pixel aperture ratio, the fractional area of a pixel that is designed to emit light. Top emitting configurations can achieve significantly higher aperture ratios than bottom emitting devices.[17] In a top emitter, the pixel address circuitry is buried in a layer beneath the OLED devices, leaving virtually the entire display surface available for emission. The design becomes critical as pixel circuitry moves from a two transistors to four, or even seven transistors per subpixel to achieve a high level of brightness uniformity across the panel.[18] High aperture ratios are desirable because target brightness specifications can be achieved at lower subpixel brightness, resulting in longer device lifetimes. However, device aperture ratio is a parameter of device design that greatly affects OLED materials patterning. Patterning methods for high aperture ratio displays must have high absolute placement accuracy and high pattern precision.

7.3.3 Pixel Array Pattern

The spatial arrangement of additive red, green, and blue primary subpixels and the design of the basic repeating unit of image information, the pixel, may be designed in a number of ways to meet certain image quality or panel cost requirements (Table 7.2, Figure 7.1). Stripe arrays are the most common pixel pattern in consumer flat panel displays today.[19] The columnar arrangement allows for crisp definition of the vertical and horizontal edges that are common in the rendering of text. Most computer monitors are based on stripe pixel arrays.

Table 7.2 Common Pixel Array Patterns

Pattern	Display Application	Comments
Stripe	Text	Most common pixel pattern; can be used for images if pixel count is sufficiently high
Delta	Images	Used for relatively low pixel count image displays in digital cameras
Mosaic	Images	More common in image sensor applications rather than displays
Pentile	Various	Variations can be used to increase apparent resolution and brightness

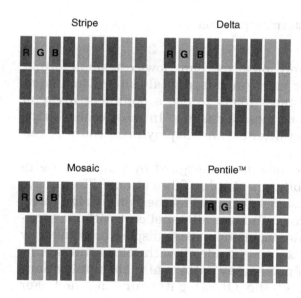

Figure 7.1 (Color figure follows p. 274.) Common pixel array patterns.

Nature abhors a straight line. Displays for viewing natural images or illustrations that are based on linear arrays may render artifacts if the pixel count is not sufficient. Curved edges can take on jagged, stair-step appearance that is jarring

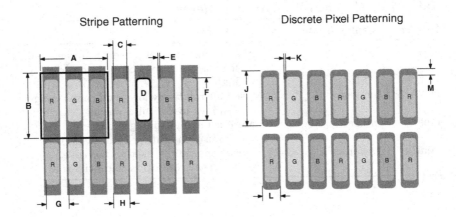

Figure 7.2   Pixel patterning parameters.

to the viewer. Relatively low pixel count, low-cost, small-format displays in consumer video, mobile, digital camera, and gaming applications are most likely to use a nonlinear array of pixels such as delta or mosaic designs. In most cases, stripe array patterns are sufficient if the display pixel count is relatively high.

Pentile™ arrays have been designed by Clairvoyante to allow for some flexibility in display fabrication.[20] The arrays can be used to increase resolution using the same design rules, to reduce display driver and overall panel cost without sacrificing apparent resolution, or to increase display brightness by adding a white subpixel (in a trade-off with color saturation).

The selection of pixel array pattern also has implications for implementation of the OLED patterning method. For example, nonlinear patterns may affect the dimensional stability of fine metal masks or the design of laser imagers.

Before embarking on a quest for the ideal patterning method for a given display, a panel maker will design a pixel array and define a number of dimensional parameters. Figure 7.2 and Table 7.3 define and illustrate the common pixel array parameters such as aperture area, pixel pitch, and stripe margin.

Table 7.3 Pixel Array Pattern Parameters

Label	Parameter
A	Pixel width (= pixel pitch)
B	Pixel height (= row pitch)
C	Subpixel width
D	Subpixel area
E	Horizontal stripe margin
F	Subpixel height
G	Subpixel pitch
H	Stripe width
J	Printed subpixel height
K	Horizontal printed subpixel margin
L	Printed subpixel width
M	Vertical printed subpixel margin
(not shown)	Angular error
3*D/(A*B)	Aperture ratio
(H-C)/2	Horizontal pixel allowance
(G-C)/2	Horizontal overlap allowance
((H-C)/2)-E	Horizontal error
(J-F)/2	Vertical pixel allowance
(B-F)/2	Vertical overlap allowance
((J-F)/2)-M	Vertical error

7.3.4 Display Format

Displays have a pixel array pattern that is typically characterized by a pixel count (e.g., 1280 × 720), a display diagonal size (e.g., 17-in. D), and a pixel pitch (e.g., 150 ppi). Early CRT monitors had a similar diagonal size, so the graphics array designations were a convenient means of addressing video resolution. The move from VGA (Video Graphics Array, 640 × 480) to XGA (Extended Graphics Array, 1024 × 768) corresponded to a move to higher resolution (pixel density) and smaller pixel size. A comparison of the various display formats for a 17-in. D panel is given in Table 7.4.

Today, flat panel displays are available in a large variety of panel sizes, from liquid crystal on silicon (LCOS) microdisplays (e.g., Brillian WUXGA LCOS display, 0.72-in. D, 1920 × 1200, 3136 ppi) to extremely large plasma display panels (PDPs) (e.g., Samsung SDI PDP, 80-in. D, 1920 × 1080, 28

Table 7.4 Resolution and Approximate Pixel Size for 17-in. D Flat-Panel Displays

Format Name	Abbreviation	Columns × Rows	Pixels (Mpixels)	Subpixels (Mpixels)	Aspect Ratio	Panel Dimensions Width (in)	Height (in)	Subpixel Dimensions Width (μm)	Height (μm)	Pixel Pitch (ppi)
Quarter Common Intermediate Format	QCIF	176 × 144	0.025	0.076	1.22	13.16	10.77	632.9	1899	13
Quarter Video Graphics Array	QVGA	320 × 240	0.077	0.230	1.33	13.60	10.20	359.8	1080	24
Full Common Intermediate Format	FCIF	352 × 288	0.101	0.304	1.22	13.16	10.77	316.5	949	27
Video Graphics Array	VGA	640 × 480	0.307	0.922	1.33	13.60	10.20	179.9	540	47
Wide Video Graphics Array	WVGA	800 × 480	0.384	1.152	1.67	14.58	8.75	154.3	463	55
Super Video Graphics Array	SVGA	800 × 600	0.480	1.440	1.33	13.60	10.20	143.9	432	59
Wide Super Video Graphics Array	WSVGA	1024 × 600	0.614	1.843	1.71	14.67	8.59	121.3	364	70
Extended Graphics Array	XGA	1024 × 768	0.786	2.359	1.33	13.60	10.20	112.4	337	75
High Definition TV 720 lines progressive	HDTV 720p	1280 × 720	0.922	2.765	1.78	14.82	8.33	98.0	294	86
Wide Extended Graphics Array	WXGA	1366 × 768	1.049	3.147	1.78	14.82	8.33	91.8	276	92
Super Extended Graphics Array	SXGA	1280 × 1024	1.311	3.932	1.25	13.27	10.62	87.8	263	96
Super Extended Graphics Array Plus	SXGA+	1400 × 1050	1.470	4.410	1.33	13.60	10.20	82.2	247	103
Ultra Extended Graphics Array	UXGA	1600 × 1200	1.920	5.760	1.33	13.60	10.20	72.0	216	118
High Definition TV 1080 lines interlaced	HDTV 1080i	1920 × 1080	2.074	6.221	1.78	14.82	8.33	65.3	196	130
Wide Ultra Extended Graphics Array	WUXGA	1920 × 1200	2.304	6.912	1.60	14.42	9.01	63.6	191	133
Quantum Extended Graphics Array	QXGA	2048 × 1536	3.146	9.437	1.33	13.60	10.20	56.2	169	151
Quantum Super Extended Graphics Array	QSXGA	2560 × 2048	5.243	15.729	1.25	13.27	10.62	43.9	132	193
Quantum Ultra Extended Graphics Array	QUXGA	3200 × 2400	7.680	23.040	1.33	13.60	10.20	36.0	108	235

Note: Approximate pixel size was estimated by geometrically subdividing panel area. Subpixel dimensions do not take aperture ratio into consideration. To calculate approximate subpixel aperture dimensions, multiply the subpixel width or height by the square root of the pixel aperture ratio.

Source: Adapted from http://www.samsungusa.com/download/semiconductors/tft_lcd/tft_fundamentals.pdf.

Table 7.5 Pixel Pitch over a Range of Flat-Panel Display Sizes

Format	Pixel Pitch (ppi)				
	2-in. D	5-in. D	17-in. D	27-in. D	40-in. D
QCIF	114	45	13	8	6
QVGA	200	80	24	15	10
FCIF	227	91	27	17	11
VGA	400	160	47	30	20
WVGA	466	187	55	35	23
SVGA	500	200	59	37	25
WSVGA	593	237	70	44	30
XGA	640	256	75	47	32
HDTV 720p	734	294	86	54	37
WXGA	784	313	92	58	39
SXGA	820	328	96	61	41
SXGA+	875	350	103	65	44
UXGA	1000	400	118	74	50
HDTV 1080i	1101	441	130	82	55
WUXGA	1132	453	133	84	57
QXGA	1280	512	151	95	64
QSXGA	1639	656	193	121	82
QUXGA	2000	800	235	148	100

ppi). Despite that graphics array acronyms have long been synonymous with display resolution, they are sufficient to describe a display — at least two of the three display format metrics (pixel count, pixel pitch, display diagonal size) are required in addition to the display aspect ratio. Table 7.5 illustrates the range of pixel pitch for a series of panels and pixel counts.

7.3.5 Substrate Size

In our laboratories, and I suspect many others, OLED research is performed on small (50×50 mm^2) glass substrates. The small size is well suited to manual laboratory cleanroom equipment such as ultrasonic cleaners, spin coaters, plasma chambers, and device test equipment. The fabrication of an OLED test coupon is done in a long sequence of batch steps, which, however arduous, can produce a reasonable number of finished devices over a period of a few days (10 to 40 devices every 2 days, in our case).

A manufacturing environment is very different — process efficiency is key. Display devices are fabricated in a sequence of batch steps and taken through the process using the largest substrate possible. The substrate, called a mother glass, is sized according to an agreed-upon industry standard so that equipment manufacturers can produce process equipment for multiple customers. The mother glass size is referred to by the generation number in a series that continually evolves toward larger glass size and equipment (Table 7.6). Figure 7.3 shows the approximate dimensions for a "6-up" array of 16-in. diagonal wide-format panels on Gen 4 mother glass. Other panelizations (allowable multiples of individual displays per mother glass) are given in Table 7.7 for Gen 4 mother glass.

The use of large substrates brings about some of the primary challenges for OLED material patterning. Some of these issues including defect density, takt-time, and absolute placement accuracy are discussed below. Currently, the target mother glass size for AMOLED fabrication is Gen 4.

Additional issues include dimensional stability and temperature control. The coefficient of thermal expansion (CTE) for display glass, a typical OLED substrate, is 3.18 ppm/°C. For every 1°C rise in temperature of a Gen 4 mother glass, the glass expands by 2.93 μm in the larger dimension (920 mm).[21] The problem is much worse for polymer substrates, with CTEs of around 18 ppm/°C.[22] Industrial processing of flat-panel display substrates is therefore done in temperature-controlled environments. Microenvironments surrounding patterning equipment may be used to regulate temperature to within ±0.05°C, if necessary.

Unlike most TFT-LCDs with color filters, AMOLEDs have a high-resolution pattern on a single substrate (an exception may be some color-by-white and color conversion AMOLEDs). This fact allows for the compensation of thermal expansion errors during the patterning process, if the process allows. For example, if the photolithographic patterning steps of an AMOLED glass substrate were performed at +1°C above

Table 7.6 Flat-Panel Display Mother Glass
Generation Standards

Generation	Width (mm)		Height (mm)	Area (m^2)	Aspect Ratio
1	270	×	360	0.097	0.75
1	300	×	350	0.105	0.86
1	300	×	400	0.120	0.75
1	320	×	400	0.128	0.80
2	360	×	465	0.167	0.77
2	370	×	470	0.174	0.79
2.5	400	×	500	0.200	0.80
2.5	400	×	505	0.202	0.79
2.5	404	×	515	0.208	0.78
2.5	410	×	520	0.213	0.79
3	550	×	650	0.358	0.85
3	550	×	660	0.363	0.83
3	550	×	670	0.369	0.82
3.25	590	×	670	0.395	0.88
3.25	600	×	720	0.432	0.83
3.25	610	×	720	0.439	0.85
3.25	620	×	720	0.446	0.86
3.5	650	×	830	0.540	0.78
3.5	680	×	880	0.598	0.77
4	730	×	920	0.672	0.79
5	1000	×	1200	1.200	0.83
5	1100	×	1250	1.375	0.88
5	1100	×	1300	1.430	0.85
5	1200	×	1300	1.560	0.92
6	1500	×	1800	2.700	0.83
7	1800	×	2100	3.780	0.86
8	2200	×	2600	5.720	0.85

the specified temperature, then the absolute size of the pattern on the substrate would be enlarged by 2.7 ppm. If OLED materials were then applied in an inkjet or laser thermal process, the fiducials on the substrate would be detected during an alignment step prior to the patterning step and the expansion could be calculated. The digital pattern for the path of the inkjet head or laser beam could then be adjusted to compensate for the error in absolute substrate pattern size. In an engineer's words, "Two wrongs can make a right."

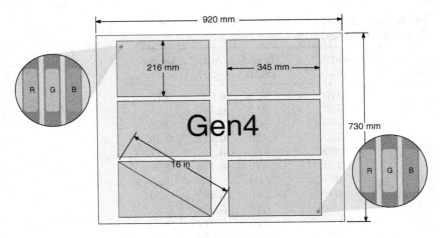

Figure 7.3 Gen 4 mother glass with 6-Up 16-in. D wide format displays.

Table 7.7 Panelization of Wide (16:9) and Standard (4:3) Displays on Gen 4 Mother Glass

16:9 Aspect Ratio		4:3 Aspect Ratio	
Display Size	Gen 4 MG Panelization	Display Size	Gen 4 MG Panelization
11	12	10	16
15	8	11	9
17	6	12	9
18	4	13	9
20	4	14	9
22	2	15	6
23	2	16	6
24	2	17	4
25	2	18	4
30	2	19	4
31	2	20	4
32	2	21	4
37	1	23	2
40	1		

7.3.6 Placement Accuracy

The smallest features (dots) that can be reproduced on a modern offset printing press are on the order of 10 μm in diameter.[23] Arrays of dots from the four inks used in color printing (cyan, magenta, yellow, and black) can be overprinted such that each dot forms a part of a "rosette" of colors. A similar statement can be made about a desktop inkjet printer. It can print dots with a volume of as little as 8 pL and produce beautiful photographic prints.[24] However, neither is likely to print features on a high-resolution substrate without significant improvements. At issue is the difference between relative and absolute placement accuracy.

For display patterning, relative placement accuracy is defined as the degree to which multiple points on a plane can be positioned with respect to each other without reference to an absolute point (or points). In an inkjet printing analogy, consider a blank piece of paper to be printed with an image. The paper has no information prior to the printing of a photographic image — hence, no reference point. The first drop of ink jets onto the paper somewhere in the vicinity of a margin corner. From that point on, each successive drop must be deposited accurately with respect to neighboring drops. The ink dots on the print may be characterized as having good relative placement accuracy on the scale of the dots because each dot is well aligned to its neighbors. However, there are also long-range distortions in the printed image that go unnoticed by an observer but would be disastrous for display patterning.

Absolute placement accuracy is defined as the degree to which multiple points on a plane can be positioned in reference to an absolute point (or points). Using the inkjet printing analogy again, consider a Gen 4 OLED mother glass with wide format 17-in. D W-UXGA panels to be printed with red, green, and blue subpixels. The subpixel electrodes must be completely covered with red-, green-, or blue-emitting material without overlapping the neighboring subpixel from one corner of the substrate to the other. Assuming that the absolute position of each subpixel anode is perfect and the anodes are each $50 \times 150 \ \mu m^2$, then a 60-μm-wide printed subpixel has a horizontal pixel allowance of only $(60 - 50)/2 = 5$ μm. If the value of the printed subpixel margin is negative, then the anode will be exposed and the subpixel will have an electrical short. If the printed subpixel margin is too great, then adjacent stripes will overlap, which may or may not cause device problems. The allowable horizontal error over the width of the mother glass is (5 μm/920,000 μm) = 5.4 ppm. For a Gen 7 mother glass with the same display format, the allowable horizontal error is less than 1.9 ppm.

OLED displays are built on photolithographically defined backplanes in which each feature has a very high positional and dimensional accuracy. Long-range distortions will cause alignment errors between the material pattern and the display elements on the mother glass. Viable patterning methods for OLEDs must have a high degree of absolute placement accuracy over the entire mother glass surface with allowable placement errors on the order of 2 ppm.

7.3.7 Takt-Time

Another important concept in flat panel display manufacturing is "takt-time," a concept taken from the Toyota Production System.[25] Takt-time (sometimes referred to as tact-time or TACT time) refers to the time allowable for completion of each step in a complex sequential process in order to meet a target production rate. The official definition is part of the "just-in-time" manufacturing method that provides a method of adjusting production rates to meet customer demand. In the flat panel display industry, the term can be defined as the inverse of the throughput for a given discrete manufacturing step. For example, if the throughput for cathode deposition on mother glass substrates is 3 per hour, then the takt-time is 20 min.

The concept is important in OLED manufacturing because it is used to balance the flow of intermediates that are produced in the overall process so that they do not accumulate along the way. In addition, it allows a manufacturer designing a process with a number of defined steps to specify the rate at which a new step must occur in order to maximize production and minimize accumulation of intermediates.

Table 7.8 OLED Pixel Patterning Defects and Possible Causes

Defect Type	Possible Cause
Misalignment	Pixel placement allowance exceeded
Overlap	Overlap allowance exceeded
Distortion	Thermal or mechanical damage
Edge defects	Edge roughness allowance exceeded
Mask defects	Material buildup on mask edge
Pinholes	Particulate contamination
Cracks	Step height limits exceeded
Particles	Poor environmental control
Dewets	Surface contamination
Cross contamination	Stray inkjet drops
Bright pixel regions	Poor layer thickness control
Dark pixel regions	Poor layer thickness control
Dark pixel edges	Non-uniform drying

7.3.8 Defect Density

From the information in Table 7.1 through Table 7.3, we can determine the number of subpixels on a Gen 4 substrate with 16-in. D wide-format panels by multiplying the panelization (6) by the number of pixels per panel (2,304,000) by the number of subpixels per pixel (3) to arrive at a total of 41,472,000 subpixels per mother glass. Assuming a liberal allowance of three nonproximate pixel defects per panel, the defect density specification is 1 defect per 6,912,000 subpixels, or 145 ppb. Given the number of possible defect types (a partial listing is shown in Table 7.8), this is an extreme challenge for any patterning technology.

7.3.9 Summary of Patterning Issues

In summary, a patterning process for AMOLED materials must meet a number of criteria to be viable in a manufacturing setting. These guidelines refer to processes that are targeted at producing large-format, high-resolution, full-color AMOLED panels.

The OLED material patterning process should be capable of the following:

- Patterning the material set of choice
- Patterning high-aperture-ratio top emitting device constructions
- Patterning the optimal pixel array patterns for the targeted OLED application
- Patterning the optimal display formats for the targeted OLED application
- Scaling to large mother glass size
- Aligning the pattern on the substrate to within the manufacturing specifications for the target OLED application
- Attaining a takt-time (or a small multiple thereof for parallel patterning approaches) that is compatible with other OLED manufacturing steps
- Producing patterned panels with an overall yield that is on par with those of competitive display technologies

7.4 ADVANCED PATTERNING METHODS

Although many patterning methods have been used in the laboratory, only three — shadow mask, inkjet, and laser thermal imaging — have succeeded in maturing to a stage where high-resolution AMOLED devices can be fabricated. These three methods are highlighted in the following discussion. Shadow mask and inkjet have dominated most work in the industry and commercial equipment is available for both. At this time, full-color OLEDs on the market are nearly all produced by the shadow mask method, although Seiko Epson, Philips, and others have demonstrated large full-color prototypes of inkjetted devices. The remaining methods are currently under development. Some, such as Laser Induced Thermal Imaging, are at a very advanced stage.

7.4.1 Shadow Mask

Shadow masks, also referred to as fine metal masks (FMMs), are used to pattern "small molecule" materials using vacuum evaporation. Typical small molecules are organic dye-like molecules that can attain an appreciable vapor pressure at temperatures well below the point of thermal decomposition. These materials include fluorescent dopants (e.g., coumarins), phosphorescent dopants (e.g., iridium complexes), and various charge transporting and blocking materials (e.g., aluminum quinolates).

In the process, a small-molecule material is placed in a crucible and heated by a surrounding oven until a desired rate of evaporation is attained. Usually, the oven temperature is adjusted to achieve a stable materials flux and then the deposition is initiated by removing a physical barrier, called the shutter, from above the source (the combination of crucible and oven). A plume of evaporated material emanates from the source and is distributed throughout the chamber, mainly in a "line of sight" manner. The materials flux is monitored during the process by sensing the frequency of an oscillating quartz crystal placed within the vapor cloud.

The technology of vacuum evaporation is mature, so there exist a wide variety of chamber vendors, chamber and source designs, and chamber configurations. Leading vendors include Ulvac (Chigasaki, Japan),[26] Tokki (Chuo-ku, Japan),[27] Doosan DND (Ansan City, South Korea),[28] and Applied Films (Alzenau, Germany).[29]

In a typical configuration, the substrate is inverted at the top of the chamber and the sources are placed in an array at the bottom. In between, a precision stencil, called the shadow mask, is positioned to physically block the material from depositing on the substrate in undesired regions. For parallax considerations, the shadow mask is placed in close proximity to the substrate surface.

In the laboratory, shadow masks can be as simple as a razor-cut piece of aluminum foil or a piece of stainless steel shim stock cut by a precision fabrication method such as electrical discharge machining (EDM). In a production setting, the masks are usually prepared by chemical etch techniques using a photopolymer mask to generate the apertures, which correspond to a single emission color on the substrate. The thin mask is then tensioned and welded to a metal frame.[30]

The patterning process begins when a substrate is brought into the vacuum chamber and optically aligned to the shadow mask using a set of fiducials on the substrate and mask. The deposition is initiated by opening the shutter above the source. Material begins depositing indiscriminately on the mask, in the apertures, and on the chamber walls. Only about 16% of the material is deposited on the shadow mask surface (including apertures), so the fraction deposited on the substrate per color is less than 6% of the evaporated material.[31]

After repeated use, the shadow mask needs to be cleaned to remove material that has built up on the surface and around the apertures. The mask is then discarded after a number of cleaning cycles.

Gravity poses another problem. Both the mask and the mother glass exhibit some degree of sag, called "vertical spread," which is proportional to the fourth power of the mother glass size.[32] For example, the vertical spread of a 1 × 1 m² mother glass (0.7 mm thickness) is 108 mm.[33] Tensioning of the glass in one dimension can compensate some of the vertical spread in the mother glass, but it generates a distortion field with reflective symmetry. The tensioning of the shadow mask on all four sides results in a different distortion pattern with fourfold rotational symmetry. Samsung has recently discussed the mathematical modeling of both types of distortion, presumably as a means of providing a geometric compensation in the shadow mask pixel pattern.[34]

The advantages of the shadow mask process include the fact that it is already in use and has been demonstrated to be capable for production of full-color OLEDs for small-format applications. The future of the technology is less certain as mother glass migrates to larger sizes and higher pixel counts. The upper limit of shadow mask usefulness (barring future technological breakthroughs) is thought to be Gen 4 or, possibly, Gen 5 mother glass.

7.4.2 Inkjet Printing

Inkjet technology has matured much in the last 10 years and become a pervasive, nearly ubiquitous patterning method in a number of industries. Printers range from desktop units given away free with a new computer purchase to industrial signage behemoths such as the Scitex Vision XLjet+, able to print ultraviolet (UV)-curable ink on 5-m-wide rolls of vinyl film at 95 m²/h with 370 dpi resolution.

Inkjet printing technology is used for (and is limited to) the patterning of soluble amorphous materials for OLEDs. For example, it has been adapted for the patterning of light-emitting polymers (LEPs) and the hole injection polymer PEDOT/PSS (poly-3,4-ethylenedioxythiophene/polystyrene sulfonate) in full-color OLEDs by several groups including those at Seiko Epson,[35] CDT/Litrex,[36] Philips,[37–39] Osram,[40] and DuPont.[41]

Successful implementations of inkjet patterning for AMOLEDs require modification of the receptor substrate. Inkjet substrates are designed to minimize cross-contamination between subpixels. A hydrophilic/hydrophobic surface patterning technique, much like that used on a printing plate, is used to confine the jetted drops. The substrate consists of an array of subpixel wells with interspersed dams or ribs. The ribs are fabricated using conventional photolithography followed by a series of surface modification steps to create hydrophobic and hydrophilic regions.

After the photolithographic patterning is complete, the exposed indium tin oxide (ITO) is treated with an oxygen plasma to improve wetting of the first jetted solution to be applied, aqueous PEDOT. The substrate is then subjected to a CF_4 plasma, which fluorinates the top surface of the patterned photopolymer layer, reducing its surface energy. The result is an array of subpixel wells with hydrophilic bottoms and low-surface-energy ribs in between. As a drop of PEDOT hits the well, it wets the bottom surface, spreads to the edges, and (if the solvent mixture has been formulated properly) dries uniformly. Drops that land partially in and partially out of a well tend to roll off the low surface energy rib and into the well.[42]

A multiplicity of challenges for the inkjet patterning method has been addressed in recent papers, including the optimization of solution properties, inkjet micropumping characteristics, and printer design. The intense research has enabled inkjet head and printer manufacturers to overcome many of the difficulties in patterning LEPs and to create viable pilot- and production-scale systems for flat-panel display fabrication.

Inkjet solution properties including viscosity, solvent composition, and evaporation rate must be adjusted with care. All types of inkjets have inherent limitations on the viscosity of fluid that may be ejected. For piezo drop-on-demand, optimum viscosity ranges from about 2 to 20 cps at operating temperature. The upper limit is set by stiffness of the pumping chamber in relation to ink properties including viscosity, bulk modulus, and the speed of sound. Viscous damping is a materials parameter related to both viscosity and operating frequency and can be used advantageously in limiting the formation of long "tails" that are common with jetted high-molecular-weight LEP solutions. Solvent composition affects both surface tension and evaporation rate. The shape of the drop is determined by its surface tension as it is ejected from the nozzle. Tailoring of the surface tension can reduce the formation of satellite drops and fluid tails during flight. Evaporation rates are critical for two reasons. First, inks with volatile solvents may readily clog nozzle openings. Second, the drying of the drops on the substrate surface determines material thickness uniformity.

Drying of inkjet drops has been studied extensively. Early efforts to form uniform coatings in subpixel wells were plagued by a number of drying artifacts including thickness non-uniformity and the so-called coffee-stain effect. These problems stem from the facts that the inks used in OLED printing are fairly dilute and the drop volume tends to be large with respect to the pixel size (Figure 7.4). The low ink concentration is dictated by viscosity requirements for proper jetting behavior. Most of these wetting and drying issues have since been resolved.

For OLED applications, pilot- and production-scale printers are available from Litrex, a jointly owned subsidiary of CDT and ULVAC, using piezoelectric drop-on-demand inkjet technology.[43] The Litrex systems consists of a series of printers, each dedicated to one material — PEDOT, red LEP, green LEP, and blue LEP. Key technology for the system was developed by

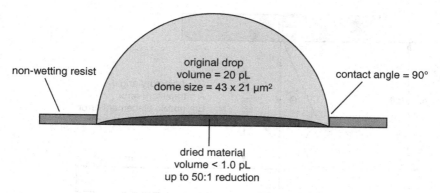

Figure 7.4 Inkjet drop volume reduction.

Spectra, an industry leader in the manufacture of high-resolution, high-precision inkjet nozzle heads.

In conjunction with Spectra, Philips Center for Industrial Technology (Philips CFT) has developed a system to calibrate heads so that drop volume variability is reduced and drop placement accuracy is increased. Figure 7.5 indicates the trajectory error and nozzle placement error components of the total drop placement error. The system works by adjusting parameters on the fly using data from optically detected drops as they exit the nozzle. Each multinozzle head can be calibrated offline using the optical system and then swapped into a production floor unit after the procedure. The technique can also be used to gauge the mean time to nozzle failure for a given jetting solution. Spectra inkjet heads can now consistently attain ultrasmall drop volumes of 12 pL ± 2% (with jet trimming), drop velocities of 8 m/s ± 5% (without tuning), and the required straightness for 65-μm pixels with a 1-mm standoff (Table 7.9 and Table 7.10).[44,45]

Perhaps the greatest leap in recent inkjet head design has just been announced by Spectra, which has developed a means of fabrication using a microelectromechanical system (MEMS) design based on a silicon wafer. The design rules and tolerances for silicon wafer fabrication are much improved from the laser-ablated nozzle approach, so nozzle uniformity,

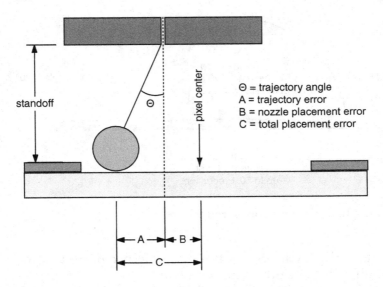

Figure 7.5 Inkjet placement error components.

Table 7.9 State of the Art Inkjet Head Characteristics

Parameter	S-Class	M-Class[a]
No. of addressable jets	128	300
Nozzle spacing	508 μm	141 μm
Drop volume	12 pL	10 pL
Adjustment range	10–12 pL	—
Drop volume variation	<2% with TDC electronics	±5%[b]
Nominal jet velocity	8 m/s	8 m/s
Spot location (all sources)	±10 μm at 1 mm	±7 μm at 1 mm
Compatibility	Solvent, UV, aqueous inks	Solvent, UV, aqueous inks
Drop velocity variation	±5% without tuning	—
Operating frequency	Up to 10 kHz	—

[a] Preliminary specifications as of October 18, 2004.
[b] Nozzles addressed with same pulse shape and size. Better performance is expected with individual nozzle addressing schemes.

nozzle number, and trajectory straightness specifications have all been improved (Table 7.9, "M-Class").[46,47] In addition, silicon nozzles will be much more resistant to chemical and

Table 7.10 Straightness Requirements for a 65-μm Pixel (1 mm inkjet head standoff)

Drop Volume (pL)	Drop Diameter (μm)	Straightness (mrad)
30	39	5.5
20	34	8
10	27	11.5
5	21	14.5

mechanical etching. Finally, the nozzles can be run at an operational frequency of as high as 40 kHz, allowing for a variation in drop volume through the use of multiple subdroplets.

There have been some impressive demonstrations of inkjet patterned AMOLEDs from Philips, Seiko Epson, and Toshiba. However, the inkjet method has not yet made its way into a production environment. One factor has been the performance of soluble OLED materials, which has lagged somewhat behind that of the small molecules (although there have been some very promising results based on dendrimers). Other challenges include nozzle mean time to failure, nozzle wear and cost, and patterning yield. Judging from the recent announcements of Litrex regarding industrial inkjet tools for the flat-panel industry that are sized for Gen 7 mother glass and the MEMS based heads from Spectra, we should see a significant number of inkjet printed OLEDs in the near future.[35]

7.4.3 Laser Thermal Printing

Since 1992, 3M has developed Laser Induced Thermal Imaging (LITI) as a high-resolution, digital patterning technique with a large number of potential applications including the patterning of OLED emitters,[48] organic thin film transistors,[49] color filters,[50] and color conversion filters, etc. (Table 7.11). Since September of 2000, 3M has partnered with Samsung SDI to jointly develop the process for AMOLEDs.

Table 7.11 LITI Attributes

LITI Attribute	Example
Patterning of materials that are incompatible with photolithography	Most organic electronic materials
Patterning of multiple co-soluble layers	OLEDs
Patterning of oriented materials	Polarizers, polarized organic emitters
Patterning of structured materials	Nanoemitters, nano- and microstructures
Patterning of biomaterials	Enzymes, cells, DNA oligomers

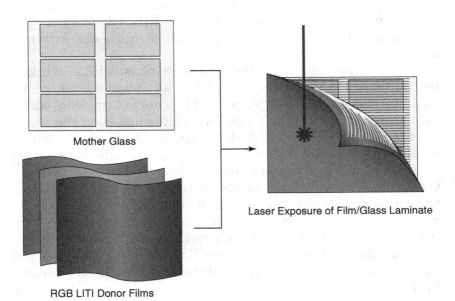

Mother Glass

Laser Exposure of Film/Glass Laminate

RGB LITI Donor Films

Figure 7.6 (Color figure follows p. 274.)   The LITI process.

LITI involves the use of a precoated donor film, a large-format laser exposure system, and a receptor (Figure 7.6). The process begins by first aligning the receptor (e.g., an AMOLED backplane) to the laser exposure system and then laminating the first of a series of precoated donor films to the receptor surface. The laser system is then used to expose the

Table 7.12 Representative LITI OLED Donor Construction

Layer	Typical Thickness (µm)	Function
Transfer layer	0.040	OLED materials
Interlayer(s)	1.0	Protective layer
LTHC	1.6	Absorptive layer
PET Substrate	75	Mechanical support

laminated assembly. During the process, the laser-exposed regions are released from the donor and adhered to the receptor. The process is repeated two or more times, depending on the OLED construction. Alignment is performed only once.

The construction of the donor film greatly influences the mode and quality of transfer. Our donor sheet is typically constructed of three coated layers on a 3-mil PET film substrate — the light to heat conversion layer (LTHC), the interlayer, and the transfer layer (Table 7.12).

The first coated layer is usually the LTHC, a photocured carbon black-based ink. Carbon black acts as a black body absorber and rapidly generates heat within the layer upon exposure. The optical density and thickness of the LTHC are important parameters in adjusting the donor heating profile, which affects imaging quality, performance, and device efficiency. Under typical conditions, the laser exposure of a standard donor film with a typical LTHC results in heating at a phenomenal rate ($\sim 70 \times 10^6$°C/s) with peak temperatures within the LTHC layer reaching from 650 to 700°C. The peak temperature within the OLED transfer layer is about 350°C, but the layer remains above 100°C for less than 1.5 ms.[51] The donor heating profile can be tailored with the use of more than one LTHC, the use of an underlayer, or the use of a gradient absorber.

A second coated layer, the so-called interlayer, is typically a photocured layer that acts to protect the transfer layer from chemical, mechanical, and thermal damage. Depending on the transfer layer requirements, the interlayer may consist of more than one layer. Laser exposure of a donor film with a carbon black-based LTHC but without an interlayer can lead to defects in which the carbon black layer ruptures and contaminates the transfer layer. The interlayer also moderates other physical defects known to arise from warping or distortion of the LTHC so that the transfer layer surface remains smooth. In addition, use of an interlayer allows for independent selection of the materials and properties for both laser heating and transfer layer release functions.

The final coated layer is the transfer layer. This layer of OLED material, commonly a doped evaporated emissive layer, is friable and readily damaged. The panel manufacturer will coat the doped emissive transfer layer on site in order to avoid damaging the OLED material during storage or while in transit.

LITI has the unique ability to pattern multiple discrete organic layers in a single step. This ability is useful, for example, in the patterning of phosphorescent OLED stacks because their optimal construction frequently requires exciton confinement and, hence, multiple charge blocking and transport layers. The entire phosphorescent stack (e.g., HTL/host:dopant/HBL/ETL) can be evaporated onto a stock roll in reverse order prior to the patterning step (Figure 7.7). During the LITI process, the entire multilayered transfer layer will be patterned on the receptor surface.

An important aspect of the LITI OLED process is the imager. The second generation of laser exposure devices is designed to pattern Gen 3 mother glass (550×650 mm) with an absolute placement accuracy of ± 2.5 μm over the imaging area. It is based on two 16-W Nd:YAG lasers operating in TEM_{00} mode and focused to a Gaussian spot of 30 by 330 μm (at $1/e^2$ intensity). The system has an optical efficiency of 50% and the power on the film plane can be varied from 0 to 16 W.

The optical design is based on the premise that transferred line edge quality and precision (e.g., edge roughness variation over the substrate surface) could be improved by increasing the slope of the laser intensity profile in the region of the transfer threshold corresponding to the line edge formation. The method of achieving this goal involves the oscillation of an elliptical beam transverse to the scan direction (dithering) (Figure 7.8 and Figure 7.9).[52] The oscillation is

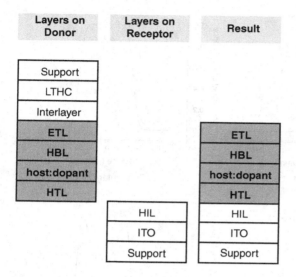

Figure 7.7 Example of multipayer transfer (transferred layers are bold).

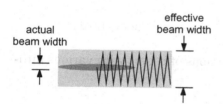

Figure 7.8 Beam cross section during transverse oscillation (dithering).

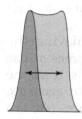

Figure 7.9 Effective width of a dithered beam.

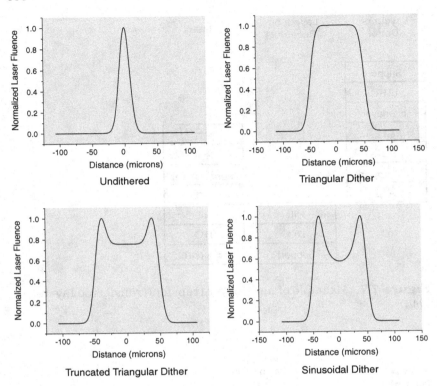

Figure 7.10 Calculated cross sections of beams with various dithering functions.

achieved via acousto-optic beam deflection. Variation of oscillation parameters such as frequency and shape (triangular, sinusoidal, etc.) allow the fluence profile to be tailored to the materials. Figure 7.10 illustrates the results of a mathematical model of the fluence profile for an undithered beam as well as a beam with three different dither functions.

The elliptical beam shape is a factor in achieving the system's high cross-scan accuracy. However, this comes at the expense of imager in-scan accuracy. The result is that the second-generation imagers are configured to write accurate line structures, but not designed to write an arbitrary pattern that requires accuracy in the scan direction (e.g., a delta subpixel configuration).

Table 7.13 LITI Compatible OLED Materials

Emission Type	Coating Method	Examples
Fluorescent	Evaporation	Alq:C545T, multilayer structures
Fluorescent	Solution	Formulated LEPs
Phosphorescent	Evaporation	Host:Ir(ppy)$_3$, multilayer structures
Phosphorescent	Solution	Phosphorescent dendrimers, MDPs

The optical design allows for continuous variation of several parameters including laser power, effective beam width, beam oscillation frequency, as well as choices for the dither function (e.g., triangular, truncated triangular, sinusoidal, etc.). Therefore, the imaging platform can expose donor films using a very large parameter space. In practice, these parameters, in conjunction with those of the donor film, are optimized to tailor the imaging process to the materials.

Laser thermal patterning is currently the only process that is compatible with all four major OLED material categories (Table 7.13). This flexibility allows for the "mixing and matching" of materials on a single substrate (laterally) or within a single subpixel (vertically). LITI allows for the use of a very broad selection of conventional OLED materials without concern for their co-solubility or method of deposition.

There are practical differences in the patterning of different OLED materials related to their bulk mechanical properties. The physical process of removing a microscopic line (the transferred stripe) from a continuous film of OLED material (the transfer layer) involves the formation of line edges defined by the points at which the laser fluence crosses the imaging threshold — effectively the border of exposed and unexposed donor film. A fine balance of adhesion and release must be achieved along the boundaries of this critical region of the donor/receptor laminate in order to create an acceptable line edge.

Our experience with the laser thermal imaging of high-molecular-weight LEPs indicates that the film strength of these materials tends to overwhelm the tearing forces at the line edge, leading to poor image quality. We have investigated several approaches to improve LITI imaging quality of LEPs. For example, reduction of the LEP molecular weight results in significant LITI pattern quality improvements. However, OLED initial device performance and, in particular, OLED device lifetime also depend on LEP molecular weight. The industry consensus is that high-molecular-weight LEPs are necessary to achieve OLEDs of acceptable performance.

As a means of imaging high-molecular-weight LEPs with acceptable image quality, we initially developed phase-separated blends of LEPs and electronically neutral polymers. The performance of these phase-separated blends was sufficient for our development partner (Samsung SDI) to produce AMOLED prototype devices.

Our current focus is the patterning of vapor-deposited phosphorescent emitting materials. This material set represents some of the highest efficiency, longest lived OLED materials that have been identified to date. We have demonstrated that these materials are compatible with the LITI process and produce devices with excellent line patterning quality.

Samsung SDI has used LITI to make a high-resolution (UXGA) 17-in. D full-color AMOLED device without the use of shadow masks (Figure 7.11, Table 7.14).[53] The laser thermal transfer technique has also been applied to successfully demonstrate the fabrication of devices based on multiple patterned layers (simultaneous or sequential). For example, we have prepared devices by simultaneously patterning a stack consisting of a hole-transport layer, an emissive layer, and a hole-blocking layer.

7.5 OTHER PATTERNING METHODS

7.5.1 Laser-Induced Dye Sublimation and Dye Diffusion

In all the patterning techniques described so far, the patterned material — be it LEP or small molecule — comprises the entire emitting layer. In the case of a host-dopant system, such as a dye-doped Alq_3 layer patterned by shadow mask, both host and dopant are patterned simultaneously. Similarly, in the laser thermal transfer of a phosphorescent doped small-molecule layer, the host–dopant system is preformed on the thermal donor and transferred as a unit.

Figure 7.11 (Color figure follows p. 274.) Samsung SDI 17-in. D LITI AMOLED.

Table 7.14 Samsung SDI 17-in. D LITI AMOLED

Display Parameter	Value
Screen size (mm)	345.6 × 259.2
Aspect ratio	4:3
Viewing angle (°)	>170
Resolution	1600 × RGB × 1200 (UXGA)
Pixel pitch (µm)	72 × 216 (118 ppi)
Emission structure	Bottom emission
Response time	<0.01
Brightness (cd/m²)	400

Two adaptations of conventional thermal printing technology take a different tack. Both laser-induced dye sublimation and laser-induced dye diffusion are techniques that may be used to pattern the dopant alone onto a substrate having a precoated layer of undoped host material. The dye is diffused into the host material either during or, in a separate step, after the process. The two techniques are similar in that they both use a donor ribbon with a dye-containing layer along with a laser imaging system to provide local heating of the dye, but the patterning mechanisms are different.

Resistive thermal head dye sublimation printing is the process that Kodak uses in its Picture Maker G3 kiosks, Professional 8500 printer, and Photo Printer 6800 to generate continuous tone photographic prints from digital image files.[54] Not surprisingly, Kodak has been active in developing an analogous technique for AMOLED patterning using lasers, rather than a resistive thermal head, to heat the donor ribbon.[55,56] As described by Kodak, the laser imager is used to expose the donor ribbon as it is in contact with the receptor. The heating process causes dopant dispersed in a layer of the donor to sublime and escape from the film, across a small gap, and onto the surface of the receiving layer. Usually, the donor ribbon incorporates the use of spacer beads that create the optimum gap between the donor and receptor. In some cases, it may be necessary to use a final heating step to redistribute the dopant from the receptor layer surface into the layer itself. Kodak has recently reported on a method to co-sublime both dopant and host from a single donor ribbon.

Laser-induced dye diffusion is also related to a color printing technique referred to as "dye diffusion thermal transfer" or D2T2. Hi-Touch Imaging Technologies offers a line of personal photo printers based on D2T2 technology using a conventional resistive head array, rather than a laser, to heat the donor ribbon.[57] For OLEDs, the method involves a dopant-containing ribbon and a laser exposure unit. However, in this case, there is no gap between the donor and the receptor. The two are held in intimate contact, usually under pressure, to allow the dopant-containing layer on the donor to form a continuous medium with the dopant-receiving (host) layer on the receptor.[58] A report on the use of this method and the sensitivity of the OLED efficiency to laser heating conditions has been made by a research group at NHK.[59]

In both methods, the dye concentration in the receptor is proportional to the laser heating of the dye layer, which can be good news or bad news. For color photos, the proportionality allows for a near continuum of color density in the print, much like the continuous tone of a silver halide color photo print. However, OLEDs have a fairly narrow optimum for dye concentration in a doped host. A variation in dopant concentration may result in an efficiency difference that will, in turn, lead to brightness non-uniformity over the display panel. Another issue with dye dopant patterning is the fact that a common host for red, green, and blue must be used. This is not the optimum case for many of the highly developed phosphorescent systems, for example, which typically require hosts with different HOMO levels and HOMO-LUMO gaps for each emission color.

7.5.2 Photolithography

Photolithography is an accepted standard for patterning in the flat-panel display and microelectronics manufacturing sector. It is a mature method with a very large infrastructure and a great number of equipment and materials suppliers. Manufacturers have developed a thorough knowledge of processing parameters as well as methods for improving process efficiency and yield and reducing process waste and cost. The method can be used to pattern complex submicron features reliably in a mass production setting. For these and other reasons, photolithography would surely be the method of choice for patterning OLED materials if it produced devices with state-of-the-art efficiency and lifetime.

The quest for photopatternable OLED materials started soon after organic electroluminescence was first demonstrated and it continues today. However, it has been intrinsically difficult for two reasons. First, OLED absorption and emission occur at about the same quantum energy as the patterning radiation, so it is difficult to selectively excite a photosensitizer without photobleaching the OLED chromophores. Second, typical organic moieties for cross-linking (e.g., acrylates) are good traps for chromophore excited states. It is difficult to completely consume the potential traps during the curing step so that they do not pose a problem during device operation.

A collaborative effort by a number of German groups recently demonstrated the potential for photolithographic patterning of OLED materials.[60,61] They employed the acid-catalyzed ring opening of oxetanes as the basis for cross-linking in LEPs. The oxetanes are not thought to quench excited state species. Red, green, and blue LEPs were constructed by combining a spirobifluorene spacer unit, a color or charge transport unit, and a cross-linkable spirobifluorenyl

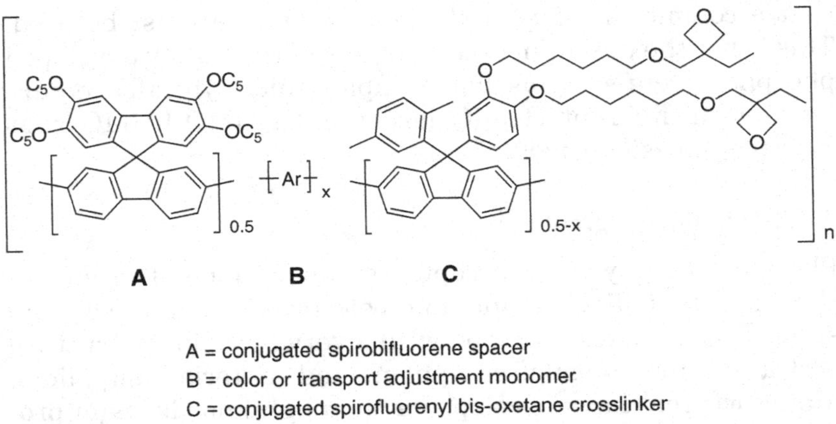
C5O
OC5
C5O
OC5
Ar
x
0.5
0.5-x
n
O
O
O
O
O
O
A
B
C
A = conjugated spirobifluorene spacer
B = color or transport adjustment monomer
C = conjugated spirofluorenyl bis-oxetane crosslinker

Figure 7.12 Photopolymerizable LEPs.

bis-oxetane cross-linking unit. Formulations of the polymer included a photoacid generator to initiate the cross-linking process upon exposure to UV radiation (Figure 7.12).

7.5.3 Other Methods

7.5.3.1 Soft Lithography

Soft lithography belongs to a family of related patterning techniques including microcontact printing,[62,63] soft contact lamination,[64] nanoimprinting,[65–67] and cold welding.[68] The basis for all is the use of a soft replicated stamp of a polysiloxane cast from a high-resolution relief pattern on a silicon wafer mold. Although there are a large number of papers in the field, there are relatively few that discuss the actual patterning of RGB emitters for OLEDs. An excellent review on the topic by a team at IBM Zurich Research Laboratory hints at the possibilities in suggesting soft lithography adaptations of conventional printing techniques such as flexography and intaglio printing.[69]

Soft lithography suffers from the same fate that flexography, screen printing, and other printing technologies in that it is based on a dimensionally unstable material embossed with a high-resolution pattern. Techniques for achieving high absolute placement accuracy will need to be developed before soft lithography will be accepted in the manufacturing arena. Until then, demonstrations of soft lithography for OLEDs may be limited to relatively small areas on small format devices.

7.5.3.2 Organic Vapor Jet Printing

Groups at both Princeton University and Kodak are developing a new means of patterning small-molecule OLED materials.[70–73] Organic Vapor Jet Printing (OVJP) is related to the general deposition technique called Organic Vapor Phase Deposition (OVPD).[74] Although OVPD is intended as a replacement for standard high-vacuum deposition methods, it must be coupled with a patterning technology (e.g., shadow mask) to enable fabrication of full-color OLEDs.

OVJP is a combination of the hot carrier gas, low pressure deposition method with a high-resolution dispensing system. It utilizes a nozzle with a small (20 μm) orifice on a scanning head and enables the printing of discrete, small areas of evaporable material. In some ways, the process has similarities to inkjet printing because a jet (in this case, a heated gas) is ejected from a micronozzle and physisorbs on the receptor surface in what resembles a jetted dot. However, the "solvent" is an inert carrier gas such as nitrogen or helium and the solute is an evaporable small molecule.

In a demonstration of the technique, the organic small molecule Alq_3 was printed onto a glass receptor surface as an array of 25-μm dots. Based on an extrapolation to a 300°C source temperature, it was estimated that the achievable rate of deposition was approximately 8000 Å/s. The authors argue that an 800 nozzle OVJP head depositing material at this rate is roughly equivalent to the printing speed of a state-of-the-art inkjet printing head. OVJP is an interesting, but immature technology for patterning volatile molecules that deserves closer inspection.

7.5.3.3 Screen Printing

Screen printing is another standard printing technique that has been applied to patterning OLED materials. In the graphic arts world, screen printing is relegated to relatively low-resolution, low-information-content graphics such as decals, signage, textiles (T-shirt prints), and promotional items.[75] The process utilizes a screen/photopolymer film laminate. The photopolymer image is developed using a standard photolithographic process, leaving regions of open screen mesh through which ink can flow. Viscous ink is applied to the screen surface and forced through the open mesh using a squeegee. The mesh size, thread diameter, and emulsion (photopolymer) thickness determine the amount of ink that is printed as the squeegee forces it through to the substrate below. In standard screen printing, the ink is too viscous to flow through the mesh unless it is forced, so the wet coating thickness is approximately the thickness of the mesh plus the thickness of the emulsion.

Screen printing has been used by a group at Siemens AG (Erlangen, Germany) to print $270 \times 270 \ \mu m^2$ pixels of an LEP and fabricate passive-matrix devices with initial performance characteristics that were on par with spin-coated devices.[76] The effect of solution viscosity and screen mesh count has been studied in the fabrication of OLEDs with screen-printed hole transport layers.[77] In 2001, the company Add-Vision changed its focus from patterning thick-film electroluminescent devices to screen-printing monochrome LEP displays for low-cost, temporary, low-usage, or disposable applications.[78,79]

Commercial screen printing is based on UV-curable inks so that multiple colors can be overprinted without disturbing the underlying images. Screen-printed OLEDs cannot easily avoid the disruption of previously coated layers. Therefore, screen printing will be used primarily for fabrication of either monochrome or zone color OLEDs.

7.5.3.4 Die Coating

Die coating is a technology that is usually applied to the continuous coating of plastic film webs.[80] It involves the flow of a low-viscosity coating solution through a manifold, to a precision slot. The fluid is metered through the slot onto the surface of a film as it passes by the coating head. Toppan

(Tokyo, Japan) has developed a die-coating method to accurately print solutions of OLED materials.[81,82] More recently, the technique has been optimized for flat-panel display coating using a glass substrate on a precision stage and moving it past a stationary coating head as a laminar flow of coating solution is applied.[83] Production units from FAS Technologies (Dallas, TX) are available for coating mother glass as large as Gen 6.[84]

Die coating is inherently a large, continuous area-coating technology. The application of the technique to flat-panel display manufacture is mainly as a surrogate for or companion to spin coating. Because the coating solution is metered accurately, the materials utilization is very high, in contrast to spin coating. The application of die coating to OLEDs is therefore limited to coating buffer layers or LEPs for monochrome displays. There is no clear path to its use in patterning full-color displays.

7.6 CONCLUSION

As OLED manufacturing technology matures, the number of viable high-resolution patterning options remains relatively small. As of today, the majority of the infrastructure is based on the tried-and-true shadow mask method. However, the limits of the method are understood today and the method cannot dominate as the trend toward both higher resolution and mother glass size continues. New technologies such as inkjet and laser thermal patterning are emerging and will attempt to prove themselves within the next few years. Other methods such as screen printing and die coating will be directed toward low-resolution information displays rather than AMOLEDs.

In summary, a number of patterning methods have been described along with a detailed explanation of AMOLED patterning issues. Table 7.15 attempts to assign a qualitative rating to the probability that a given patterning method can achieve the goals set forth in the section "Summary of Patterning Issues." The assessments are not firm and should be

Table 7.15 Qualitative Assessment of OLED Patterning Method Attributes

	Material Type	Aperture Ratio	Device Design	Pixel Pattern	Display Format	Substrate Size	Placement Accuracy	Takt-time	Defect Density
Shadow mask	Fair	Fair	Good	Good	Fair	Poor	Fair	Fair	Poor
Inkjet	Fair	Poor	Good	Good	Fair	Good	Fair	Good	?
LITI	Good	Good	Good	Fair	Good	Good	Good	Good	?
Dye sub/dye diffusion	Fair	Good	Good	Good	Good	Good	Good	Good	?
Photolithography	Poor	Good	Good	Good	Good	Good	Good	Good	Good
Soft lithography	Poor	N/A	N/A	N/A	N/A	N/A	N/A	N/A	N/A
OVJP	Fair	?	?	?	?	?	?	Good	?
Screen printing	Fair	N/A	N/A	N/A	N/A	Good	Poor	Good	N/A
Die coating	Fair	N/A	N/A	N/A	N/A	Good	N/A	Good	N/A

Note: N/A indicates that the method is not currently compatible with large area AMOLED device production. Either the method has not yet been demonstrated in the fabrication of an AMOLED or there are known issues with large format printing or high resolution patterning.
? indicates that the patterning method could be assessed for the attribute, but those data have not yet been reported.

used with caution since the technologies for patterning are evolving rapidly and continuously.

ACKNOWLEDGMENTS

I acknowledge many colleagues, associates, and co-workers that have helped in discussing and proofreading this chapter, including Dr. Vadim Savvate'ev, James M. Nelson, Dr. Sergey Lamansky, Bradley Zinke, Dr. Linda Creagh (Spectra), Dr. Eliav Haskel (Philips), and others. The LITI section of the chapter represents the work of a large number of people at 3M (St. Paul, MN), Sumitomo/3M (Sagamihara, Japan), and Samsung SDI (Yongin City, Korea). In particular, I acknowledge all my colleagues, team leaders, and managers in each location — Dr. William A. Tolbert, Rick L. Neby (St. Paul), Dr. Takashi Yamasaki, Keizo Yamanaka (Sagamihara), and Dr. Seong Taek Lee, Dr. H.K. Chung (Yongin City). In addition, I thank Covion Semiconductors GmbH for its assistance and for providing many OLED materials.

REFERENCES

1. Alternative Display Technology Report: OLED Technology, Display Search, Austin, TX, 2002, table 7.3.

2. Alternative Display Technology Report: OLED Technology, Display Search, Austin, TX, 2003, table 7.4.

3. "Patterning of OLED material on a volume production scale with high throughput, high yield, and scalability to large substrates is clearly the next major challenge facing the SM OLED industry," Alternative Technology Display Report: OLED Technology, Display Search, Austin, TX, 2003, 115.

4. Kashiwabara, M. et al., Advanced AM-OLED display based on white emitter with microcavity structure, *SID Dig.*, 1017, 2004.

5. Ghosh, A.P. et al., Color changing materials for OLED microdisplays, *SID Dig.*, 983, 2000.

6. Tang, C.W. and VanSlyke, S.A., Organic electroluminescent diodes, *Appl. Phys. Lett.*, 51, 913, 1987.

7. Burroughes, J.H. et al., Light-emitting diodes based on conjugated polymers, *Nature,* 347, 539, 1990.

8. Strohriegl, P. and Grazulevicius, J.V., Charge-transporting molecular glasses, *Adv. Mater.,* 14(20), 1439, 2002.

9. Chen, A.C.A et al., Organic polarized light-emitting diodes via Förster energy transfer using monodisperse conjugated oligomers, *Adv. Mater.,* 16(9–10), 783, 2004.

10. Markham, J.P.J. et al., Highly efficient solution-processable dendrimer LEDs, *SID Dig.,* 1032, 2002.

11. Tokito, S., Suzuki, M., and Sato, F., Improvement of emission efficiency in polymer light-emitting devices based on phosphorescent polymers, *Thin Solid Films,* 445, 353, 2003.

12. Baldo, M.A. et al., Highly efficient phosphorescent emission from organic electroluminescent devices, *Nature,* 395, 151, 1998.

13. Baldo, M.A. et al., Very high-efficiency green organic light-emitting devices based on electrophosphorescence, *Appl. Phys. Lett.,* 75(1), 4, 1999.

14. Kwong, R.C. et al., High operational stability of electrophosphorescent devices, *Appl. Phys. Lett.,* 81(1), 162, 2002.

15. Sumitomo, CDT link in PLED, *Chem. Wk.,* 166(8), 22, 2004.

16. Lee, C.J. et al., On the problem of microcavity effects on the top emitting OLED with semitransparent metal cathode, *Phys. Status Solidi A,* 201(5), 1022, 2004.

17. Lu, M.-H. et al., High-efficiency top-emitting organic light-emitting devices, *Appl. Phys. Lett.,* 81(21), 3921, 2002.

18. Lee, J.-H. et al., A new a-Si:H TFT pixel design compensating threshold voltage degradation of TFT and OLED, *SID Dig.,* 264, 2004.

19. Based on the author's informal survey of every flat panel display on the showroom floors of three large (St. Paul, MN) electronic stores (BestBuy, Ultimate Electronics, CompUSA). The only non-stripe patterns were found on a small number of digital video camera and digital still camera viewfinder displays.

20. http://www.clairvoyante.com.

21. Corning® EAGLE2000™ AMLCD Glass Substrates Material Information, MIE 201, Corning, NY, 2002.

22. DeLassus, P.T. and Whiteman, N.F., Physical and mechanical properties of some important polymers, in *The Wiley Database of Polymer Properties*, John Wiley & Sons, New York, 1999.

23. Wadle, H. and Blum, D., *Heidelberg Introduction to Screening*, Heidelberg Druckmaschinen, Heidelberg, 2002.

24. http://www.hp.com/oeminkjet/reports/techpress_11.pdf.

25. Linck, J. and Cochran, D.S., *The Importance of Takt Time for Manufacturing System Design*, SAE International Automotive Manufacturing Conference, Detroit, MI, 1999.

26. http://www.ulvac.com/flatpanel/oled.asp.

27. http://www.tokki.co.jp/eng/products.html.

28. http://www.doosandnd.com/.

29. http://www.appliedfilms.com/display.htm.

30. Kang, C.H. and Kim, T.S., Mask for evaporation, mask frame assembly including the mask for evaporation, and methods of manufacturing the mask and the mask frame assembly, U.S. Patent Appl. 2003/0221613 A1.

31. For a 1×1 m^2 substrate suspended 1 m above a source, the fraction of the material deposited from the source onto the substrate may be approximated by assuming uniform material distribution within a hemisphere. The fractional area of the substrate within the hemisphere 1 sr/(2π sr) = 0.16.

32. Meda, G., Optimal support of plates to minimize sag, Corning Technical Information Paper, Report L-5822 MAN, Corning, NY, date unknown.

33. Meda, G., Support designs for reducing the sag of horizontally supported sheets, Corning Technical Information Paper, Report L-5701 MAN, Corning, NY, 2003.

34. Chung, H.K., LITI technology for high definition and large size AMOLED fabrication, presented at the FPD International Exhibition, Session E-6, Yokohama, Japan, 2004.

35. Shimoda, T. et al., Inkjet printing of light-emitting polymer displays, *MRS Bull.*, 28(11), 821, 2003.

36. Edwards, C. et al., Precision industrial ink jet printing technology for full color PLED display manufacturing, in *International Display Manufacturing Conference (IDMC) Proceedings*, Taipei, 2003.

37. Fleuster, M. et al., Mass manufacturing of full color passive-matrix and active-matrix PLED displays, *SID Dig.*, 1276, 2004.

38. van der Vaart, N.C. et al., Towards large-area full-color active-matrix printed polymer OLED television, *SID Dig.*, 1284, 2004.

39. de Gans, B.-J., Duineveld, P.C., and Schubert, U.S., Inkjet printing of polymers: state of the art and future developments, *Adv. Mater.*, 16(3), 203, 2004.

40. Gupta, R. et al., Ink jet printed organic displays, *SID Dig.*, 1281, 2004.

41. MacPherson, C. et al., Development of full color passive PLED displays by inkjet printing, *SID Dig.*, 1191, 2003.

42. Sirringhaus, H. et al., Active matrix displays made with printed polymer thin film transistors, *SID Dig.*, 1084, 2003.

43. http://www.litrex.com/Products.htm.

44. Creagh, L.T., McDonald, M., and Letendre, W., Progress in non-contact precision deposition for manufacturing FPDs, presented at USDC Flexible Electronics Conference, Phoenix, 2004.

45. Creagh, L.T., personal communication.

46. Menzel, C., Bibl, A., and Hoisington, P., MEMS solutions for precision micro-fluidic dispensing applications, presented at Imaging Science & Technology Non-Impact Printing 20 Conference (NIP20), Salt Lake City, UT, November 2004.

47. Letendre, W. and Brady, A., Advances in piezoelectric micro-pump precision deposition using silicon nozzles, presented at Imaging Science & Technology Non-Impact Printing 20 Conference (NIP20), Salt Lake City, UT, November 2004.

48. Wolk, M.B. et al., Thermal transfer element and process for forming organic electroluminescent devices, U.S. Patent 6194119.

49. Wolk, M.B. et al., Thermal transfer element for forming multi-layer devices, U.S. Patent 6144088.

50. Wolk, M.B. et al., Methods for preparing color filter elements using laser induced transfer of colorants with associated liquid crystal display device, U.S. Patent 5521035.

51. Calculated values for the 3M Alpha imager using two lasers, 16 W/laser, scan velocity = 12.3 m/s, threshold dose = 650 mJ/cm². Thomas R. Hoffend, Jr., personal communication.

52. Fillion, T., Savikovsky, A., and Ehrmann, J.S., Method and apparatus for shaping a laser-beam intensity profile by dithering, U.S. Patent 6496292 B2.

53. Lee, S.T. et al., A novel patterning method for full-color organic light-emitting devices: Laser Induced Thermal Imaging (LITI), *SID Dig.*, 1008, 2004.

54. http://www.kodak.com/US/en/corp/pressReleases/pr20040217-03.shtml.

55. Burberry, M.S. et al., Fabrication of electroluminescent display devices using radiation-induced transfer of organic materials, U.S. Patent 6610455 B1.

56. Tutt, L.W., Culver, M.W., and Tang, C.W., Laser thermal transfer donor including a separate dopant layer, U.S. Patent Appl. 2004029039 A1.

57. http://www.hitouchimaging.com/Products/PhotoPrinters.asp.

58. Wu, C.-C. et al., Finite-source dye-diffusion thermal transfer for doping and color integration of organic light-emitting devices, *Appl. Phys. Lett.*, 77(6), 794, 2000.

59. Tanaka, I. et al., Selective heat-transfer dye diffusion technique using laser irradiation for polymer electroluminescent devices, *Displays*, 23, 249, 2002.

60. Müller, C.D. et al., Multi-color organic light-emitting displays by solution processing, *Nature*, 421, 829, 2003.

61. Pogantsch, A. et al., Photochemical patterning approaches for multicolor polymer light emitting devices, *Mater. Res. Soc. Symp. Proc.*, 771, 307, 2003.

62. Advincula, R. et al., Micro-contact printing and the electrochemical cross-linking approach to conjugated polymer ultrathin films and devices, *Polym. Prepr.*, 44(2), 309, 2003.

63. Liang, Z. et al., Micropatterning of conducting polymer thin films on reactive self-assembled monolayers, *Chem. Mater.*, 15, 2699, 2003.

64. Lee, T.-W. et al., Organic light-emitting diodes formed by soft contact lamination, *Proc. Natl. Acad. Sci. U.S.A.*, 101(2), 429, 2004.

65. Koide, Y. et al., Hot microcontact printing for patterning ITO surfaces. Methodology, morphology, microstructure, and OLED charge injection barrier imaging, *Langmuir,* 19, 86, 2003.

66. Wang, J. et al., Direct nanoimprint of submicron organic light-emitting structures, *Appl. Phys. Lett.*, 75(18), 2767, 1999.

67. Cavallini, M., Murgia, M., and Biscarini, F., Direct patterning of tris-(8-hydroxyquinoline)-aluminum (III) thin film at submicron scale by modified micro-transfer molding, *Mater. Sci. Eng. C*, 19, 275, 2002.

68. Kim, C. and Forrest, S.R., Fabrication of organic light-emitting devices by low-pressure cold welding, *Adv. Mater.,* 15(6), 541, 2003.

69. Michel, B. et al., Printing meets lithography: soft approaches to high-resolution patterning, *IBM J. Res. Dev.*, 45(5), 697, 2001.

70. Shtein, M. et al., Direct, mask- and solvent-free printing of molecular organic semiconductors, *Adv. Mater.*, 16(18), 1615, 2004.

71. Shtein, M., Forrest, S.R., and Benzinger, J.B., Device and method for organic vapor jet deposition, U.S. Patent Appl. 2004/0048000 A1.

72. Marcus, M.A. et al., Device for depositing patterning layers in OLED displays, U.S. Patent Appl. 2004/0056244 A1.

73. Marcus, M.A. et al., Depositing layers in OLED devices using viscous flow, U.S. Patent Appl. 2004/0062856 A1.

74. Shtein, M. et al., Material transport regimes and mechanisms for growth of molecular organic thin films using low-pressure organic vapor phase deposition, *J. Appl. Phys.*, 89(2), 1470, 2001.

75. McSweeney, T.B., Screen printing, in *Coatings Technology Handbook,* Satas, D., Ed., Marcel Dekker, New York, 2001, chap. 15.

76. Birnstock, J. et al., Screen-printed passive matrix displays based on light-emitting polymers, *Appl. Phys. Lett.*, 78(24), 3905, 2001.

77. Jabbour, G., Radspin, R., and Peyghambarian, N., Screen printing for the fabrication of organic light-emitting devices, *IEEE J. Selected Top. Quantum Electron.*, 7(5), 769, 2001.

78. http://www.add-vision.com/index.html.

79. Victor, J.G., Wilkinson, M., and Carter, S., Screen printing light-emitting polymer patterned devices, U.S. Patent Appl. 6605483 B2.

80. Lippert, H.G., Slot die coating for low viscosity fluids, in *Coatings Technology Handbook,* Satas, D., Ed., Marcel Dekker, New York, 2001, chap. 11.

81. Shimizu, T. et al., Coating apparatus for cost-effective manufacture of organic electroluminescent devices with high luminance and good durability, Japanese Patent Appl. 2004139814 A2.

82. McCormick, F.B. et al., Large area organic electronic devices having conducting polymer buffer layers and methods of making same, PCT Appl. WO 2001018889 A1.

83. Gibson, G.M. et al., Moving head coating apparatus and method, U.S. Patent 6540833 B1.

84. http://www.fas.com/advantage6.htm.

8

AMOLED Display Pixel Electronics

MATIAS TROCCOLI, MILTIADIS K. HATALIS,
AND APOSTOLOS T. VOUTSAS

CONTENTS

8.1 INTRODUCTION

Liquid crystal displays (LCDs) have now been on the market for a number of years; however, recent enhancements on organic light-emitting devices (OLEDs) offer a new alternative to the flat panel display industry. OLEDs offer improved luminous efficiency, high brightness, and excellent viewing angle. In addition, OLED displays have thinner profiles, weigh less, and consume less power than LCDs. Other interesting aspects of this new technology are the flexible display and flexible electronics product solutions it offers. Because of their emissive nature, OLED displays do not require a backlight, which makes this technology a lower-cost and more suitable alternative to flexible display applications. OLED displays facilitate the implementation of displays on plastics and metal foils, which make for more rugged and lighter products. Furthermore, when using flexible substrates, these displays could ultimately be fabricated using cheaper processing techniques such as roll-to-roll processing.

According to recent market trends, in order for OLED displays to be commercially feasible, they will have to meet or exceed the performance of LCD displays and do so at a competitive cost. Consumers are not usually concerned about what type of technology is inside a product or how it works; performance and price are the key factors that ultimately attract buyers. However, several technological challenges remain for OLED displays to overtake the next generation display market. Some such challenges pertain to OLED fabrication and performance (i.e., lifetime, color), and others pertain to pixel circuits and display drivers (i.e., pixel output uniformity, yield).

Display drivers are divided into two different addressing schemes: active matrix addressing and passive matrix addressing. The active matrix pixel circuitry is classified according to the type of silicon thin film used to form the thin film transistors (TFTs): amorphous silicon or polycrystalline silicon.

In the following sections, we concentrate our discussion on the electronic aspects of OLED displays. We compare active matrix and passive matrix addressing schemes and discuss the different types of silicon thin film technologies currently used. Because the focus of our attention lies on active matrix OLED (AMOLED) displays that are driven with TFTs, we show various pixel circuits examples designed for either amorphous or polysilicon technologies. In general, pixel circuitry addressing can be classified as voltage or current addressing techniques and examples of both schemes are shown, as well as a brief description of digital addressing techniques. Finally in the technology section we show some physical layout designs examples for commonly used pixel circuits as well as some basic processing steps for TFT fabrication.

8.2 GENERAL CIRCUIT CONSIDERATIONS

8.2.1 OLED Characteristics

For circuit designers it is important to be able to represent OLEDs with an appropriate electric circuit model that can be used in the design process of TFT pixel circuits. For this reason we begin by discussing how OLEDs behave from an electrical point of view. We need to study their voltage and current behavior, which can ultimately be approximated with standard circuit elements such as resistors, sources, etc.

It is important to define what physical parameters control the brightness level for any type of lighting element. The

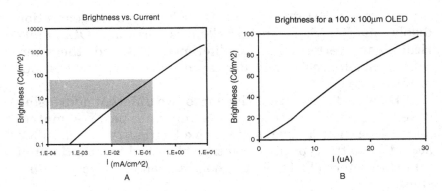

Figure 8.1 OLED brightness curves: (A) normalized per area, (B) for an OLED 100 × 100 μm.

first demonstration of efficient light emission for OLEDs was accomplished in small-molecule organic materials;[1] however, OLEDs are made of a variety of organic thin films. They are mainly classified into two groups: small-molecule and polymer OLEDs.[2] Their structure consists of a series of thin films that usually include electron and hole injection layers, electron and hole transport layers, and an emissive layer. (In general, polymer OLEDs require only a hole transport layer and an emissive layer.) The anode (metal that makes contact with the hole injection layer) is usually made of a high-work-function metal such as transparent indium-tin oxide (ITO) while the cathode is made by a low-work-function metal. When the holes and electrons recombine in the emissive layer, light is produced. For this reason, brightness is directly related to the OLED current. Figure 8.1 shows this relationship between current and brightness for a typical device.[3] As we can see from the graph, the dependence is almost perfectly linear. The shaded area on plot A corresponds to the typical operation area for an OLED. We define the turn-on voltage of the device as the voltage needed for the device to reach a brightness level of 1.0 cd/m^2, which for our example device corresponds to approximately 0.5 mA/cm^2 or 0.5 μA for a 100 × 100 μm OLED. (It is important to point out that sometimes the electric

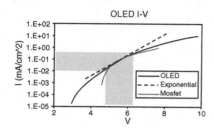

Figure 8.2   OLED *I–V* curve.

turn-on voltage for an OLED can be different from the light turn-on voltage.)

To improvise an appropriate OLED electrical model, we first have to analyze its current-voltage relationship. As we can see from Figure 8.2, inside the normal operating values of our sample device, the OLED current (solid black) approximates an exponential function of the voltage (dotted line). When designing pixel circuits it is convenient to approximate that relationship inside that voltage range with a diode connected transistor.[4–7]

Even though the *I–V* curve for a diode-connected transistor is a quadratic relationship, it provides relatively good accuracy for our circuit simulations. In order for the electrical model to capture transient responses for the OLED, we will add a capacitor in parallel to the diode-connected transistor. This capacitor is the effective parasitic capacitance of the organic layers, which for a 100×100 μm OLED is in the order of 1 to 10 pF. We can examine the transient response of the OLED by driving it with a step input (i.e., a step current source). Initially, the input current is divided between the transistor and the capacitor. According to Figure 8.3, I_d is the diode current (which in effect controls the brightness) and I_c is the capacitor current. The latter charges the capacitor until it reaches a steady-state voltage (the OLED voltage), at which time all of the current flows through the transistor. Analytically, we can solve the corresponding differential equation for this circuit as follows:[4–7]

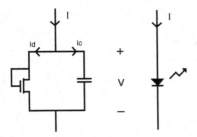

Figure 8.3   OLED electrical model.

$$C\frac{dV(t)}{dt}+\frac{\beta}{2}V(t)^2 = I ,$$ (8.1)

where

$$\beta = C_0\mu\frac{W}{L}$$

For a MOSFET in saturation:

$$Id = \frac{\beta}{2}\left(V_{gs}-V_{th}\right)^2$$ (8.2)

and when diode connected (assuming turn-on voltage, $V_{th} = 0$ for simplicity):

$$V_{gs} = V_{ds} = V(t) \quad \text{and} \quad Id = \frac{\beta}{2}V(t)^2 .$$ (8.3)

The solution to Equation 8.1 is

$$V(t) = V(\infty)\left[\frac{V(\infty)\tanh\dfrac{t}{\tau}+V(0)}{V(\infty)+V(0)\tanh\dfrac{t}{\tau}}\right].$$ (8.4)

For $V(0) = 0$, Equation 8.4 reduces to

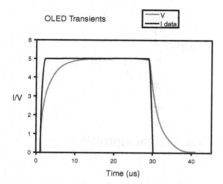

Figure 8.4 OLED transient characteristics.

$$V(t) = V(\infty) \cdot \tanh\frac{t}{\tau}, \tag{8.5}$$

where

$$\tau = \frac{C}{\sqrt{\frac{\beta}{2}Id}} \quad \text{and} \quad V(\infty) = \sqrt{\frac{2I}{\beta}}. \tag{8.6}$$

A graphical solution to this equation (OLED transient simulation) is shown in Figure 8.4. As we can see in this plot, the settling time for the OLED voltage could be quite significant. As we will see later on, this will be an important factor in pixel circuit topology because OLED voltage transients should not affect the current driving capabilities of the current source inside each pixel.

Another important consideration that has to be accounted for when designing for OLED displays is the lifetime of the device. This is perhaps the number one problem to be addressed over the next few years to make this technology more competitive with the well-performing LCD industry. Recently, there has been a significant improvement in OLED lifetime. However, even though the lifetime of a device could be on the order of 10,000 h, devices can lose up to 10% of their

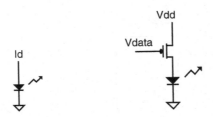

Figure 8.5   Passive and active addressing configurations.

intensity in just a few hundred hours.[3] These figures are not yet up to commercial standards, although some small OLED displays are being fitted into commercial products. Different compensation schemes have been proposed to correct this problem, but as we show in following section, the pixel complexity can increase dramatically.

8.2.2   Passive Matrix and Active Matrix Addressing

As we mentioned above, displays can be addressed with two different addressing schemes, passive and active matrix addressing.[7–12] Figure 8.5 shows the schematic configurations for passive and active addressing. Passive matrix displays apply a voltage directly across the lighting element, whereas active matrix displays use a data voltage to control a current through the LED. The fundamental difference between both techniques relates to the fraction of time that a given pixel is allowed to emit light. When one chooses to address a display passively, only the pixels in one row are turned on at any instant in time. On the other hand, pixels in active matrix displays emit light at all times (except when no light is desired). The difference in design approaches results in radically different power efficiencies, brightness, lifetime, and fabrication complexity between displays.

Passive matrix addressing exploits the fact that the human eye image-sensing cycle is much longer than the refresh rate of the display. By rapidly turning on (or off) one line at a time, and by repeating this process many times per second (usually 60 times per second) the human eye is not able to capture the rapid on/off transitions of the pixels, and a still image is perceived. To achieve this, the light-emitting element has to be turned on at high brightness levels because the human eye averages that given intensity over time until the OLED is turned on again. For example, a single row could be turned on at a very bright light level, but if it remains at this state for only one fraction of the time (depending on the number of rows) a much lower light level is perceived. This means that high currents and high voltages are needed to control each pixel in a passive addressed display. Consequently, the power efficiency of the display is poor while the high current levels during operating cause reduced lifetime.[3] Active matrix displays are significantly more power efficient since every pixel is kept on at the appropriate brightness level for the entire frame cycle.

Figure 8.6 shows the IBM standard addressing timing pulses for a 640 × 480 (VGA) operating at 60 Hz. A typical VGA display has 640 columns and 480 rows and its refresh rate is 60 Hz. This means that the entire array is addressed 60 times per second (i.e., frame time is 1/60 ms and the row time is approximately 32 μs). Timing diagrams change for different displays of different sizes and resolution. Currently, displays that run at higher frequencies are being introduced to the market since they reduce eye fatigue.

Figure 8.7 shows a simplified diagram of control signals for both passive and addressing schemes for three rows and three columns in a display. We can clearly notice how in a passive display only one line is turned on at any instant in time (different brightness levels for each pixel are determined by the duration of the current pulses). On the other hand, pixels in an active display are always on. In active matrix displays, when an address line is activated in a row (pixel switches are on), each pixel on that row reads the data from each corresponding voltage data line (V_{data}). Once the address signal voltage is deactivated, that information is stored and the process is repeated in the rest of the rows. The data information is stored in each pixel in the form of a capacitor voltage. This capacitor voltage controls the gate of a transistor that behaves as a current source feeding the OLED.

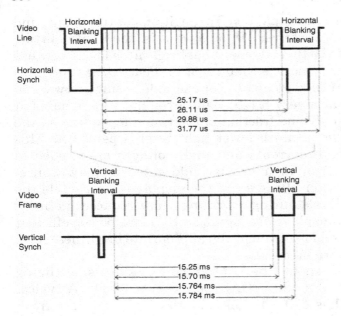

Figure 8.6 IBM standard addressing timing for VGA displays at 60 Hz.

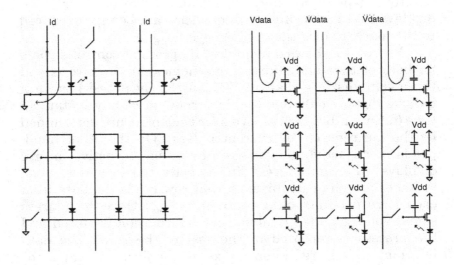

Figure 8.7 Passive (left) and active (right) addressing schemes.

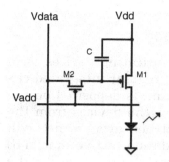

Figure 8.8 Simple active matrix pixel circuit.

Figure 8.8 shows a typical active matrix pixel circuit consisting of a control switch (M2), a current source (M1), and a storage element (C). The size of the capacitor has to be very carefully selected to allow proper operation. It has to be large enough to keep the voltage stored for the entire row time (charge in the capacitor can leak through the control switch or through the capacitor itself). On the other hand, it has to be small enough to allow the TFT gate-source voltage (Vgs) to reach its required value within the addressing time. As we see later on, when programming a pixel brightness level with a reference current instead of a data voltage, a small current (few micro amps) has to charge that capacitance. Therefore, the proper sizing of the storage capacitance value is critical.

Evidently, it is much simpler to address a display passively, as no transistors are required to control passive pixels. However, despite the need of TFTs to drive active matrix pixels, these displays present a series of advantages that far outweigh any drawbacks that the more elaborate display fabrication might present. Such advantages become more apparent in large displays where high power efficiency, lifetime, brightness, and color definition surpass that of passive displays (significant voltage drops can be seen across a large passive array due to the high currents that flow through the display lines). For these reasons the display industry is steering toward active matrix displays, especially when high-definition, large-area displays are targeted.

8.2.3 Active Matrix OLED

8.2.3.1 Amorphous and Polysilicon TFT

When fabricating TFTs, instead of fabricating devices on a single-crystal silicon substrate like in regular integrated circuit (IC) technologies, a silicon thin film is deposited on top of a thick dielectric layer that isolates the devices from the substrate. Silicon islands are then etched away, which will ultimately provide the physical foundation (active material) for the transistors to be formed. Drain and source for the transistors will be defined once the gate dielectric and gate terminal are formed. For this reason, many of the electric characteristics of TFTs depend on the properties of the silicon thin film. Immediately after depositing the silicon, we have two options: leave the amorphous silicon as it is (i.e., hydrogenated amorphous silicon) or crystallize it (usually done through thermal or laser annealing). During crystallization, grains of different sizes are formed resulting in polycrystalline silicon (polysilicon).

To the circuit designer, different silicon technologies offer different device characteristics that can dramatically change the operation of circuits. Some of these properties are summarized in Table 8.1. Amorphous silicon transistors present electrical challenges such as lower mobility (up to three orders of magnitude lower), threshold voltage drift over time, and the lack of p-type devices. In the case of the amorphous devices, threshold voltage variations happen when their duty cycle is close to unity (typically in active matrix display pixels,

Table 8.1 Summary of TFT Properties

	Amorphous Silicon	Polysilicon	Single-Crystal Silicon
Grain size	Amorphous	0.5–5 µm	Single grain
Threshold voltage uniformity	Good	Fair	Excellent
Threshold voltage drift	High	Low	None
Mobility	<1	100–500	>500
Mobility uniformity	Good	Fair	Excellent
Devices	n-MOS	CMOS	CMOS

transistors are turned off only when no light is desired). OLED driving transistors are then operating in saturation mostly; in a DC state that leads to drift of the threshold voltage.[12] However, if a good pixel circuit design that compensates for these disadvantages is used, then the overall fabrication cost of the display could be lowered (since crystallization could be an expensive processing step).

Polysilicon TFTs have several performance advantages when compared to amorphous silicon transistors. These devices can be fabricated as either both PMOS or NMOS, with carrier mobility exceeding 100 cm²/Vs and threshold voltage drifts that are not as pronounced. However, variation in grain size and orientation make it difficult to maintain uniform threshold voltage and mobility values across the array.

Single-crystal silicon devices have the best performance of all TFT devices, but the processing techniques used to obtain single-crystal islands could be very expensive and also limit the size of an array to the size of a silicon wafer. AMOLEDs have been demonstrated with all three different types of silicon devices.

In Figure 8.9 we can see the transfer response of poly-silicon devices. To produce well-performing, high-resolution displays, the on-current vs. off-current ratio for the addressing transistors should be of several orders of magnitude, while

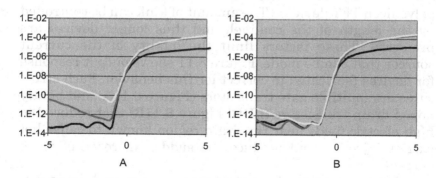

Figure 8.9 Single (A) and double (B) gate NMOS polysilicon TFT transfer characteristics at $V_{ds} = 0.1$, 2.5, and 5. (Courtesy of Display Research Laboratory, Lehigh University, Bethlehem, PA.)

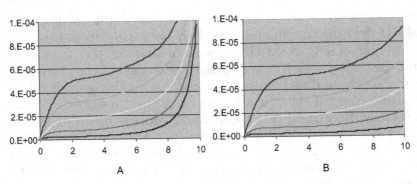

A   B

Figure 8.10 Effect of length on the output response of polysilicon TFT, W/L = 1: (A) L = 8 μm; (B) L = 16 μm. (Courtesy of Display Research Laboratory, Lehigh University, Bethlehem, PA.)

the driving transistors should have high mobility and low threshold voltage. (Notice the difference in leakage current at high Vds between single- and double-gate transistors. As we see later on, this will prove to be very useful in circuit implementations.) Figure 8.10 shows output response plots for long and short devices. We can see that the saturation region of polysilicon TFT devices is not entirely flat (which corresponds to a finite output impedance) and we can also notice a sudden increase of drain current at high drain voltages (in part due to the early Kink effect that occurs in polysilicon TFT devices.) The amount of kink can be controlled by the length of the channel, and thus longer devices are preferred. These factors limit the quality of the current sources that can be made. Figure 8.11 shows output response for devices fabricated to correct for this behavior. Such structures are multiple gate transistors (Figure 8.11A) and lightly doped drain transistors (LDD; Figure 8.11B) that reduce the high electric fields near the drain region known to cause hot carrier injection and produce the sudden increase of drain current.

As we can see, TFTs present many choices and challenges to the circuit designer, and it is for this very reason that many pixel circuits have been proposed for both amorphous and

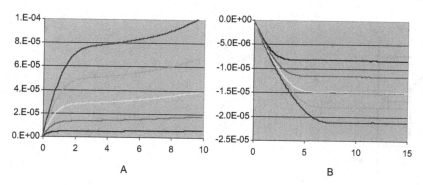

Figure 8.11 Output response of polysilicon TFT: (A) triple gate; (B) LDD PMOS. (Courtesy of Display Research Laboratory, Lehigh University, Bethlehem, PA.)

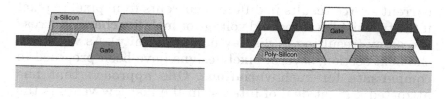

Figure 8.12 Amorphous and polysilicon TFT.

polysilicon devices to compensate for such problems (Figure 8.12).

8.2.3.2 Voltage and Current Programming

Traditionally, LCD displays have been voltage addressed. This means that the brightness level is transmitted to each individual pixel with a voltage data line. This voltage is stored in a capacitor that determines the current value of the drive TFT and this in turn determines the level of emitted light. In the previous section we presented some of the problems that frequently arise when fabricating TFT circuitry. Changes in the LED threshold voltage and in the transistor electrical characteristics result in non-uniform display brightness.

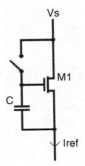

Figure 8.13   Current memory cell.

When using voltage programming (as shown in Figure 8.8), the voltage applied to the gate of the TFT acting as a constant current source results in different currents from pixel to pixel if the transistor's threshold voltage or mobility changes across the array (resulting in display non-uniformity). As we see in the next section, several pixel designs have been proposed to compensate for such variations. One approach that has attracted a great deal of interest in the last few years is the current addressing technique. When a pixel is directly addressed with a current instead of a voltage, the TFT threshold voltage and mobility variations do not affect the LED brightness. This is done by forcing a data current through the OLED driving transistor during programming, and therefore the voltage on its gate adopts the required value to keep the current the same after programming. Figure 8.13 shows a typical circuit implementation for a current memory or current copy cell. This circuit has been used in a variety of analog circuits including high-resolution digital to analog converters, precision current sources, and analog memories.[14] The circuit operates as follows: when the switch is closed and a data current, I_{ref} is forced through the driving transistor M1, a voltage is established on its gate to allow that current to flow entirely through M1 (once the capacitor voltage settles). When the programming current, I_{ref} is removed, and the switch is opened, the capacitor C holds the gate voltage constant, and

M1 now operates as a current source. It is clear that the current provided by M1 after programming (when the switch opens) is equal to the forced current I_{ref} (during programming), and mobility or threshold voltage variations of M1 will not affect this current.

When the control switch is closed, the current memory cell has the same topology as that of the OLED model shown in Figure 8.3. Therefore, the transient characteristics for this circuit are also described by the differential equation (Equation 8.1), and to compute the transient times for the gate voltages we can use Equation 8.6. Note that when programming a pixel with a voltage that is applied directly to the capacitor, the time it takes for the capacitor to reach the correct data value is very small (only a function of the current capabilities of the data voltage source and line capacitances). However, when a pixel is addressed with a reference current of a few micro amps, the charging times could increase significantly and we need to pay close attention to the transient response. Equation 8.4 is the solution to the differential equation. In this function, $V(t)$ is the gate voltage stored in the pixel capacitor. This capacitor will hold the gate of the driving transistor at the appropriate voltage during the entire frame cycle:

$$V(t) = V(\infty)\left[\frac{V(\infty)\tanh\dfrac{t}{\tau} + V(0)}{V(\infty) + V(0)\tanh\dfrac{t}{\tau}}\right],$$

which for $V(0) = 0$ and programming current I_{ref} reduces to

$$V(t) = \sqrt{\frac{2I_{ref}}{\beta}} \cdot \tanh\frac{t}{\tau}. \tag{8.7}$$

This hyperbolic charging function has a time constant τ at which the final voltage is at 76% of its final value, and it reaches 99.5% of its final value at three time constants. For a VGA display, the row time is 25 µs. If we use this value in Equation 8.6 we can estimate the value of the capacitance that the programming current is capable of charging during the row time:

$$C = \tau \sqrt{I_{ref} \frac{C_0 \mu}{2} \frac{W}{L}}. \qquad (8.8)$$

This capacitance has to account for parasitic capacitance in the data line as well as input capacitance for each pixel. Sometimes these values are too large and a pre-charging time with higher data currents is required to charge the line. These long transients are perhaps one of the greatest drawbacks in the implementation of current programmed pixels.

In addition to voltage and current pixel programming, other active addressing schemes have been proposed that control brightness levels digitally instead of using analog voltages.[8,9,12] In these methods, known as digital programming, the brightness levels are determined either by the duration of time in which a pixel is programmed on or off (time ratio gray-scale, or TRG) or by the size of the OLED that is turned on or off (area ratio gray-scale, or ARG). In the former digital addressing method, the driving transistor is turned either fully on or fully off. The different brightness levels for each pixel are determined when a frame is divided into subframes of different duration. The latter schemes drive pixels that are physically divided into subpixels of different area. Brightness levels are determined by the total area of OLED being turned on (or off). These schemes present low display brightness non-uniformity because current levels are not determined by transistor parameters, but instead by OLED film uniformity (current-day OLEDs film thicknesses present good uniformity throughout the entire array).

8.2.3.3　AMOLED Pixel Circuit Design Guidelines

In the previous sections we have presented the different technologies and addressing techniques that a circuit designer has to choose from. Given these alternatives, we can now discuss certain considerations that must be kept in mind when designing OLED pixel circuits. Some of the most important ones are display brightness uniformity, power consumption, aperture ratio, yield, cost, and integration with driving circuitry.

If we look at the saturation current equation for a typical MOSFET we notice that high mobility values are not necessary to reach the required currents needed to achieve the brightness levels suggested in Section 8.2.1. From Equation 8.9, we can properly size the transistor length and width given required drive current capabilities and carrier mobility.

$$Id = \frac{1}{2} \cdot C_0 \cdot \mu \cdot \frac{W}{L} \cdot \left(Vgs - Vth\right)^2 . \tag{8.9}$$

When using amorphous devices with effective mobilities on the order of 1 cm²/Vs, Equation 8.9 shows that appropriate current levels can be achieved with wide enough devices (usually in the order of 100 μm). However, there has been an attempt in the recent years to integrate more and more circuitry in the same display substrate (i.e., scan and data drivers). This requires better performance TFTs that can only be fabricated using polysilicon technology. Devices with good mobility are needed to create buffers (used to provide more current capability), and p-type transistors are needed to create CMOS implementations of registers, Op-Amps, and D/A converters that have far lower power consumption.

On the other hand, both technologies present variations in transistor parameters, and circuits have been proposed to compensate for these variations in both voltage and current mode addressing. However, when implementing these designs in amorphous silicon, many times the number of transistors and/or lines increases rapidly due to the unavailability of p-type devices. This means that the effective area available to place the OLED decreases when we add more devices and control lines. Furthermore, line crossings increase the possibility of short circuits between them, and a larger number of transistors also lowers the yield of the overall display (due to higher numbers of device failure). It is also important to point out that the drift over time of the properties of amorphous devices is much greater than that of polysilicon TFTs. In the next section we show some of the circuits that compensate for device parameter changes and make for more uniform display brightness.

Other types of design considerations concern OLED processing techniques and issues. For example, OLEDs are most frequently used with their cathode connected directly to ground. This is because the cathode is the last step in the fabrication of the OLED. As we saw before, the OLEDs voltage could take a long time to settle, so if we use an n-type device to control the current, its gate voltage would include the voltage across the OLED. This means that any changes in the OLED voltage or long settling times would affect the transistors' Vgs and consequently the current that it provides. For this reason, it is convenient to use a p-type device to create the current source since the transistor's source terminal is connected to the supply voltage instead of the OLED.

Finally, good aperture ratio is perhaps one of the most important goals in the design of any TFT pixel circuit. We define the aperture ratio as the fraction of area available for OLED fabrication with respect to the total pixel area. Any devices such as transistors or capacitors, as well as control lines, occupy a portion of the pixel that in most applications cannot be used to emit light. Displays with good aperture ratio offer good resolution and good brightness levels. Recent advances in top emitting OLEDs that do not need to be implemented in an open pixel area present an alternative solution to the more conventional bottom-emitting OLEDs. By applying planarization layers on top of the pixel circuitry, OLEDs that emit light away from the pixel (instead of through the pixel) can be fabricated utilizing most of the pixel area and thus achieving high aperture ratios.

8.3 AMOLED TFT PIXEL CIRCUIT EXAMPLES

Several pixel circuits have been proposed to drive TFT displays. They vary in performance and complexity, from the simple 2TFT voltage–addressed pixel, to the more complex optical feedback circuits.[8,9] The main motivation behind most designs is to be able to obtain good display performance (high

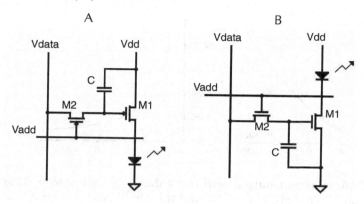

Figure 8.14 2TFT voltage–addressed pixel circuit: (A) p-MOS; (B) n-MOS implementations.

brightness levels and good uniformity) while reducing the number of devices and control lines.

Simulation programs are usually used to study the performance of the different circuits. Figures of absolute normalized current variations are reported over the entire data range as well as variations in output current as a function of threshold voltage and mobility variations.

8.3.1 Voltage Addressed Pixel Circuit with 2TFTs

The most commonly known pixel circuit is the 2TFT circuit.[10,11] In Figure 8.14, both NMOS and PMOS implementations of this circuit are shown. The PMOS architecture is usually preferred due to OLED processing issues and performance, particularly if a polysilicon TFT technology is used. It has been noticed that when connecting the OLED to the power supply (NMOS implementation) the ITO layer has to be deposited on top of the organic material degrading the lifetime of the OLED.[12] Another reason PMOS devices are preferred is because voltage transients across the diode do not affect the current output of a p-type current source (this current is controlled by the TFT gate-source voltage: $V_{data} - V_{dd}$).

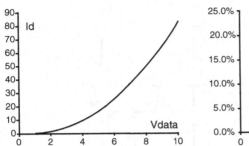
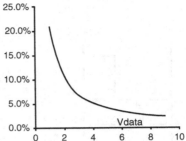

Figure 8.15 Current output and normalized variation over data range for mobility variations of 4% and threshold voltage variations of 9%.

Both circuits operate in the same fashion with M1 behaving as a voltage controlled current source, and M2 as the control-addressing switch. This means that there are three lines going into each pixel (V_{data}, V_{add}, and V_{dd}). This circuit does not compensate for any TFT device parameter variations.

This circuit works as follows: an analog voltage is programmed in the data line during addressing time (when M2 is turned on by the address line). This voltage activates M1 allowing a controlled current to flow through it. When the addressing period is over, M2 is turned off and C stores the data voltage. This allows M1 to keep supplying the correct current during the remaining of the frame time.

The plots in Figure 8.15 show simulations of the output current vs. data voltage as well as variations in current output for a 10% threshold variation and 4% mobility variation. It is easy to observe from the MOSFET current equation that variations in the current output with mobility non-uniformity have the same normalized effect over the entire range of data. That is, at a first level approximation, if mobility varies by 4%, the output current will vary 4% at any data value (since the mobility is a first-order coefficient in the MOSFET current equation). We can obtain the theoretical variation[14] by solving the difference equations for mobility (8.10) and threshold voltage (8.11) at different gate bias. (First-level approximation is done with TFTs working on saturation.)

$$\partial I = \partial \mu \cdot k \cdot \left(Vgs - Vt\right)^2 \qquad (8.10)$$

$$\partial I = \partial Vt \cdot 2 \cdot k \cdot \mu \cdot \left(Vgs - Vt\right). \qquad (8.11)$$

If we normalize for the absolute value of current and notice that $\partial \mu = \mu \cdot \xi_\mu$ and $\partial Vt = \overline{Vt} \cdot \xi_{Vt}$ (where μ, ξ_μ and $\overline{Vt}$, ξ_{Vt} are the average and the percentage mobility and threshold voltage variations, respectively), we obtain

$$\frac{\partial I}{I} = \xi_\mu \qquad (8.12)$$

for mobility changes of ξ_μ percent, and

$$\frac{\partial I}{I} = 2 \cdot \frac{\overline{Vt} \cdot \xi_{Vt}}{Vgs - \overline{Vt}} \qquad (8.13)$$

for threshold voltage changes of ξ_{Vt} percent.

As we can see the output uniformity is a function of the device properties, which in turn depend entirely on the quality of the fabrication process. Other factors such as Kink effects, also contribute to display non-uniformity to a great degree, and this is why long or multiple gate devices are preferred for the driving TFT.

8.3.2 Voltage Addressed Threshold Voltage Compensation Circuit

Another family of pixel circuits, used to compensate for threshold voltage variations only, is the threshold voltage read circuit.[7,9,15] These pixel designs actually measure the threshold voltage of the driving TFT and add it to the data voltage. In this fashion, the current value has a certain degree of immunity to the threshold voltage in each pixel. They operate on the principle illustrated in Figure 8.16. Looking at this plot, we can see how when a MOSFET is diode connected and its gate charge is allowed to discharge until no more current flows ($I_d = 0$), the final voltage on the gate will approximate its threshold voltage.

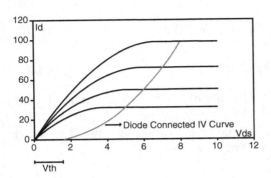

Figure 8.16   Output curves for MOSFET.

The following circuit[8,9] is a good example of how complicated pixel circuits can become when trying to compensate
for device parameter variations. Clearly, the large number of
devices and control lines does not make it a good candidate
for practical applications. However, it shows that compensation can be achieved. This circuit compensates only for threshold voltage variations, and it requires more than simply an
addressing and running mode of operation; it needs a threshold voltage read period.

The circuit works as follows: during threshold voltage
read time, M3 is on and the capacitor C2 is allowed to discharge through M1 (M2 and M4 are off). When the voltage
on the gate of M1 reaches its threshold voltage, the discharge
stops. This value is retained in C2. When in addressing time,
M2 is turned on and the data voltage is allowed to charge up
C1. At this time, the gate of M1 sees the data voltage and the
threshold voltage added together. Every individual pixel will
perform this operation; therefore, current is only a function
of V_{data} all across the array and independent on the threshold
voltage of individual TFTs (Figure 8.17).

Several variations of this circuit have been published with
varying numbers of transistors and control lines.[7,20] It is important to notice that the addition of an extra addressing period
during which the threshold voltage is measured makes for a
more complex driving scheme that could not be implemented

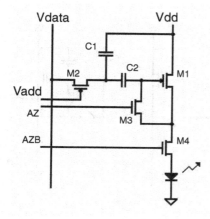

Figure 8.17 Threshold voltage compensation circuit.

with off-the-shelf drivers. Despite the introduction of an extra TFT, extra line, and extra capacitor, the above circuitry compensates only for threshold voltage variations, remaining sensitive to mobility variations. A pixel circuit that compensates for both mobility and threshold voltage variations is discussed in the next section.

8.3.3 Current Copy Pixel Circuit with 4TFTs

The 4TFT current copy circuit[8,9] is perhaps a good compromise between performance and complexity. As mentioned before, current programming a pixel eliminates the dependency on device parameters and this particular circuit achieves a good degree of uniformity without adding any extra lines. However, it relies heavily on the current source characteristics of the TFT. That is, it works better when the saturation region of the *I–V* curve is very flat, showing high output impedance. Long, LDD, or multiple-gate devices are thus recommended.

This circuit works as follows: during addressing time, Figure 8.18 (M2 and M3 on, M4 off), a reference current is sinked through I_{data}. Since M1 is diode connected, a corresponding voltage is established on its gate to allow that cur-

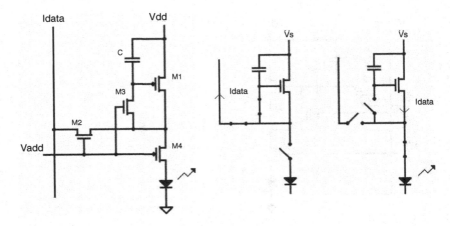

Figure 8.18   Pixel schematic, addressing and operating modes.

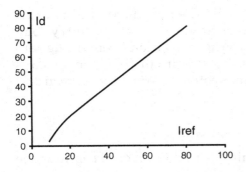

Figure 8.19   Driving vs. programming current.

rent to flow. When M2 and M3 are turned off and M4 is turned on, C holds the programmed voltage on the gate of M1 allowing that current (I_{data}) to now flow through the diode.

Figure 8.19 shows the ideal relationship between programming current and data current (i.e., data current, I_d equal to programming current, I_{ref}). This circuit follows the addressing transients discussed earlier. These transients depend heavily on the mobility of the transistors, so even though similar circuits have been implemented with NMOS amorphous devices,[16–19] it

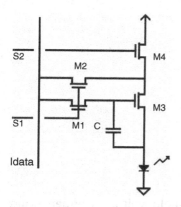

Figure 8.20 Standard current addressed pixel implemented in amorphous silicon.

has the potential to run at faster frame rates with polysilicon devices that present higher mobilities.

It is worth observing that the transient analysis discussed in an earlier section applies to this pixel circuit. When addressing a pixel with a current, the line capacitance tends to slow the addressing time by a considerable amount. It has been shown that adding a "dummy pixel" as a sample and hold block that can be used to charge up the line instead of operating an OLED can solve this problem. (This effectively adds a current memory cell that drives the column over an extended time.)

8.3.4 Amorphous Silicon Current Addressed Pixel

The standard current addressed pixel discussed in Section 8.3.3 has also been implemented in amorphous silicon.[16–20] However, the absence of p-type devices in amorphous silicon technologies makes its implementation somewhat more difficult. For this reason, an extra control line was required to direct the current. This circuit is shown in Figure 8.20.

In this variation of the circuit, transistors M1 and M2 are on during addressing time and M4 is kept off. The result is that M3 is diode connected and the addressing current is sourced into the pixel to program M3. Following addressing,

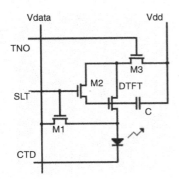

Figure 8.21 Amorphous silicon threshold voltage compensation circuit.

M2 and M1 are turned off and M4 is turned on, allowing M3 to now sink the data current from the power supply.

8.3.5 Amorphous Silicon Threshold Voltage Compensation Circuit

This circuit[21] is a variation similar to the circuit discussed previously, but it is meant to run with amorphous silicon devices (Figure 8.21). Its main purpose is to compensate threshold voltage shift over time of the driving TFT due to gate bias stress. Just like the previous circuit, it requires extra addressing lines as well as a pre-charging or threshold voltage read period.

The circuit works as follows: during pre-charging SLT and CTD are high, preventing current to flow through the OLED. At this point the source of DTFT is at V_{data}. Following pre-charging, the threshold voltage read cycle starts by turning off M3 allowing the gate of DTFT to discharge (notice how DTFT is diode connected) until it settles at $V_{th} + V_{data}$. The capacitor holds this voltage constant during operation at which time TNO returns high while SLT and CTD return low.

8.3.6 Amorphous Silicon Voltage Addressed Circuit

This circuit[22] was specifically aimed at amorphous silicon technology (Figure 8.22). It was designed to compensate for

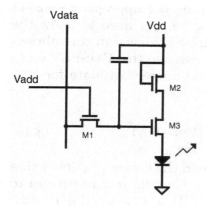

Figure 8.22 Amorphous silicon voltage addressed pixel.

device parameters shifts that are present in amorphous devices. It is a simple three-transistor design that utilizes the normal number of control lines. However, it does not compensate for parameter variations from pixel to pixel.

It works as follows: during addressing time M1 allows the data voltage to be stored in the capacitor. This controls the current through M3. M2 behaves as an active resistance that compensates for current drifts. If the current through M3 changes due to its threshold voltage drift, the voltage of the active resistor will change in a compensating manner. For example, if the current goes down, the voltage across the active resistor would go down as well, allowing a higher current to flow back through the OLED (because the capacitor voltage stays constant).

8.4 PIXEL LAYOUTS AND FABRICATION

8.4.1 Pixel Layouts and Device Sizing

In this step of the display design process all the circuit considerations come together to form an appropriate physical realization of the pixel. Aspects such as capacitor value, transistor electrical characteristics, parasitics, and most importantly aperture ratio are determined at this moment.

384 *Troccoli, Hatalis, and Voutsas*

We might begin by determining the appropriate size of the driving transistor. To do this, we first need to know the typical value of mobility that can be achieved in the fabrication lab. Once this is known, we can use the MOSFET transistor equations to compute a first-degree estimate for W/L ratios.

$$Id = \frac{1}{2} \cdot C_0 \cdot \mu \cdot \frac{W}{L} \cdot \left(Vgs - Vth\right)^2 . \qquad (8.14)$$

The maximum required current and the lowest possible value of mobility should be used to solve for W/L. It is important to determine whether self-aligned, LDD, or gate overlap devices are needed at this point, because different devices could take up different areas. Furthermore, amorphous TFTs have much lower mobility; so larger transistors will be needed to provide the same current.

The size of the switching transistor is not as relevant as that of the driving transistor. However, its off characteristics are very important. A low leakage device is needed to prevent the capacitor to discharge (a multiple gate transistor usually reduces this leakage current). To calculate the size of the capacitor, we must decide how many bits of color gray scale is the display going to provide and what the refresh rate of the display will be. For example, if we need 3 bits of resolution, then that corresponds to 8 gray scale levels. Then, if the data voltage range is 4 V, 0.5 V separate each gray scale level. This means that in one refresh cycle the capacitor voltage cannot change more than 0.5 V. To calculate the capacitance needed to accomplish this we can use the constituent law for capacitors:

$$I = C \frac{\partial V}{\partial t} . \qquad (8.15)$$

We can replace I with the leakage current of our switch, ∂V, by the gray scale level, and ∂t by the frame time. After obtaining the appropriate capacitance value, we can solve for the capacitor physical dimensions using the following equation:

$$C = \frac{K_{OX} \cdot \varepsilon_0}{x_0} \cdot A, \qquad (8.16)$$

where x_0 is the thickness of the oxide layer and A is the capacitor area.

Many more layout design rules should be followed to complete the layout of the pixels. Factors such as spacing between layers, minimum contact hole sizes, minimum line thickness, and many others will ultimately determine yield of the overall display. Finally, there are some general conventions that should be followed. For example, addressing lines are usually horizontal lines, while power and data lines are vertical. Also, the ITO region (which will ultimately determine the light-emitting area) should be placed as close to the x–y center of the pixel as possible to try to avoid line patterns in the overall display.

As examples for pixel layouts, Figure 8.23 shows possible physical implementations of the traditional 2TFT and the 4TFT current copy pixels from Figures 8.14 and 8.16, respectively. The total size of this pixel is 250×250 µm. Note that to maintain similar capacitor and ITO sizes, the 4TFT-circuit layout implementation has a reduced number of gates in the switch (which increases its leakage current) and a reduced number of contact holes (which reduces the yield of the overall display). This is one simple example of the trade-offs the designer has to account for when opting for different pixel circuits.

As the number of transistors increases and size of the pixel is reduced, the effective light-emitting area is reduced as well. For this reason there has been an increased interest in fabricating top emitting OLEDs instead of the more common bottom emitting devices. In this way, most of the pixel area can be used as the light-emitting area, increasing resolution and reducing power consumption.

8.4.2 TFT Fabrication

We now briefly discuss an example of the TFT fabrication process. Glass and quartz have traditionally been the most

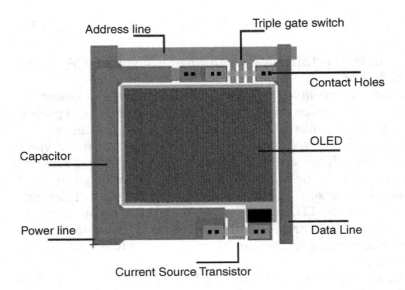

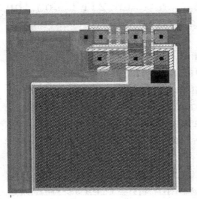

Figure 8.23   2TFT (top) and 4TFT (bottom) pixel layouts.

widely used substrates. However, to exploit the full potential of TFT OLED displays it has been of great interest to use flexible substrates such as plastics[22,23] and flexible metal

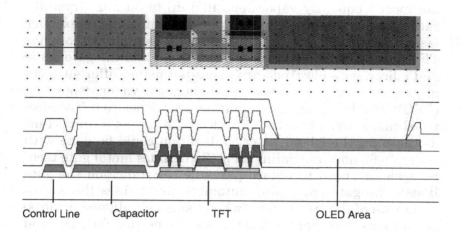

Control Line   Capacitor   TFT   OLED Area

Figure 8.24 Cross section (vertical dimensions are not to scale).

foils.[14,24,25] These substrates can be used to fabricate more rugged displays or used for specific purposes such as wearable or conformable electronics or lightweight displays. Furthermore, flexible substrates offer the alternative of roll-to-roll processing, which could ultimately reduce the cost of production by a considerable amount. Displays made of polymer or plastic substrates can be transparent and therefore compatible with already used light-emitting techniques. On the other hand, metal foils substrates could yield better-performing more uniform devices as they are compatible with high-temperature CMOS processing techniques that yield better-quality films.

Figure 8.24 shows the cross sections for typical TFT structures. Usually when fabricating TFT displays,[10,11,14] we begin by growing a thick passivation/isolation layer on top of the substrate. The thickness of this passivation layer can determine many important aspects of the circuit performance such as parasitic capacitance with the substrate. A thicker oxide reduces this capacitance and also prevents contaminants from the substrate to diffuse to the TFT devices.

Silicon island preparation comes next. The silicon films are most frequently vapor-deposited up to 50 nm. Crystallization can be achieved through thermal annealing or excimer laser annealing (ELA). At this point, the sample is ready for TFT formation. The islands are patterned and the gate oxide and silicon gate electrode are deposited using different techniques such as plasma-enhanced chemical vapor deposition (PECVD) or low-pressure chemical vapor deposition. The electrical characteristics of the gate electrode are not as important as those for the island, so crystallization could be done with a simple furnace-annealing step. (Sometimes metal gate technologies are used to reduce parasitic resistance on the gate lines). The gate patterning automatically defines the source and drain of the transistors (when using self-aligned devices) or an extra lithographic mask would be needed for LDD and gate overlap devices. Immediately following the gate patterning, the drain and source are doped at the appropriate concentration levels and the sample is heated to the appropriate activation temperature. Typically, we would finish the transistor formation at this point, but recently there has been an increased interest in improving device uniformity and reducing series resistance. It has been shown that silicidation of the drain and source achieves both these goals. The silicides can be formed by sputtering nickel and then followed by vacuum annealing. A thick passivation oxide is deposited to cover the TFT, which protects the transistor from forming any shorts with control lines. Contact holes are opened and metallization follows. Data lines and power lines are usually made with aluminum, but more than one metal such as nickel and titanium are added to prevent the formation of defects such as hillocks. After another thick passivation layer (many times a double layer of PECVD oxide is used) a spin on glass (SOG) is performed to achieve a planarized surface for the deposition of the ITO. Contact holes are opened at this point and the ITO is sputtered and patterned. A second SOG is spun on the wafer and cured. Finally, a large contact area is opened, which provides the region for the formation of the OLED. This is the actual area that will emit light.

8.5 SUMMARY

In the previous sections, we have discussed several different aspects that engineers account for when designing OLED displays. Factors such as OLED electrical characteristics, addressing scheme, and fabrication technologies help determine the appropriate choice of driving circuits and devices that would achieve the desired level of performance.

At the present time, active matrix LCD displays dominate the flat panel display market, due to advances in TFT technology and manufacturing processes. Positive results in OLED science and technology as well as higher-performance thin film devices such as polysilicon TFTs increase the range of possibilities of the flat panel display industry. Large-area, high-resolution, slim displays are now the focus of attention of major consumer product companies and active matrix OLED displays have the potential to claim a portion of the flat panel display market currently being dominated by the LCD industry.

REFERENCES

1. C. Tang and S. VanSlyke, *Appl. Phys. Lett.*, 51, 913, 1987.

2. E. Forsythe, Operation of organic light emitting devices, *SID Dig.*, M5, 2002.

3. V. Christou et al., *22nd International Display Research Conference Proceedings*, 5-1, SID-France, 2002, 89.

4. M. Jac Junsky, *Trans. Electron Devices*, 46(6), 1146, 1999.

5. R. Hattory, *IEICE Trans. Electron.*, E83-C, 779, 2000.

6. R. Hattory, AM-LCD'01, 2001, 223.

7. R. Dawson, *IEDM 1998*, 32, 875, 1998.

8. I. Hunter et al., OLED addressing, presented at 40th SID Meeting, Vol. 1, Sem. M-6, 2002.

9. D. Fish, *SID Dig.*, 33(2), 968, 2002.

10. M. Hatalis et al., Cockpit displays IV, *SPIE Proc.*, 3057, 277, 1997.

11. M. Hatalis et al., Flat panel display technology, *SPIE Proc.*, 3636, 22, 1999.

12. D. Pribat, *Thin Solid Films*, 383, 25, 2001.

13. R. Baker, CMOS Circuit Design and Simulation, IEEE.

14. M. Troccoli, Mat. Res. Soc. Symp. Proc. Vol. 769, H3.7.1, Materials Research Society, 2003.

15. J. Goh, *Electron Device Lett.*, 23(9), 544, 2002.

16. J. Kanicki, *Proc. of Asia Display*, 2001, 315.

17. Y. He, *Electron Device Lett.*, 21(12), 590, 2000.

18. Y. Hong, *Proc. of Asia Display*, 2001, 1443.

19. Y. Hong, International Symposium Digest of Technical Papers, 4.5, SID, 2003.

20. J. Goh, *SID Dig.*, P.72(1), 494, 2003.

21. J. Kim, *SID Dig.*, 33(1) 614, 2002.

22. S. Wagner et al., *Symposium on Amorphous and Micro-Crystalline Silicon Technology*, MRS Proceedings, Vol. 467, 1997, 843.

23. D. Stryahilev, *J. Vacuum Sci. Technol. A*, 20, 1087, 2002.

24. T. Afentakis, Polysilicon TFT AMOLED on thin flexible metal substrate, *SPIE Proc.*, 2003.

25. P. Kazlas et al., *22nd International Display Research Conference Proceedings*, 14-3, SID-France, 2002, 259.

9

Past, Present, and Future Directions of Organic Electroluminescent Displays

TAKEO WAKIMOTO

CONTENTS

9.1 INTRODUCTION

The cathode ray tube (CRT) has played an important role in the electronic industry ever since its birth. For more than 100 years, especially with their use as television screens, CRTs have dominated the field of displays. However, there is a strong need to develop a new kind of display technology that can help provide an innovative human–machine interface in the multimedia culture. Recently, information display devices have attracted much attention and have led to an emphasis on lighter, smaller, and, in some cases, portable devices. Conventional CRTs have many disadvantages, such as their heavy weight, large volume, and high power consumption, that preclude their use in portable applications. Many new technologies have been recently developed for flat panel displays (FPDs), including liquid crystal displays (LCDs), plasma display panels (PDPs), and electroluminescent displays (ELDs) based on light-emitting diodes (LEDs).

Application-oriented display types have also been developed to address the various needs of multimedia applications. These include head-mounted and three-dimensional (3-D) displays in search of realism and portable displays such as personal digital assistants (PDAs), mobile telephones, and projection displays for large screens. The requirements for displays suitable for multimedia applications are numerous,[1] including high display quality, both for liveliness or more realism. The market demands displays that are available any time, anywhere, and to everybody. There is therefore a need for displays that meet the following requirements:[1] (1) high-resolution pixels, (2) large screen size, (3) 3-D features, and (4) portability.

In 1987, Tang and VanSlyke[2] reported electroluminescence (EL) from organic light-emitting diodes (OLEDs) based on multilayer organic thin films with high efficiency and brightness. Since this demonstration, many studies have been conducted on OLED devices based on thin films of small-molecules. OLEDs are generally characterized by the material used in the emissive layer: either small molecules such as aluminum (III) tris-(quinolin-8-olato) (Alq_3) or polymers such as poly(*para*-phenylenevinylene) (PPV). OLED devices offer many advantages for flat panel display applications:

1. Very thin solid-state device (less than 300 nm thick)
2. Light weight
3. High luminous power efficiency
4. Fast response time that makes animations and motion crisp and entertaining
5. Wide viewing angle without brightness or image loss (170+ degrees)
6. Self-emitting, which eliminates the need for a backlight illumination source
7. Color tuning throughout the entire visible spectrum for full-color displays

An unlimited variety of organic fluorescent and phosphorescent dyes and semiconductors can also be synthesized, in comparison with the more limited range of useful inorganic compounds.

OLEDs that have these merits in addition to their temporal stability are expected to have many applications. Devices that are very thin and lightweight and have low power consumption are especially suitable for portable equipment (e.g., wireless phones, PDAs, view finders for digital cameras) and other portable imaging devices. Monochrome, multiple-color, and full-color display technologies using OLEDs are presented in this chapter, as are future directions relating to applications based on the needs in the multimedia era.

9.2 PRESENT STATE OF OLED PRODUCTS

9.2.1 Monochrome Dot-Matrix Display

Pioneer Corporation (herein after referred to as "we") demonstrated a dot-matrix display at the Japan Electronics Show in 1995.[3] Then we manufactured this type of display as a character broadcasting display by frequency modulation for the car stereo in 1997. These panels were passive matrix types whose anode columns and cathode rows are intersected. Passive matrix displays have no switching elements controlling individual pixels. Pixels are turned on and off by biasing the row and column electrodes. The luminance of the OLEDs is proportional to the injected current in the device. The current flow through a selected row is necessarily pulsed to a level that is proportional to the total number of rows in the display. For example, for a display with 64 addressable rows and an average luminance of 100 cd/m^2, the operating voltage and current density corresponding to a maximum luminance of 6400 cd/m^2 high compared with DC devices, leading to high power consumption and limited lifetime. However, the target for this type of passive dot-matrix display is to realize low-cost displays because of its simple structure. Figure 9.1 shows a photograph of the world's first ever OLED display.

Figure 9.1   The first OLED display, which appeared in 1997.

9.2.1.1   Cathode Micropatterning Method[4]

First, we introduce the cathode micropatterning method, which is the key to fabricating passive matrix displays. We have to pattern the top electrodes on the organic semiconductor to fabricate the passive matrix display. One method to pattern the cathode is to use the integrated shadow mask to deposit metals on the organic layer. However, it is difficult to obtain high-resolution patterns using this method. Another method is to use a conventional chemical etching process with organic solutions. We have developed a new, solvent-free method for micropatterning metal OLED cathodes, which is illustrated in Figure 9.2. This process consists of three main steps. First, the cathode separator is formed on a glass substrate that has prepatterned indium tin oxide (ITO) anodes. The ITO anodes are formed in stripes (width 340 µm, gap 30 µm) made by photo-etching the ITO. Then, a hole-injection layer, a hole-transport layer, an emitting layer, an electron-transport layer, and a metal cathode film are deposited on the glass substrate. Organic layers are deposited obliquely to the substrate surface so that the deposited material can surround the root of the separator. Finally, the metal cathode is deposited perpendicularly to the substrate surface with a smaller angle than the taper angle of the cathode separators. Through this process, the cathode separators electrically insulate the adjacent metal cathodes. Figure 9.3 shows scanning electron

1. Forming cathode separators

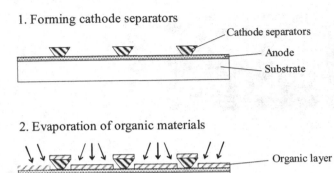

2. Evaporation of organic materials

3. Evaporation of cathode metal

Figure 9.2   Schematic illustrations of the cathode micropatterning process.

| 120 sec | 150 sec | 180 sec |

Figure 9.3   SEM photographs of the cathode separator using different developing times.

microscope (SEM) photographs of the cathode separator for different developing times. The cathodes are made in 300-μm-wide stripes, separated by 30-μm gaps.

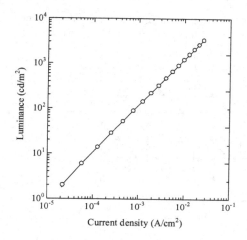

Figure 9.4 The luminance-current density characteristics of an OLED cell.

Table 9.1 Specifications of the Dot-Matrix OLED Display

Item	Specifications
Luminance (cd/m^2)	100
Driving duty ratio	1/64
Frame rate (Hz)	150
Contrast ratio	>100:1 (at 500 lux)
Power consumption (W)	0.5 (panel only)

9.2.1.2 The Cell Characteristics and Specifications of the Dot-Matrix Display[5]

The luminance-current density characteristics of the cell that constitutes the product display are shown in Figure 9.4. The luminance and luminous power efficiency at a current density of 1 mA/cm² are quite high, and are 160 cd/m² and 12 lm/W, respectively. A developed dot-matrix display consists of monochrome green 256 × 64 dots. The display system is made of a panel, a driving circuit, and an interconnecting board. Representative specifications of this display are given in Table 9.1.

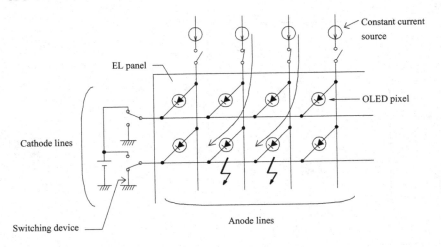

Figure 9.5 A conceptual schematic diagram of the driving circuit for the OLED display.

9.2.1.3 Driving Circuit for the Dot-Matrix Display[5]

An anisotropic conductive film connects electrodes distributed at the outer edges of the panel to a flexible printed circuit (FPC). The FPC is connected to the driver circuit board via a connector. A conceptual schematic of the driving circuit for the display is shown in Figure 9.5. Each pixel is represented by a diode whose current density-drive voltage characteristics are shown in Figure 9.6. An independent, constant current source is allocated to every anode line. A switching device is placed for every cathode line, and is used for the selection of a certain potential. Each cathode is switched every 100 ms sequentially, and one picture is formed every 6.7 ms. The cathode line of the pixel to be selected and the anode line are connected to the ground line and the constant current source, respectively. EL takes place at a specific luminance according to a preset current value. A dedicated IC has been developed as a current source using 8-pin configuration. The gray scale of this display can be varied in three levels (0, ½, 1) via pulse width modulation.

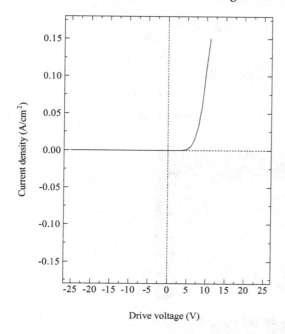

Figure 9.6 The current density-drive voltage characteristics of an OLED cell, which constitutes the pixel for a monochrome display.

9.2.1.4 Stability of the Dot-Matrix Display[5]

9.2.1.4.1 Storage Lifetime of OLEDs

Because OLEDs are very vulnerable to moisture and oxygen, they need to be sealed. After fabrication of the OLED cells, glue is applied to the surrounding of the display area of the panel and the OLED active area is sealed by the metal can. During this process, N_2 gas is trapped in the gap between the panel substrate and the backing capsule. Barium oxide (BaO) is also added as a desiccant. This combination of insert gas and desiccant effectively prevents the growth of dark spots (non-emitting areas).[6] Figure 9.7 shows a schematic diagram of an encapsulated OLED and photographs of the emitting area in the presence or absence of the BaO after 500 h of storage at 60°C and 95% relative humidity (RH). As can be

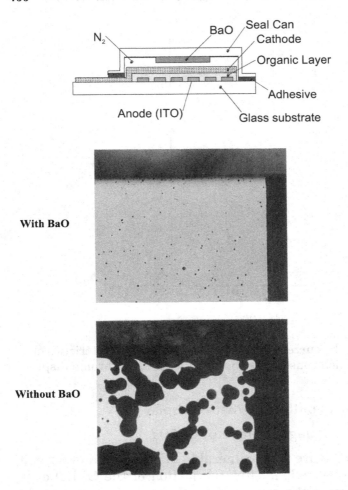

Figure 9.7 A schematic diagram of an encapsulated OLED and photographs of the emitting areas in the presence and absence of the BaO desiccant after 500 h of storage at 60°C and 95% relative humidity.

seen from this figure, the growth of dark spots was minimal in the presence of the BaO desiccant. The performance of the panel was also evaluated at 85°C for a period of 500 h. Figure 9.8 shows the luminance-current density characteristics of a cell, which was constructed using the same materials as the

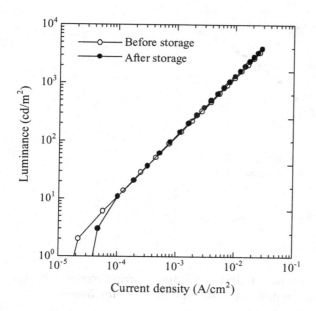

Figure 9.8 The luminance-current density characteristics of an OLED cell, which was constructed using the same materials as the dot-matrix panel before and after storage at 85°C for 500 h.

dot-matrix panel before and after storage for 500 h. Figure 9.9 shows the luminance-drive voltage characteristics of the same cell before and after storage for 500 h. Figure 9.10 shows photographs of the dark spots observed in the OLED cell before and after the storage test.

9.2.1.4.2   Driving Lifetime of OLEDs[4]

OLED device lifetime was once a big problem, because it was believed that the organic molecules degrade during the injection of high current. The half-life ($\tau_{1/2}$) of an OLED is defined to be the amount of time required for the luminance to drop to half of its initial value while being driven under constant current. The lifetime of the OLEDs based on our product has been extended to $\tau_{1/2} > 10{,}000$ h at an initial luminance of 300 cd/m² at DC constant current. A half-life $\tau_{1/2} \sim 4000$ h is

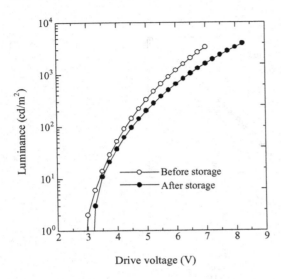

Figure 9.9 The luminance-drive voltage characteristics of the same OLED cell before and after storage at 85°C for 500 h.

reasonable for an automotive application. This would correspond to about 100,000 km mileage, a good indicator of the average lifetime of a car used in Japan. The lifetime performance of the OLED cell is shown in Figure 9.11 at room temperature and 60°C. The lifetime of the cell, which was constructed using the same materials as the dot-matrix panel, has been extended to more than 7000 h of half luminance from an initial luminance of 200 cd/m². The driving condition was 1/64 duty pulse of AC constant current.

9.2.1.5 High Contrast Display

It is difficult to obtain high contrast with OLEDs, because of light reflection from the metal cathode, which functions as a good mirror. Therefore, an actual display panel uses a polarizing plate, which enhances the contrast ratio but halves the luminance. Figure 9.12 indicates the fundamental principle to obtain high contrast using a polarizing plate.

A

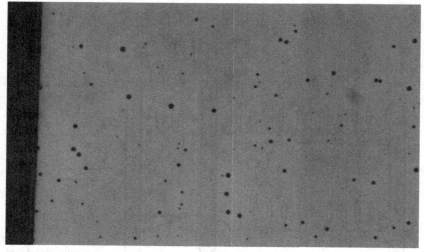

B

Figure 9.10 Photographs of dark spots observed in the OLED cell before (A) and after (B) storage for 500 h at 85°C.

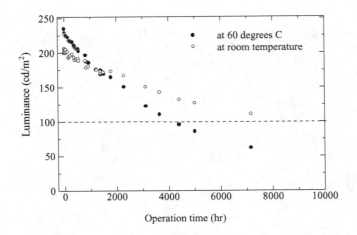

Figure 9.11   The lifetime performance of an OLED cell at room temperature and 60°C (AC drive, 1/64 duty, constant current).

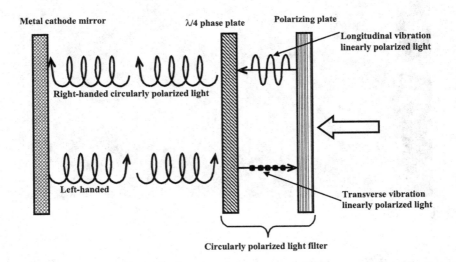

Figure 9.12   The fundamental principle to obtain high contrast using a polarizing plate.

Figure 9.13 Photograph of the multiple-color display, which has yellow and orange icons in addition to the blue and green passive matrix displays.

9.2.2 Multiple-Color Displays

In 1998, we started manufacturing multiple-color displays for car stereos. Figure 9.13 shows a photograph of the multiple-color display, which has yellow and orange icons implemented in a base blue and green passive matrix display. OLED provides full-motion, 3-D gradient images to give consumers better display interaction. In addition, this multicolor display provides a very wide viewing angle, without loss of visibility under sunlight or a bright environment. A driver and passengers can put graphical and animated information on the screen, making it much easier to read and interpret quickly their input.

The benefits of multicolor OLED technology are many:

1. Fast refresh rate for vivid 3-D images
2. Wide viewing angle with high visibility under any brightness condition
3. A self-emitting device, OLED eliminates the need for back-lighting
4. Ultrathin panel suitable for use in a limited, narrow space

TDK Corporation provided a white dot-matrix display and icons with color filters for car audio use in 2000. In the same year, we provided the multiple-color display for a wireless phone. This was the beginning of applications for portable OLED displays.

9.3 FULL-COLOR DISPLAY

The next product target was a full-color display. Efforts toward producing a full-color OLED display have continuously made steady progress. Prototypes have been demonstrated or reported since 1995 by a number of research organizations taking different fabrication approaches. For example, one approach adopted by TDK Corp. in 1995 uses white OLEDs with color filter arrays such as LCD panels. A second method uses blue OLED cells for light source with fluorescent color arrays named color-changing medium (CCM) to change blue into red or green fluorescent colors.[7] A third approach employs patterned lateral red, green, and blue (RGB) emitters.[8] Other methods such as stacked RGB emitters,[9] patterns by photo-bleaching adopted by Kido et al.,[11] and ink-jet printing[10] have been applied. All these approaches offer advantages and disadvantages. Figure 9.14 shows several methods that can realize full-color displays. In this section, some of the methods to fabricate full-color OLEDs are discussed.

9.3.1 White OLED Backlight with Color Filter

One of the first approaches was to use a white OLED cell as a backlight. Color filters on the glass substrate, such as those used for LCDs, could be employed to filter the RGB subpixels from the white backlight. Several techniques to make white OLEDs have been proposed. Kido et al.[11] used RGB dyes dispersed in the polymer poly(N-vinylcarbazole) (PVK). Sato[12] developed devices with multiple layers of blue-green- and orange-emitting materials. The filters can be patterned on a separate plate using photolithography, and the white OLED is directly fabricated on the substrate.

Use of color filters on the glass substrate is conventionally practiced in the LCD industry. There is no need to pattern each one of the pixels other than the OLED white backlight. However, one disadvantage of this type of patterned RGB OLED display is a poor luminous power efficiency due to absorption of the light by the color filters.

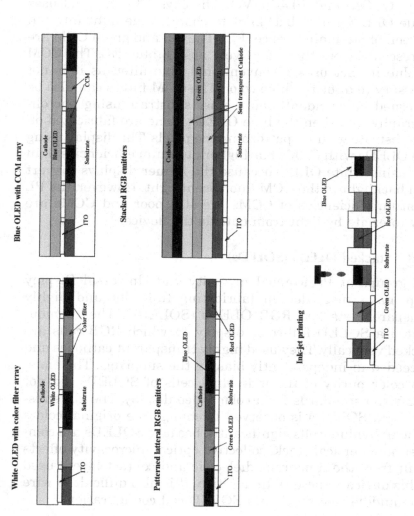

Figure 9.14 (Color figure follows p. 274.) Methods used to pattern OLEDs for RGB emission and full-color displays.

9.3.2 Blue OLED Backlight with the CCM

A group at Idemitsu Kosan has studied the CCM approach. The group has demonstrated color 5.2-in. quarter video graphics array (QVGA) and 10.4-in. VGA displays. This method uses a blue OLED cell as backlight to change blue light into red or green photoluminescence (PL). The red and green CCM are fluorescent layers that effectively absorb blue EL. The CCM for blue in fact uses a conventional blue filter, as it is not necessary to alter the blue color. The CCM filters can also be patterned independently on a glass substrate using photolithography, and then the blue OLEDs alone are fabricated on the substrate without patterning the pixels. The display using blue OLEDs with CCMs has higher quantum efficiencies than that using white OLED because the former displays convert from backlight to the CCM fluorescent light. However, the PL quantum efficiencies of CCMs are still poor and CCMs are easy to excite by light from outside the device.

9.3.3 Stacked OLEDs (SOLEDs)

The groups at Princeton University and Universal Display Corp. have succeeded in fabricating full-color and highly transparent stacked RGB OLEDs (SOLEDs). They demonstrated a SOLED full-color display, in which RGB cells are stacked vertically. They used highly transparent cathodes and succeeded in independently biasing the subpixels. The emission color purity of the individual cells of SOLEDs did not satisfy the standards for a color video display. The EL emission from SOLEDs is observed to change the original colors of the individual cells significantly, because SOLEDs are combined in a vertical stack; collective optical microcavity effects result from the numerous dielectric and contact layers used in this device structure. In addition, it is also difficult to wire some middle electrodes in a SOLED cell configuration.

9.3.4 Ink-Jet Printing Method

Light-emitting polymers are easy to coat on a large-area substrate as they can be processed from liquid solution. However, it is difficult to integrate RGB-emitting polymers on the pixels separately with high resolution using conventional coating techniques. To resolve this problem, ink-jet printing technology has been used to pattern three different RGB colors to make pixels. The ink-jet printing method, once viable, is expected to be very successful due to its presumably low cost of manufacturing displays. Ink-jet technology is quite established and has been successfully used for color printers. It has the potential for direct patterning of the emitting layers and large-area processing with high resolution and lower cost. However, it is difficult to control the film thickness distribution in the pixel. This means that the emission in the cells is not uniform, and hence EL colors may change. Seiko-Epson and CDT groups developed a 2-in. full-color display using thin film transistors (TFTs) in 2000.[13]

9.3.5 Patterned Lateral RGB Emitters[7]

We prototyped a 5.2-in. QVGA passive-matrix OLED display using the cathode micropatterning method developed earlier in 1998. This was coupled with an RGB emitters' selective deposition method using an automatic high-accuracy shadow mask moving mechanism in vacuum.

9.3.5.1 Substrate Structure of the Display

The basic structure is identical to the commercialized monochrome dot-matrix display. However, due to the smaller pixel size and the higher duty ratio needed to address 240 lines, we appended the auxiliary metal anodes to reduce the electric resistance of the ITO anodes. The auxiliary metal anodes are placed on top of the ITO anodes and are 10 μm wide. They result in a reduction of the anode resistance to 1/60 of its previous value.

9.3.5.2 Selective Deposition of RGB Organic Materials

When RGB emitters are being patterned by using a precision shadow mask and a high-accuracy moving mechanism mask, it is important to maintain as small as possible a gap between the substrate and the shadow mask. This is particularly important when a guest-host system is used[14] as the emitting layer because the two organic materials are co-evaporated from two different crucibles. For example, let the distance between a host material crucible and a dopant material counterpart be 10 cm, and the distance from the crucibles to the substrate be 1 m. In this case, the pattern of the dopant material will shift by about 5 µm from the pattern of the host material, even if the thickness of the mask were negligible and that the gap between the substrate and the shadow mask was only 50 µm. Furthermore, it is difficult to keep a uniform gap with a thin shadow mask, as it tends to bend easily. In addition, materials damage may occur from touching the shadow mask because RGB emitters are sequentially deposited.

This problem was solved by maintaining a gap of 5 µm between the substrate and the shadow mask, using cathode separators as stoppers for the shadow mask. A schematic illustration of the emitters' selective deposition process is shown in Figure 9.15. Selective deposition of RGB materials using cathode separators was successful, not only as their original intended function, but also as protectors against shadow mask settings. In addition, the process minimizes

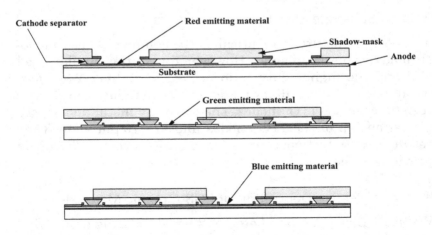

Figure 9.15 Schematic diagram depicting the emitter selective deposition process.

exposure to ambient atmosphere, which may result in material or device degradation. Hence, an automatic high-accuracy shadow mask moving mechanism was developed inside a vacuum chamber to deposit RGB organic emitting materials. For the cathode metal deposition, another shadow mask was used, which was also automatically changed in the vacuum chamber.

9.3.5.3 Driving Method of a Full-Color OLED Display

The same driving scheme used for monochrome dot-matrix OLED displays was adopted here (successive line scanning and constant current driving). As OLEDs have very fast response time, very high momentary luminance is required. The duty ratio was reduced by dividing the anodes at the center. Consequently, this display was driven in both directions (up and down) at the same time sequence by the duty ratio of 1/120, even though there are 240 cathodes. Cathodes were driven simultaneously from both sides, because of relatively large current flows in the cathodes, which cause a large voltage drop. The OLED cells were driven for a length of time whose duration was in rigorous proportion to the signal magnitude. This technique is known as pulse-width modulation (PWM). This system has 64 (6 bits) gray scale for each one of the RGB signals. The display therefore can represent 262,144 colors. Figure 9.16 shows a luminance curve as a function of the PWM pulse width. The display exhibits good linearity to the pulse width.

9.3.5.4 Optimization of the OLED Cell Structure[15]

In Section 9.3.5.3., we described how we developed a full-color OLED display by adopting the selective deposition method to pattern RGB emitters. Individual RGB cells are patterned laterally to form RGB emitters. Therefore, we can fabricate these cells keeping the optimal conditions for the cell structure to obtain pure RGB colors, the highest luminance efficiency, and the longest lifetime. The thickness of the organic layers and ITO within each individual, laterally patterned RGB cell can also be optimized. The thickness of the organic layers and that of the ITO are very important parameters to

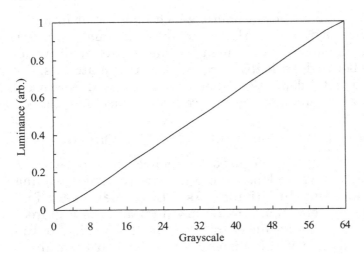

Figure 9.16 Luminance curve as a function of pulse width.

control the J-V-L characteristics of the cells for a full-color
display. When varying the thickness of the layers in the OLED
cell, we have to consider mainly two characteristics of the EL
cell. One is the electronic characteristic of the cells such as
the driving voltage; this depends on the resistance of the
organic layers and the ITO. The other is the optical characteristic of the cells as well as interference effects in the cells.

It is known that the EL spectrum outside a glass substrate is changed by the influence of the light reflected from
the cathode mirror.[2,16,17] There are many reports of very narrow EL spectra achieved by inserting a high-reflective-coeffi-
cient mirror between the ITO and the glass substrate that
works as a microcavity structure.[18,19] The EL spectra of the
RGB cells depend on the PL of the emitter layer and the
optical interference filter characteristics of the cells.[20,21] The
color of a pixel is characterized by its CIE coordinates, using
a universally accepted system developed by the Commission
Internationale de l'Eclairage of France. We generally notice
only the reflection by the cathode mirror because there are
no other obvious mirrors except the cathode in the cells. However, there is a large change in the refractive index at the

Table 9.2 Refractive Indices of Layers in an OLED

Layer	at 460 nm	at 520 nm	at 625 nm
Organic	1.83*	1.76*	1.72*
ITO	2.00	1.94	1.83
Glass	1.53	1.52	1.52

* Central value.

boundary between the glass and the ITO, especially in the blue spectral region (see Table 9.2). Light reflected by the cathode mirror at this boundary affects both the EL spectrum and the external EL quantum efficiency.

Figure 9.17A shows the cell structure seen from the electronic point of view, while Figure 9.17B shows the same structure seen from the optical point of view. If we focus on optical distances between each of the boundaries, the following three boundaries are dominate optical interference phenomena (see Figure 9.17B):

1. The boundary between the hole-transport layer (HTL) and the emitting layer (EML)
2. The boundary between the organic layer and the cathode mirror
3. The boundary between the glass substrate and the ITO anode.

The refractive indices of each layer of the RGB cells, at PL peak wavelengths, are shown in Table 9.2. In the red-emitting spectral region, the film thickness of the composition layers affects only slightly the optical characteristics of the cells, because the differences of refractive indices of these layers at boundaries are small. However, large refractive index steps emerge in green and the blue color wavelength region at the glass/ITO boundary. We took advantage of reflection not only at the cathode mirror but also at the glass/ITO boundary to adjust the emission color and cell EL efficiencies.

The actual emission zone of the cell depends on the cell structure and the organic materials used. This emission zone occurs within in a few hundred angstroms of the HTL/EML

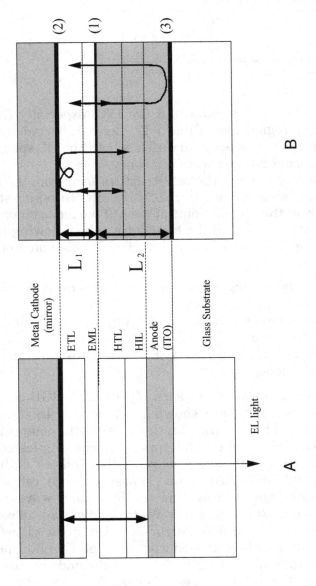

Figure 9.17 A cell structure is seen from (A) the electronic and (B) the optical perspectives.

boundary. For simplicity, we will assume that the emission site is located only at the EML/HTL boundary, and that the emitted light is reflected only once at the cathode mirror and the glass/ITO boundary. The entire reflection and interference phenomena need to be taken into account due to the multiple optical pathways of reflected lights at all boundaries in the cells. However, we regard the multiple optical pathways as single reflections, because the reflection coefficient of the boundaries is less than 20% except that of the cathode mirror.

We calculate the conditions for an optical distance L_1, which is taken from the cathode mirror to the boundary (1), and a distance L_2, which is taken from the boundary (1) to the boundary (3) to enhance the interference effects. Generally, an optical length L is given by

$$L = \sum_i n_i d_i,$$

where n_i is the refractive index and d_i is the thickness of layer i. The conditions to enhance the interference of the wavelength λ in the normal direction to the glass substrate are given by

$$L_1 = \tfrac{1}{4}(2m+1)\lambda$$

$$L_2 = \tfrac{1}{2}(m+1)\lambda, \quad m = 0,1,\ldots.$$

Figure 9.18 shows the EL spectra of the green light-emitting cells for different HTL thicknesses. In this chapter, the term "EL spectrum" is defined as that measured from the normal direction to the glass. This figure tells us that the peak wavelengths of the EL spectra gradually change. Therefore, we optimized each of the RGB cells paying special attention to the multiple reflections at the multilayers' boundaries and succeeded in improving the color purity, enhancing the

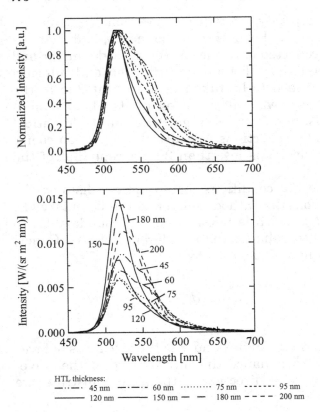

Figure 9.18 EL spectra of the green light-emitting cells varying only the HTL thickness.

external quantum efficiency, and reducing the display power consumption. The viewing angle dependence of the RGB color cells was small, as shown in Figure 9.19. The figure indicates that the CIE coordinates of the RGB cells were measured from perpendicular to 60° or 80° to the glass substrate. The panel, which optimized the thickness of each layer by using the interference effects, did not result in unnatural color change from any angular direction of view. Good color purity for all emission, virtually identical to the NTSC standard CIE coordinates, was obtained. Table 9.3 shows the EL characteristics of the optimized RGB cells for the full-color display.

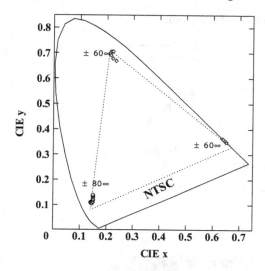

Figure 9.19 Viewing angle dependence of the RGB cells in the CIE coordinates.

Table 9.3 EL Spectral Characteristics of the Optimized RGB Cells for the Full-Color Display

EL Color			Blue	Green	Red
Peak wavelength (nm)			460	520	625
External QE (%)*			4.7	4.1	2.7
Luminance efficiency*			3.9	16.1	2.6
This display	CIE	x	0.14	0.22	0.65
		y	0.11	0.70	0.35
NTSC standard		x	0.14	0.21	0.67
		y	0.08	0.71	0.33

* = @ 300 cd/m².

9.3.5.5   Specifications for a Full-Color OLED Display

Full-color OLED displays were developed whose specifications are listed in Table 9.4. Figure 9.20 and Figure 9.21 show a photograph of a video rate image and a micrograph picture of the pixel arrangements, respectively. Figure 9.21 shows precise emission from each color by virtue of our RGB emitters' selective deposition method. We can see two pixels at

Table 9.4   Specifications of the Full-Color
OLED Display

Item	Specifications
Display size (in.)	5.2
Pixel number	320 (×3) × 240
Aperture ratio (%)	63
Gray scale	64 each
Luminance (cd/m²)	150
Driving duty ratio	1/120
Power consumption (W) @ 150 cd/m² (in white peak)	6

Figure 9.20   A photograph of a video rate.

each color, but actually they form one pixel, because the auxiliary metal anode crosses the center of one pixel. This display
has good color purity for all emissions and compares well to
those obtained by the NTSC standards. Using this display,
we can obtain high-resolution video rate pictures similar in
quality to that of a CRT with fast response time, good CIE
coordinates, and a wide viewing angle.[22]

Figure 9.21 (Color figure follows p. 274.) A micrograph picture of the pixel arrangements.

9.4 FUTURE APPLICATIONS AND FUTURE DIRECTIONS OF OLEDS

9.4.1 OLEDs for High-Picture-Quality Display

Super-high-resolution and large-screen-size OLEDs are needed to achieve a high-quality display. The goal for the resolution of an electronic display is to match the resolution of the human eye (80 µm) at a range of 30 cm. This translates into roughly 10 million pixels, although it depends on the screen size of the display and the viewing distance.[1] Applications such as a 40- to 50-in. display for high-definition television (HDTV) viewing and the giant screen display represented by a projector require high picture quality. The realization of large-screen OLED displays will be a challenge for many reasons, including the difficulty of manufacturing. A sensible approach is to aim initially at matching LCD screen sizes. The resolution of an LCD is around 4-in. QVGA for PDAs and

15-in. XGA for full-color PC monitors. This level of resolution can be easily achieved by using a metal shadow mask. A limit of the lateral patterning of RGB emitters, made using metal shadow mask, is about a pitch of 40 to 50 μm between cells, or the resolution limit of one pixel (RGB set) is about 120 to 150 μm pitch. The pitch corresponds to that of a 15-in. SXGA display. However, the larger the glass substrate, the more difficult it becomes to obtain the fine pixels due to the waving of the metal mask because it is difficult to keep the distance between the substrate and the metal shadow mask fixed over large areas. If even higher resolution is required, the white backlight of OLEDs or blue OLED using the CCM method has an advantage over the patterning method for full-color displays. Because the technique of placing a color or CCM filter on the substrate is commonly practiced in the LCD industry, OLED cells are finally made by finely patterning the ITO layer on the substrate.

Therefore, it is easy to make fine-pitch pixels by using a monochrome white or blue OLED. At present, the only methods for producing high-resolution full-color displays are ink-jet printing, white backlights, and CCM methods. Furthermore, to drive a large-size or fine-pitch display of passive matrix, a high drive voltage and a high current are necessary. In a passive matrix OLED display, the column lines provide the data signal, and the row lines are addressed one at a, very short, time. The current flow through a selected row line requires a peak instantaneous current more than the current level proportional to the total number of row lines in the display. While an active matrix drive is indispensable in the large-size or fine-pitch display, amorphous silicon TFT, which are used in LCDs, are not ideal to drive OLEDs because of its low carrier mobility. On the other hand, low-temperature polysilicon (p-Si) TFTs on a large substrate have been made and used for LCDs. Eastman-Kodak and Sanyo demonstrated active-matrix OLED displays based on p-Si.[23] OLED displays based on p-Si have some advantages compared to the passive matrix OLED display, which include low power consumption and long driving lifetime.

9.4.2 OLEDs for Displays with High Liveliness

For liveliness, or more realism, both large-screen and 3-D technologies are necessary. One approach to get a large screen is to use a projector. Seiko Epson and Toyota Central R&D laboratories demonstrated a prototype of a projection display using OLEDs for each RGB light source.[24] When an OLED is used as a light source, the system can use a small optical system without a backlight, which is different from one using RGB emission from a backlight white source. However, this type of display requires a very high luminance OLED. IBM Research and eMagin Corporation have developed a micro OLED display on a Si chip intended for direct view applications.[25] This display is capable of rendering very high resolution images with a pixel density of 740 pixels/in., in a VGA (640 × 480 pixels) display. By using this microdisplay, a variety of optical techniques can be used to enlarge this small-size image. The technology might find a full-size HDTV screen at home, or a handheld computer screen that might be seen through a viewing lens. This technology has the potential that it can be used as a head mounted display for computers and other future applications.

9.4.3 OLEDs for Portable Applications

Small-size OLED displays are best suited for portable applications, such as cellular phones, PDAs, view finders for cameras, etc. The important criteria for OLEDs used in portable applications are that they have low power consumption and be very thin and lightweight. Power consumption is especially for notebook computers, PDAs, and cellular phones, in which batteries are required to provide long driving time.

9.4.3.1 Low-Power-Consumption Display

Several approaches have been suggested to reduce the device power consumption. This is equivalent to obtaining higher OLED efficiencies via optical optimization, an efficient driving scheme, and improved emitting materials.

1. We can improve the EL emission output efficiency from the glass substrate to the air by studying and optimizing the optical properties. For example, the groups at Kyushu University and Matsushita Electric Works, Co. Ltd. demonstrated that the insertion of a silica aerogel layer with a refractive index $n <$ 1.03 between the glass substrate and the ITO anode contributed to a large increase in the device light output-coupling efficiency.[26] They observed the disappearance of waveguided light, which shows up as edge emission on the glass substrate and succeeded in increasing the device out-coupling efficiency by more than 60%.

2. OLEDs can be driven by a DC constant current supply by using active matrix driving scheme. In an active-matrix OLED display, each individual pixel can be switched on or off within a frame of time. As a result, the device does not suffer from the limitations of the resolution or the high instantaneous current requirements of a passive-matrix design. Consequently, the active-matrix drive can reduce the current density and driving voltage for the cells, leading to longer lifetimes.

3. Another way to reduce the power consumption is to increase the luminous power efficiency of the OLED cells. Increasing the conductivity of the individual organic layers in the cell is one approach. The conductivity (σ) is defined by the following formula:

$$\sigma = ne\mu,$$

where n the density of carriers, e the elementary electric charge, and μ the carrier mobility.

Doping donors or acceptors into the organic layer (Lewis acid–doped hole-transport layer and lithium-doped electron-transport layer)[27,28] can increase the carrier density. The existence of an inherent upper limit to the internal quantum efficiency (QE) of OLEDs has been widely argued, because charge carrier combination on molecules composed of localized π-conjugated systems produces both emissive singlet- and non-emissive triplet-excited states. According to the simplest quantum mechanical assumption, the ratio of production of singlet- and triplet-excited states is 1:3, leading to an upper limit for the internal QE of 25%.[29,30] If we assume the device light output coupling factor to be around 20%, which can be derived from simple classical optics, the upper limit of external QE is estimated to be around 5%.[31] Recently, Princeton University has demonstrated a high external EL QE of 8.0% using a phosphorescent dye as a triplet emissive center.[32] The reported value is much higher than the upper-limit of the external QE of 5% based on fluorescence and gives a key indication for the existence of low triplet–singlet branching ratio.

9.4.3.2 Ultrathin and Lightweight Displays

Although the thickness of the OLED panel is determined almost entirely by the thickness of the glass substrate, the thickness of the "seal can" used to protect the organic layers contributes to the overall display thickness. This can be critical if the device is used for portable applications. Display manufacturers have applied a protective film instead of a seal can to reduce the overall display weight and thickness. Moreover, extensive ongoing research focuses on the use of plastic substrates, similar to LCD panels, for flexible OLED displays.[33,34] The high luminous power efficiency was enhanced by using phosphorescent materials.[35,36] A SiN film was also examined as a protective film for sealing.[37]

9.4.4 Phosphorescent Dyes as the Triplet Emitter

In 1999, Baldo et al.[32] demonstrated that platinum complexes are reasonably efficient triplet emitters and reported an external EL QE of 4%, comparable to the best external EL QE values measured for OLEDs using singlet emitters. They have also used a highly phosphorescent iridium-complex with a phosphorescence QE of 0.4 at room temperature. Figure 9.22 shows the basic OLED cell structure and the organic materials

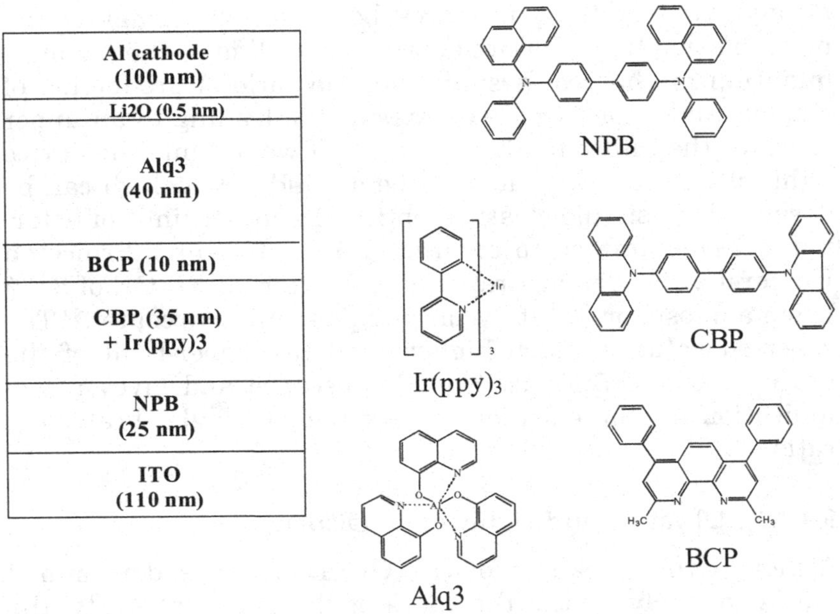

Figure 9.22 The OLED cell structure and the materials used in this work.

used. In this cell, the emitting layer consists of the phosphorescent guest emitter, tris(2-phenylpyridine)iridium ($Ir(ppy)_3$) doped in a host 4,4'-N,N'-dicarbazol-biphenyl (CBP). A film of 4,4'-bis[N-(naphthyl)-N-phenyl-amino]biphenyl (NPB) serves as the hole-transport layer and another film of 2,9-dimethyl-4,7-diphenyl-1,10-phenanthroline (BCP) is used as the hole-blocking layer. We tried to optimize the device structure by changing the layer thickness.[35] Table 9.5 shows the OLED characteristics as a function of the NPB and the CBP layer thicknesses with a device thickness of fixed to 60 nm. Based on this study, a 30-nm-thick NPB layer gave the highest external EL QE. The EL spectra, measured at 2.5 mA/cm² as a function of the NPB and CBP thickness, are shown in Figure 9.23. The EL spectra of the cells are almost unaffected by the ratio of NPB to CBP thickness as the total device thickness is held constant. Moreover, the spatial emission patterns of

Table 9.5 OLED Device Characteristics as a Function of NPB and CBP Film Thicknesses: ITO(110 nm)/NPB (60-x nm)/CBP(x nm)+Ir(ppy)$_3$(8.7 wt%)/BCP(10 nm)/Alq$_3$ (40 nm)/Li$_2$O/Al

CBP Thickness (nm)	Luminance (cd/m²)	Voltage (V)	External EL QE (%)	Power Efficiency at 100 cd/m² (lm/W)
10	1060	5.44	11.76	38.2
20	1264	5.4	13.92	46.4
30	1356	6.17	14.93	43.4
40	1310	7.2	14.45	37.3
60	836	9.49	9.27	22.2

Note: Luminance, voltage, and quantum efficiency measured at 2.5 mA/cm².

these cells were nearly identical and very close to Lambertian. The correction factor for the EL QE was calculated to be 0.95. Table 9.6 shows the corrected external EL QE as a function of NPB and CBP film thicknesses. Based on the corrected values of the EL QE, the apparent external EL QE estimated to be 14.8% drops to an external EL QE of 14.0%.

The initial device lifetime measurements were not encouraging. We studied ways to increase the device driving lifetime,[36] varying the guest molecule concentration, inserting a copper phthalocyanine (CuPc) layer between the anode (ITO) and the hole-transport layer (NPB) to prevent driving voltage from rising during constant current operation, and exchanging the BCP hole-blocking layer with ((1,1'-biphenyl)-4-olato)bis(2-methyl-8-quinolinolato N1,O8)aluminum (BAlq).

We combined these three results to see if we can achieve a longer operation lifetime. We also optimized the optical length of the cells, paying special attention to keep the optimal thickness using 50-nm-thick ITO at the anode side. This optimized cell has the following configuration:

ITO(50 nm)/CuPc(25 nm)/NPB(45 nm)/CBP(35 nm) + Ir(ppy)$_3$/BAlq(10 nm)/Alq3(40 nm)/Cathode

The driving decay curve of the cell using an emitting layer consisting of 2.9 wt% Ir(ppy)$_3$ dopant is shown in Figure 9.24. The half-life decay time is 3300 h, measured at an initial

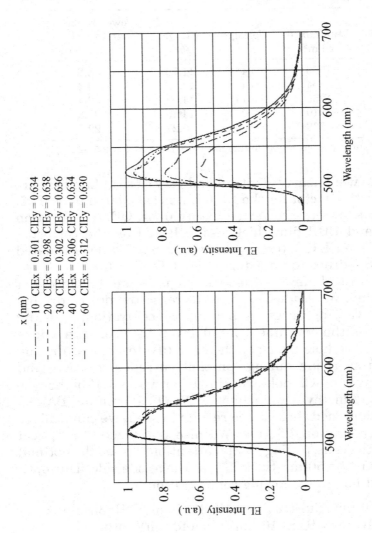

Figure 9.23 The EL spectra as a function of NPB and CBP thickness measured at 2.5 mA/cm^2: ITO(110 nm)/NPB(60-x nm)/CBP(x nm) + Ir(ppy)$_3$(8.7 wt%)/BCP(10 nm)/Alq$_3$(40 nm)/Li$_2$O/Al.

Table 9.6 Corrected External EL QE as a Function of NPB and CBP Thickness: ITO(110 nm)/NPB (60-x nm)/CBP(x nm) + Ir(ppy)$_3$(6.5 wt%)/BCP (10 nm)/Alq$_3$(40 nm)/Li$_2$O/Al

CBP Thickness (nm)	Corrected Value at 1.25 mA/cm^2	Uncorrected External QE (%)	Corrected External QE (%)
10	0.952	11.40	10.81
35	0.947	14.80	14.00

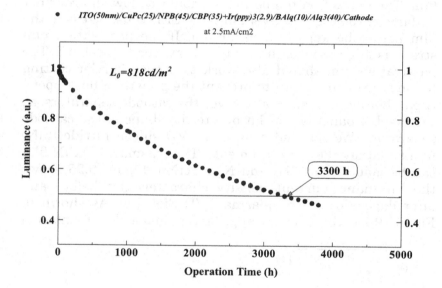

• *ITO(50nm)/CuPc(25)/NPB(45)/CBP(35)+Ir(ppy)3(2.9)/BAlq(10)/Alq3(40)/Cathode*

at 2.5mA/cm2

$L_0 = 818cd/m^2$

3300 h

Figure 9.24 The operation lifetime of an optimized OLED cell for durability and emitting efficiency, measured under continuous constant-current driving of 2.5 mA/cm^2.

luminance of 818 cd/m^2. This high-durability device, whose dopant concentration is 2.9 wt%, exhibited external EL QE of 8.9% and a luminance of 818 cd/m^2 at a current density of 2.5 mA/cm^2. A half decay lifetime ≥ 25,000 h is extrapolated for an initial luminance of 100 cd/m^2 using constant current driving.

9.4.5 Passivation Film for OLEDs[37]

The OLED cells are very vulnerable to moisture and oxygen so that they need to be sealed from ambient atmosphere. A sealing canister with ultraviolet (UV) adhesive was used under N_2 atmosphere to protect the OLED from moisture and oxygen. BaO was added as a desiccant. A passivation film is a thinner alternative and could be used to fabricate a thinner and lighter panel. The passivation film excels in radiating heat and also acts as a heat sink.

There are a number of requirements for a passivation film. The deposition temperature should be low, in order not to damage the organic layers in the OLED cells. Stress to the film has to be kept to a minimum. If the passivation film stress is high, the organic layer may easily peel off. The passivation film should also work as a good barrier coating to moisture and oxygen to prevent the growth of dark spots. A passivation film must also cover the cathode separators, as the display panel is not flat due to the shape of the cathode separators. We selected a plasma CVD silicon nitride (SiN) film to satisfy these requirements. The plasma CVD SiN film can be made from SiH_4 and N_2 reaction. Figure 9.25 shows the luminance-current density characteristics before and after deposition of the plasma CVD SiN film. As shown in Figure 9.25, there is no apparent damage due to plasma

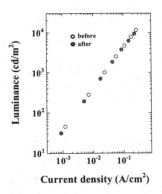

Figure 9.25 The luminance-current density characteristics before and after deposition of a plasma CVD silicon nitride film.

Table 9.7 Preparation Conditions of
the Plasma CVD SiN Film

SiH_4 flow rate	10 SCCM
N_2 flow rate	200 SCCM
RF power density	0.05 W/cm
Substrate temperature	100°
Pressure	0.9 torr
Frequency	13.56 MHz
Deposition time	60 min
Deposition rate	50 nm/min
Film thickness	3000 nm

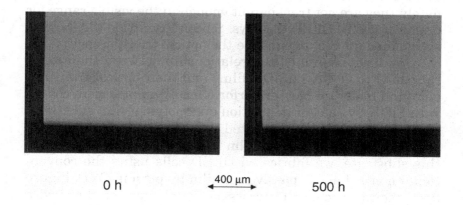

0 h   ←400 μm→   500 h

Figure 9.26 The dark spot growth before and after storage for 500 h at 100°C.

deposition in the luminance-current density characteristics. There was no change in the transmission peaks after storage at 60°C at 95% relative humidity. A SiN film was deposited from a SiH_4 and N_2 gas mixture and used as a passivation film for the OLED device. Table 9.7 lists the preparation conditions used for the plasma CVD SiN film. We also investigated the effect of using a SiNx passivation film. Figure 9.26 shows the dark spot formation before and after storage for 500 h at 100°C. Similar phenomena occur in a seal made by the sealing canister approach. The OLED characteristics remained almost identical and the dark spots growth were negligible for practical use after storage at room temperature, 60°C at 95% relative humidity, and at 100°C.

9.4.6   OLEDs on Polymer Film Substrates[34]

OLEDs are beginning to compete seriously with LCDs and may eventually replace them for flat panel displays. OLEDs on polymer substrates[38–39] may extend their applications in the field of electronic displays.

The key issue to achieve polymer film-based OLEDs is to use substrates that are impermeable to moisture. SiN passivation films work well as a moisture barrier for OLEDs. Unfortunately, they are not transparent enough in the visible range for practical use in OLED displays. Silicon oxynitride (SiON) was substituted for SiN to enhance the optical transparency.

We have examined the relationship between the atomic ratio of O to O + N in SiON films and their optical transparency and moisture barrier performance. Samples were coated with SiON by sputter deposition on a polymer film substrate. The SiON films were deposited using various ratios of O/(O + N). The thickness of each film is approximately 200 nm. On this substrate, we fabricated OLED cells using the conventional method and a passivation film by plasma CVD. Figure 9.27 shows a schematic cross-sectional view of the device structure. Each device has a square-shaped emissive area formed by the intersection of the anode and the cathode, each consisting of 2-mm-wide stripes. When moisture creeps into

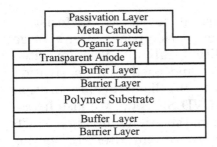

Figure 9.27 The schematic cross-sectional view of a device.

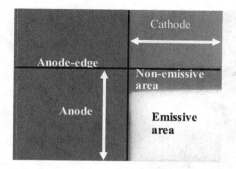

Figure 9.28 The growth of non-emissive areas from the anode edge.

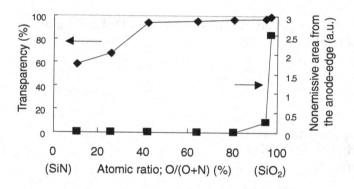

Figure 9.29 The optical transparency and the moisture resistance of SiON films as a function of the O/O + N ratio.

a device through a substrate, non-emissive areas grow from the anode edge as shown in Figure 9.28. We evaluated the resistance of a device to moisture using the growth of dark spots that appeared in devices following 500 h of storage under 60°C and 95% relative humidity.

Figure 9.29 shows the optical transmission of the SiON films and their moisture immunity as a function of the O/O + N ratio. The transparency was >90% for an O/(O + N) ratio >40%. Good moisture immunity was obtained when the O/(O + N) was <80%. It is also worthwhile to note that non-emissive areas grew rapidly when the O/(O + N) ratio was >90%.

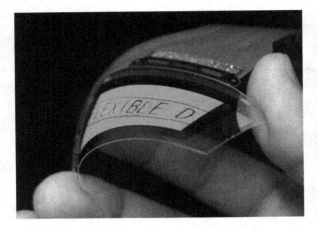

Figure 9.30 A photograph of a passive matrix OLED display on a polymer substrate.

Table 9.8 Specifications of a Passive Matrix OLED Display on a Polymer Substrate

Color	Monochrome
Active area	22.4 * 44.8 mm
Resolution	64 * 128 dots
Thickness	0.2 mm
Weight	1 g (with IC)
Supply voltage	12 V

By optimizing the O/(O + N) in the SiON barrier film, we could fabricate OLEDs on polymer substrates with high transparency in the visible and reasonably good protection from moisture. Figure 9.30 shows a photograph of a prototype passive matrix OLED display on a polymer substrate. Table 9.8 shows the specifications of this display.

REFERENCES

1. T. Tani, Display sentangijutu [in Japanese], Kyoritsu Shuppan Co., Ltd., 1998.

2. C.W. Tang and S.A. VanSlyke, *Appl. Phys. Lett.*, 51, 913, 1987.

3. R.F. Service, *Science*, 273, 878, 1996.

4. K. Nagayama, T. Yahagi, H. Nakada, T. Tohma, T. Watanabe, K. Yoshida, and S. Miyaguchi, *Jpn. J. Appl. Phys.*, 36, L1555, 1997.

5. T. Wakimoto, R. Murayama, K. Nagayama, Y. Okuda, H. Nakada, and T. Tohma, *SID 96 Dig.*, May, 849, 1996; T. Wakimoto, H. Ochi, S. Kawami, H. Ohata, K. Nagayama, R. Murayama, Y.Okuda, H. Nakada, T. Tohma, T. Naito, and H. Abiko, *J. SID*, 5, 235, 1997.

6. S. Kawami, T. Naito, H. Ohata, and H. Nakada, *Extended Abstr. 3, 45th Spring Meeting,* Japan Society of Applied Physics and Related Societies, Tokyo, 1998, 1223.

7. C. Hosokawa, E. Eida, M. Matsuura, F. Fukuoka, H. Nakamura, and T. Kusumoto, *SID 97 Dig.*, 1037, 1997.

8. S. Miyaguchi, S. Ishizuka, T. Wakimoto, J. Funaki, Y. Fukuda, H. Kubota, K. Yoshida, T. Watanabe, H. Ochi, T. Sakamoto, M. Tsuchida, I. Ohshita, and T. Tohma, *Extended Abstr., 9th International Workshop on Inorganic and Organic Electroluminescence*, 1998, 137; S. Miyaguchi, S. Ishizuka, T. Wakimoto, J. Funaki, Y. Fukuda, H. Kubota, K. Yoshida, T. Watanabe, H. Ochi, T. Sakamoto, M. Tsuchida, I. Ohshita, and T. Tohma, *J. SID*, 7, 221, 1999.

9. P.E. Burrows, V. Khalfin, G. Gu, and S.R. Forrest, *Appl. Phys. Lett.*, 73, 435, 1998.

10. T. Shimoda, S. Kanbe, H. Kobayashi, S. Seki, H. Kiguchi, I. Yudasaka, M. Kimura, S. Miyashita, R.H. Friend, J.H. Burroughes, and C.R. Towns, *SID 99 Dig.*, May, 376, 1999.

11. J. Kido, K. Hongawa, K. Okuyama, and K. Nagai, *Appl. Phys. Lett.*, 64, 815, 1994.

12. Y. Sato, T. Ogata, S. Ichinosawa, and Y. Murata, *Synth. Met.*, 91, 103, 1997.

13. O. Yokoyama, S. Seki, K. Morii, H. Kobayashi, M. Kimura, S. Inoue, S. Miyashita, T. Shimoda, J.H. Burroughes, C.R. Towns, and R.H. Friend, *Proc. 6th Int. Display Workshop*, Kobe, 2000, 239.

14. C.W. Tang, S.A. VanSlyke, and C.H. Chen, *J. Appl. Phys.*, 3610, 65, 1987.

15. Y. Fukuda, T. Watanabe, T. Wakimoto, S. Miyaguchi, and M. Tsuchida, *Synth. Met.*, 111–112, 1, 2000.

16. K. Amemiya, N. Tanaka, Y. Yonemoto, and M. Manabe, *The Abstract for the 38th Annual Spring Meeting of Japan Applied Phys. Soc.*, 1991, 1085.

17. T. Tsutsui and K. Yamamoto, presented at SPIE Conference on Organic Light-Emitting Materials and Devices, San Diego, CA, July 1998, 2.

18. T. Nakayama, Y. Itoh, and A. Kakuta, *Appl. Phys. Lett.*, 63, 594, 1993.

19. N. Takada, T. Tsutsui, and S. Saito, *Appl. Phys. Lett.*, 63, 2032, 1993.

20. P.E. Burrows, V. Khalfin, G. Gu, and S.R. Forrest, *Appl. Phys. Lett.*, 73, 435, 1998.

21. T. Granlund, L.A.A. Pettersson, M.R. Anderson, and O. Inganas, *J. Appl. Phys.*, 81, 8097, 1997.

22. K. Ziemelis, *Nature*, 399, 408, 1999.

23. G. Rajeswaran, M. Itoh, M. Boroson, S. Barry, T.K. Hatwar, K.B. Kahen, K. Yoneda, R. Yokoyama, T. Yamada, N. Komiya, H. Kanno, and H. Takahashi, *SID 2000 Dig.*, May, 974, 2000.

24. S. Tokito and Y. Taga, *Proceedings of the 9th International Workshop on Inorganic and Organic Elecroluminescence*, 1998, 1.

25. K. Pichler, W.E. Howard, and O. Prache, *Proc. SPIE*, 3797, 258, 1999.

26. T. Tsutsui, M. Yahiro, K. Yokogawa, K. Kawano, and M. Yokoyama, *Extended Abstract of 2000 MRS Fall Meeting*, 2000, JJ4.4.

27. J. Kido, J. Endo, K. Mori, and T. Matsumoto, *Extended Abstract of 1999 MRS Fall Meeting*, 1999, BB2.5.

28. J. Kido and T. Matsumoto, *Appl. Phys. Lett.*, 73, 2886, 1998.

29. W. Helfrich and W.G. Schneider: *J. Chem. Phys.*, 44, 2902, 1966.

30. T. Tsutsui and S. Saito, in *Intrinsically Conducting Polymers: An Emerging Technology*, M. Aldissi, Ed., Kluwer Academic, Dordrecht, 1993, 123.

31. T. Tsutsui, *MRS Bull.*, 22, 39, 1997.

32. M.A. Baldo, S. Lamansky, P.E. Burrows, M.E. Thompson, and S.R. Forrest: *Appl. Phys. Lett.*, 75, 4, 1999.

33. P.E. Burrows, G. Gu, V. Bulovic, S.R. Forrest, and M.E. Thompson, *IEEE Trans. Electron. Dev.*, 44, 1188, 1997.

34. A. Sugimoto, A. Yoshida, T. Miyadera, and S. Miyaguchi, *Proceedings of the 10th International Workshop on Inorganic and Organic Elecroluminescence*, 2000, 365.

35. T. Watanabe, K. Nakamura, S. Kawami, Y. Fukuda, T. Tsuji, T. Wakimoto, S. Miyaguchi, M. Yahiro, M.-J. Yang, and T. Tsutsui, *Synth. Met.*, 122, 203, 2001.

36. T. Watanabe, K. Nakamura, S. Kawami, Y. Fukuda, T. Tsuji, T. Wakimoto, and S. Miyaguchi, *Proc. SPIE*, 4105, 175, 2000.

37. H. Kubota, S. Miyaguchi, S. Ishizuka, T. Wakimoto, J. Funaki, Y. Fukuda, T. Watanabe, H. Ochi, T. Sakamoto, T. Miyake, M. Tsuchida, I. Ohshita, H. Nakada, and T. Tohma, *J. Luminesc.*, 87–89, 56, 2000.

38. G. Gustafsson, Y. Cao, G.M. Treacy, F. Klavetter, N. Colaneri, and A.J. Heeger, *Nature*, 357, 477, 1992.

39. G. Gu, P.E. Burrows, S. Venkatesh, S.R. Forrest, and M.E. Thompson, *Opt. Lett.*, 22, 172, 1997.

10

Organic Electroluminescent Devices for Solid State Lighting

ANIL R. DUGGAL

CONTENTS

10.1 INTRODUCTION

As demonstrated throughout this volume, over the last decade tremendous strides have been made in the science and technology of organic electroluminescence. Most of this progress has been fueled by interest in developing flat panel displays based on the technology. If this rate of progress can be sustained into the next decade, organic electroluminescence technology has the potential to exert an impact not only on displays, but also on general lighting applications. In particular, a large-area white-light-producing organic light-emitting device (OLED) could potentially provide a solid state diffuse light source that could compete with conventional lighting technologies in performance and cost. In this chapter, we briefly discuss the solid state lighting vision and then describe the goals, challenges, and recent progress in developing OLEDs for general illumination.

The vision of solid state lighting has largely been driven by the desire to reduce energy consumption.[1] Lighting consumes ~20% of the electricity generated in the U.S.[2] Incandescent and fluorescent lighting technologies together account for about 80% of this energy usage. Incandescent bulbs represent the oldest and least-energy-efficient lighting technology currently in use (~15 lm/W). However, they provide the highest illumination quality light. In the U.S., they are mostly sold into the residential market where consumers prefer the light quality and low initial cost and are less concerned with energy efficiency or cost of energy. In contrast, linear fluorescent bulbs are the most efficient (~100 lm/W) of the available high-illumination-quality light sources. They are sold primarily into the commercial and industrial markets where it is understood that the cost of energy is a key component of the overall cost of light.

Both incandescent and fluorescent technologies are relatively mature and so scientific breakthroughs resulting in large improvements in energy efficiency are not expected. In contrast, both inorganic light-emitting diode (LED) and OLED technologies continue to exhibit exponential progress in performance. Today, red LEDs, which have had the longest tenure as a technology, have demonstrated a wall-plug efficiency (optical watts out divided by electrical watts in) of ~45%.[3] This is almost twice the wall-plug efficiency of a fluorescent bulb (~25%). However, LEDs were not seriously considered as a potential general illumination source until the development in the 1990s of the GaN material system, which enabled efficient blue, and ultimately white, LEDs.[4] The enormous rate of progress with the GaN system combined with the demonstrated capability of other inorganic material systems sparked the idea that a solid state light source could potentially provide light with efficiency greater than standard light sources. In fact, white inorganic LED devices have already been demonstrated with higher efficiency than incandescent bulbs.[5]

Worldwide effort in developing OLEDs did not start until the 1990s, and so organic electroluminescent technology is not as mature as its inorganic cousin. However, the tremendous rate of scientific and technical progress in OLED electroluminescence combined with its emerging viability as a commercial display technology has led to active consideration of OLEDs as potential solid state light sources. In fact, in terms of application potential, OLEDs nicely complement inorganic LEDs. Inorganic LEDs are bright point sources of light and so are naturally suited to applications such as spot or task lighting that require spatial control over the illuminating beam. Today incandescent or high-intensity discharge sources are typically used for such applications. In contrast, OLEDs represent a diffuse source of light and so are naturally suited to large-area general-lighting or signage applications where today fluorescent lamps are typically utilized. An additional reason for interest in OLEDs for solid state lighting is the potential for extremely low cost "newspaper-like" processing because the active organic layers are amorphous and hence do not require the precise epitaxial growth control needed to fabricate inorganic LEDs. As described in detail in Section 8.2.5, a technology with high efficiency without low cost will not be a viable lighting technology and hence this aspect of OLED technology is an important potential advantage.

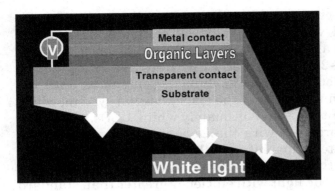

Figure 10.1   Vision for an OLED-based solid state light source.

Given that OLEDs naturally provide diffuse lighting, the ultimate solid state lighting goal is to develop the technology to the point where it is superior to fluorescent technology. One possible vision, illustrated in Figure 10.1, would be a mechanically flexible lighting sheet fabricated in a low-cost roll-to-roll process that could be conformed to any surface and provide high-efficiency illumination. In the next section, the formidable challenges that need to be overcome to achieve this solid state lighting vision are described.

10.2   REQUIREMENTS FOR AN OLED LIGHT SOURCE

The potential for OLEDs to be mechanically flexible and conformable would allow new form factors that would make an OLED lighting technology attractive for many applications that are less demanding than general illumination. In fact, even without mechanical flexibility, certain specialty lighting applications are technically viable even with the current state of technology development. These applications are not yet commercially viable only due to the relatively immature manufacturing infrastructure for OLEDs. As the current OLED display market grows and this manufacturing infrastructure matures, specialty lighting products are likely to become commercially available.

In the following we concentrate on the technology requirements for the ultimate solid state vision where OLEDs displace fluorescents as the premier diffuse lighting technology. The key performance attributes that need to be addressed are illumination quality, efficacy, luminance (brightness), lifetime, and cost. In the following, the current capability of fluorescent technology in these areas is reviewed and relevant goals for OLED technology are proposed.

10.2.1 Illumination-Quality Light

In order for a light-emitting device to be acceptable as a general illumination source, it clearly must provide light that is "pleasing" for the average human being. Through long experience, the lighting industry has found that the primary attributes that define a "pleasing" or high-illumination-quality light source are the color of the light and its color-rendering capability.

10.2.1.1 Color

For general illumination, it has been found that the preferred white light source colors are those of a blackbody emitter with a temperature between 2800 and 6500 K. Hence the color of a light source is typically characterized in terms of its color temperature. Color temperature can be related to the CIE x, y chromaticity coordinates typically used by the display industry by calculating the blackbody spectrum $BB(\lambda)$ at a specific temperature using Equation 10.1:

$$BB(\lambda) = \frac{c_1}{\lambda^5 \left(e^{\frac{c_2}{\lambda T}} - 1 \right)}, \qquad (10.1)$$

where $c_1 = 3.7418 \times 10^{-16}$ (W/m^2) and $c_2 = 1.4388 \times 10^{-2}$ (m K), and then calculating the x, y coordinates from the spectrum in the usual manner.[6,7] Doing this for every color temperature maps out the "blackbody locus" in x, y space as depicted in Figure 10.2. Note that, in lighting parlance, a lower color temperature corresponds to a "warmer" light due to the rel-

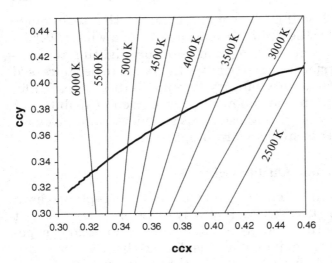

Figure 10.2 Blackbody locus (bold curve) in the CIE *x,y* color space along with lines of constant correlated color temperature (light curves).

atively higher red component in the spectrum and, conversely, a higher color temperature corresponds to a "colder" light due to the relatively higher blue spectral component.

Only a few light sources have a spectrum that closely approximates Equation 10.1. Examples include a candle with a color temperature of ~1900 K, an incandescent bulb (~2800 K), and average daylight (~6500 K). Other sources such as fluorescent and metal-halide aim to have color temperatures on the blackbody locus but have spectra that do not in any way resemble that of a blackbody at the same temperature.

If the *x,y* coordinates of an illumination source do not exactly sit on the blackbody locus, the source is characterized in terms of its correlated color temperature or CCT. The CCT is simply defined as the color temperature that most closely matches the color of the source. There is an accepted method[7] to determine lines of constant correlated color temperature in *x, y* space; Figure 10.2 depicts such lines at 500-K increments. Although Figure 10.2 depicts lines of constant CCT extending far from the blackbody locus, it is important to note that, for customer acceptance, light sources need to be as close to the blackbody locus as possible. The x, y color coordinate space is not linear in terms of color differences and so, in terms of x, y units, the maximum acceptable distance from the blackbody locus depends nonlinearly on color temperature and is discussed in terms of MacAdam ellipses, which are beyond the scope of this chapter.[7] Nevertheless, a good rule of thumb is that the color of an illumination source should be within 0.01 x, y units from the blackbody locus. This corresponds approximately to the four-step MacAdam ellipse standard employed by the lighting industry.

10.2.1.2 Color Rendition

After color, the most important aspect determining the illumination quality of a light source is its color rendition, i.e., how "true" different colored objects appear when illuminated by the source. As might be expected, this depends in a complicated fashion on the spectral power distribution of the light source, the spectral reflectance of the colored object of interest, and human perception. Color rendering needs to be understood separately from color temperature because there is not a one-to-one correspondence between the color coordinates and the spectral power distribution of a light source. Thus, two light sources with different spectral power distributions can have the same color temperature, but will, in general, have different color-rendering properties.

There is a generally accepted, quantitative measure of color rendition called the color-rendering index (CRI). It is based on the concept that, for every color temperature, there is a standard illuminant that provides ideal color rendition. This standard illuminant may or may not be physically realizable but its spectral power distribution is well defined. For color temperatures below 5000 K, the standard illuminant is defined to be a blackbody radiator. For higher color temperatures, the standard illuminant is referred to as "standard illuminant D," which is meant to approximate the various phases of daylight. The method for calculating the spectral power distribution for standard illuminant D at different color temperatures can be found in standard texts.[6,7]

For a given light source, the CRI attempts to quantify how different a set of test colors appears when illuminated by the source compared to when the same test colors are illuminated by the standard illuminant with the same correlated color temperature. The appearance of an illuminated test color can be quantified in terms of its color coordinates — calculated from the test color's reflectance spectrum weighted by the spectral power distribution of the light source. The difference in the rendering of the test color between the standard illuminant and the light source in question can then be quantified in terms of the difference in color coordinates (ΔE) obtained with the two sources. The actual calculation of ΔE involves various color space transformations, which are beyond the scope of this chapter.[6,7] The final calculation of CRI then involves calculating ΔE for eight test colors with well-defined reflectance spectra and then averaging according to the following definition:

$$CRI = R_a = 100 - 4.6 \sum_{i=1}^{8} \frac{\Delta E_i}{8}, \qquad (10.2)$$

where the index refers to the different test colors. We can see that the highest possible CRI value is 100, and this occurs when there is no difference in color rendering between the light source and the standard illuminant. An example of such a light source is the incandescent lamp. When a color rendering difference exists, ΔE is greater than zero and the CRI is less than 100. Note that negative CRI values are possible. Achieving illumination-quality white light generally requires a CRI value of 80 or greater.

It is important to realize that, since the reflectance spectra of the relevant test colors and the spectral power distributions of the standard illuminants are all tabulated, the CRI for a light source can be calculated as soon as its spectral power distribution is measured. Figure 10.3 shows the spectra and resultant CRI values for three commercially available

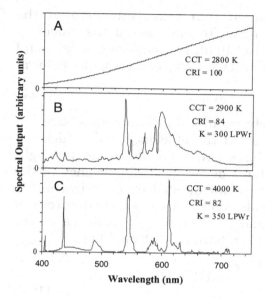

Figure 10.3   Spectra of three conventional light sources: incandescent (A), ceramic metal halide (B), and fluorescent (C).

light sources that provide illumination-quality light. Note that the spectra for the fluorescent and ceramic metal halide sources contain well-defined peaks and look dramatically different from the incandescent spectrum and yet still achieve respectable CRI values.

OLEDs have enormous flexibility in spectral tuning through varying the chemical structure of the emissive species, mixing different emissive species, modifying device structure, etc. Hence, we would expect that achieving high-illumination-quality light should not be a major hurdle in the development of an OLED light source. Indeed, as is discussed in Section 10.3, various demonstrations of high-illumination-quality OLEDs have already been achieved.

10.2.2   Efficiency

The efficiency metric for a light source is the luminous efficacy with units of lumens per input electrical watt (LPW). This is a photometric measurement that takes into account the human eye response to light. To understand the efficiency challenges for any potential lighting technology, it is useful to divide the luminous efficacy (ε) into two terms as follows:

$$\varepsilon = K\eta .$$ (10.3)

Here K, which is properly termed the "luminous efficacy of radiant flux," is a measure of the efficiency with which the eye responds to the radiant power emitted from the light source and η is the radiant or "wall-plug" efficiency of the light source (radiant power out divided by electrical power in). The first term depends solely on the spectral output of the light source, $S(\lambda)$, and the spectral dependence of the photopic eye sensitivity to radiation, $V(\lambda)$, as follows:

$$K = K_m \frac{\int V(\lambda)S(\lambda)d\lambda}{\int S(\lambda)d\lambda} ,$$ (10.4)

where K_m = 683 lumen per radiant watt (LPWr). The separation of terms in Equation 10.3 allows separate analysis of the power efficiency and the spectral efficiency of a light source for illumination.

It is straightforward to develop an intuitive understanding of K values by examining the eye sensitivity curve $V(\lambda)$ and the overlap between $V(\lambda)$ and the spectral output of light sources. Figure 10.4 provides such a comparison for two hypothetical incandescent sources with different color temperatures. We can see that a relatively small portion of the spectral output from these sources overlaps the eye sensitivity curve and hence both K values are relatively low. This is the main reason that incandescent bulbs are relatively inefficient. We can also see that the higher color temperature spectrum has a larger relative overlap with the eye sensitivity curve and hence a higher K value. This, to a large extent, explains the higher efficiency of the higher temperature halogen bulbs compared to standard incandescent bulbs. The K values for fluorescent and ceramic metal halide light sources are shown

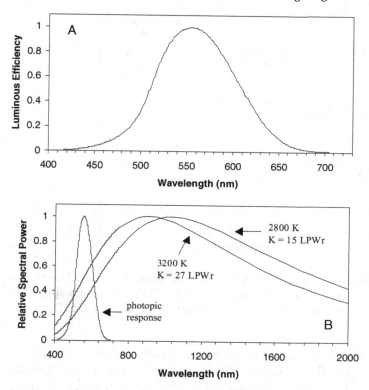

Figure 10.4 Photopic eye response to radiation (A) and comparison between photopic eye response curve (spectral luminous efficiency) and the spectral power distribution of two hypothetical incandescent sources with different color temperatures (B).

in Figure 10.3. Note that these values are more than a factor of 10 higher than incandescent values because most of the radiation generated is in the visible range.

Ideally, to maximize the energy efficiency of a light source, we would aim to maximize K by appropriately adjusting the spectral output to maximize overlap with the eye response curve. However, there is a clear trade-off between doing this and achieving high-illumination-quality light. This is because attaining a given color temperature while maximizing K is best achieved by designing a source that has

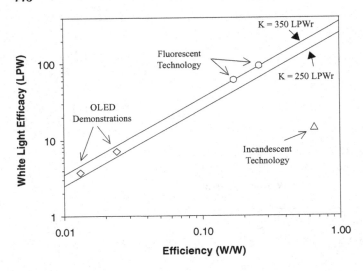

Figure 10.5 White light efficacy vs. wall plug efficiency. The region between the K = 350 LPWr and K = 250 LPWr lines corresponds to the expected achievable efficacy of illumination-quality OLEDs.

spectral components as close to the peak of the eye response curve as possible with as little spectral power as possible outside of this region. In general, such a spectrum would have very poor color-rendering properties. For example, K values above 500 LPWr for white light on the blackbody locus are possible, but these have negative CRI values. Much effort has gone into modeling this trade-off between K and CRI to try to design optimal spectral power distributions for illumination-quality white light.[8–10] In fact, the spectra of current fluorescent lamps are the result of such modeling combined with the capability of today's lamp phosphors.

For the relatively broad spectra typical of OLEDs, it is likely that optimal output spectra will have K values in the range of 250 to 350 LPWr. Figure 10.5 shows the expected linear relationship between luminous efficacy and power efficiency for these two K values. Depending on the details of how the white-light is generated, it is likely that the performance of illumination-quality OLEDs will lie somewhere between these two lines. Also shown on Figure 10.5 are the data points for various conventional light sources and recent illumination-quality OLED demonstration devices. The recent OLED points correspond to the device performance with an output brightness or luminance of 1000 cd/m^2, which, as discussed in the next section, is in the range required for lighting. We can see that these exhibit luminous efficacies of less than 10 LPW and power efficiencies of less than 3%. To achieve luminous efficacies on par with today's fluorescent technology the OLED power efficiencies will have to improve to between 15 and 25% to match compact fluorescent bulbs at ~60 LPW and to between 25 and 35% to match linear fluorescent bulbs at ~90 LPW.

The most efficient OLEDs today are typically green-emitting devices that utilize phosphorescence to harvest both singlet and triplet excitons. A typical high-performance wall-plug efficiency number for these devices at a luminance of 1000 cd/m^2 is 14%.[11] If this performance could be translated to an illumination-quality white device, we can see from Figure 10.5 that the luminous efficacy could range between 35 and 50 LPW depending on the details of the white-light generation process. These values are significantly better than possible with incandescent technology and are close to matching the efficacy values of compact fluorescent technology. Thus, a key challenge that needs to be overcome is to bring the efficiency performance of illumination-quality white OLEDs up to that demonstrated with green OLEDs.

As discussed in Section 10.1, to truly revolutionize lighting, OLEDs will have to demonstrate higher performance than linear fluorescent technology. This will require a doubling of the current best efficiency demonstrated today by OLED technology. Achieving this will likely require new types of device designs specifically tailored for lighting that have not yet been explored due to the OLED application emphasis to-date on flat-panel displays.

10.2.3 Luminance

The objective of a light source is to deliver lumens to an area. For a potential OLED area light source such as depicted in

Figure 10.1, a given lumen output can be accomplished either by utilizing a relatively small area OLED operating at high luminance or by utilizing a larger-area OLED operating at lower luminance. In addition, it is likely that different applications will have different luminance needs. Thus, there is not a definitive luminance requirement for an OLED light source. However, there are definitely technical trade-offs with luminance that need to be understood. One trade-off is that between luminance and life. It is well known that the lifetime of an OLED decreases as its operating luminance is increased. The details of this trade-off are not well understood and, at present, there does not appear to be a general function true for all OLEDs but, in many cases, the lifetime decreases faster than linearly with luminance.[12] This argues for designing lower-luminance OLEDs for lighting. On the other hand, lower luminance implies that more OLED area will be necessary and this affects the cost to achieve a certain lumen output. Thus, cost considerations, discussed in detail in Section 10.2.5, argue for higher-luminance OLEDs.

A meaningful luminance target that can guide OLED technology development for lighting can be found by examining modern lighting installations where fluorescent bulbs are aimed at the ceiling and the diffuse reflection from the lighted ceiling area provides light for the room. Here the lighting designer often tries to maximize the lighted ceiling area while simultaneously minimizing the glare that can occur when an area is too bright relative to its surroundings. Such considerations lead to an optimal area luminance of 850 to 1200 cd/m^2.[13] Thus, a reasonable luminance target for OLED technology development for lighting is 1000 cd/m^2. Note that this luminance target is approximately an order of magnitude greater than what is necessary for typical display applications and approximately an order of magnitude less than the surface luminance of a typical linear fluorescent lamp.

As a point of reference, it is useful to determine how much area would be required from an OLED light source to deliver the lumen output of a common recessed fluorescent ceiling fixture. The combined output from the lamps of a modern, single ballast, four-lamp fixture is ~9000 lm of light.

Typically, 30% or more of this light is then lost due to inefficiencies in the fixture. Note that an OLED operating under the glare threshold would not require any extra fixturing or luminaire design to reduce glare and so would not be subject to this inefficiency. Based on these numbers, and assuming Lambertian emission, ~2 m^2 of an OLED operating at 1000 cd/m^2 would be required to provide the same lumen output as a modern fluorescent system. As discussed above, larger areas at lower luminance or smaller areas at higher luminance could be utilized depending on other technical considerations.

10.2.4 Lifetime

The required lifetime for an OLED light source clearly depends on application and the performance of current lighting products. Incandescent technology exhibits the shortest lifetime of less than 1000 h. Thus, a first general-lighting entry product for an OLED would likely have to match this performance. However, to achieve the solid state lighting vision, OLEDs need to displace fluorescent technology and hence must at least match the fluorescent lifetime performance. The operating lifetime of linear fluorescent lamps — defined as the time at which 50% of the lamps from a large sample population have failed — is greater than 15,000 h. During this lifetime, the actual light output decreases, but typically by less than 20%.

The lifetime of an optimized OLED is not limited by a catastrophic failure where light emission abruptly ceases, but rather by a continuous decrease in light output with time. Hence, OLED lifetime is typically characterized in terms of time to a given luminance reduction at a fixed current or voltage. Thus, an appropriate operating lifetime goal for OLEDs that would be comparable to fluorescent performance would be a time to 80% luminance of 15,000 h. As a comparison, white OLEDs today have times to 50% luminance of ~20,000 h at display luminance levels and of ~1,000 h at 850 cd/m^2.[14] This performance is close to adequate for an initial entry lighting product, but a factor of ~20× improvement in lifetime will be necessary to achieve parity with fluorescents.

Another important lifetime performance factor is color stability. Ideally, there should not be a noticeable color change during operating lifetime. Quantifying this requirement involves specifying Macadam ellipses in x, y chromaticity space but, as in Section 10.2.1, a good rule of thumb is that the color needs to remain stable to within ~0.01 x, y units. This can be an issue if the white light is generated by mixing different dye materials into the active regions of an OLED to attain the necessary spectral output. In particular, if the individual lifetime of each dye material under operating conditions is different, then the color of the summed output can shift significantly before the total lumen output decreases below the lifetime threshold level. As is discussed in Section 10.3, this effect can be mitigated through designing device structures and/or materials that do not exhibit this differential aging effect.

10.2.5   Cost

OLED technology can potentially provide new form factors that have not been possible with conventional lighting technologies. In particular, OLEDs could provide a thin, mechanically flexible lighting "sheet" that could be wrapped around objects to provide ambient lighting, decoration, delineation, or some combination thereof. This capability, in the hands of capable designers, could change the way we think about lighting. It is likely that consumers would be willing to pay a premium to access such new designs. This is undoubtedly true for numerous existing and future specialty lighting applications. However, the vast majority of fluorescent bulbs are sold into commercial and industrial markets for the lighting of spaces such as factories, stores, office buildings, etc. Here, lighting design potential is very much a secondary consideration in evaluating a new lighting technology. Instead, the primary consideration is the overall cost-of-light — i.e., the first cost to buy the lighting product plus the operating costs, installation and maintenance costs, etc. For the commercial and industrial markets, it is generally the case that a new lighting technology will not be adopted unless it provides a reduction in the overall cost of light within 2 years. Hence, to displace fluorescent technology, the cost goal for an OLED solid state light source needs to be consistent with this reality.

If we assume that maintenance and installation costs are comparable among lighting technologies, then a cost comparison need only take into account the initial cost of the light source and the cost of electricity to run it. In this case, the cost as a function of time to provide 1 kilolumen of light is simply given by the following equation:

$$Cost(t) = Source\_Cost +$$

$$\frac{1000L \times Duty\_Cycle \times Electricity\_Cost \times t}{\varepsilon \times Fixture\_Efficiency} \quad (10.5)$$

Values appropriate for today's fluorescent technology, along with the total cost for a 2-year operating period are given in Table 10.1. Note that for an OLED with an efficacy of 120 lm/W and no fixture loss, the source cost needs to be below ~$7/kilolumen to provide a total cost of light that is competitive with today's fluorescent technology. From Equation 10.5, it is clear that if the efficacy were increased even further, more could be charged for the OLED. However, the obvious limit here is 100% efficiency, which, for an efficient white light spectrum, corresponds to an efficacy of ~350 LPW. Table 10.1 shows that, even at this performance level, the source cost needs to be less than $11/kilolumen to be competitive. This is a direct consequence of the current market requirement that costs be comparable within 2 years.

The cost numbers in Table 10.1 utilize the given assumptions about electricity cost, duty cycle, etc. and ignore other installation and maintenance costs that might be different for a new OLED technology. Hence, they should be used only as an order of magnitude guide for technology development. They can be converted to a cost per area for an OLED once a luminance specification is fixed. As an example, if we assume a luminance of 1000 cd/m^2, and assume Lambertian emission, then $7/kilolumen corresponds to $22/m^2. This cost is considerably lower than the current cost of OLED displays, which

Table 10.1 Calculation Using Equation 10.5 and Assuming a Duty Cycle of 50% and an Electricity Cost of $0.076/kWh

Source Type	Source Cost ($/kilolumen)	Efficacy (LPW)	Fixture Efficiency (%)	Duty Cycle	Electricity Cost ($/kWh)	2 Year Total Cost ($)
Fluorescent	2.40	94	70	50	0.076	12.52
OLED 1	6.97	120	100	50	0.076	12.52
OLED 2	10.62	350	100	50	0.076	12.52

Note: The fluorescent source cost includes the cost of the ballast.

is typically measured in dollars per square inch, but it should be achievable if the promise of low-cost continuous roll-to-roll processing can be realized.

10.3 RECENT PROGRESS: ILLUMINATION QUALITY LIGHT

The OLED technical community has only recently become aware of the solid state lighting opportunity, and so research efforts aimed at overcoming the challenges outlined in the last section are only just beginning. In this section we concentrate on the research area that has received the most attention so far — the generation of illumination-quality white light. We first outline the various methods that have been formulated to generate white light using OLEDs, and then highlight the recent work aimed at optimizing illumination-quality.

OLEDs possess unique flexibility in terms of choice of materials and choice of device design and, as a result, numerous examples of white light generation have been demonstrated. The first examples were published in the mid-1990s. Kido and co-workers[15,16] at Yamagata University first showed that white light could be generated by doping various dyes into a polymer layer or in different layers of a small-molecule device.[17] Since then, numerous variations on this theme have been demonstrated. Small-molecule examples include devices with a subset of dyes in one layer combined with dyes in separated layers[18] and the inclusion of blocking layers between separated dye-doped layers to better control exciton formation and energy transfer.[19-21] Many examples of white light generation using these device structures and incorporating new materials have been reported.[22-25] For polymer devices, presumably due to the difficulty in fabricating multiple layers, white device demonstrations have focused on single emission layers, either continuing the early dye doping efforts[26] or in conjugated polymer blends.[27,28] Other approaches to white light generation that have been demonstrated include the utilization of emission from excited states involving two identical (excimer emission) or different (exciplex emission) molecules[29-31] and the use of microcavities to enhance specific spectral components selectively.[32,33]

Most of these demonstrations have not specifically addressed the lighting application needs and so it is difficult to *a priori* choose the approach with the greatest potential. One requirement that, of course, must be kept in mind is the ability to provide both the correct color temperature and high CRI. It is likely that most of these approaches, given the right material set, should be able to accomplish this. Another requirement is that the color, and hence the spectrum, not change as the luminance of the device changes. Many of the approaches outlined above, where exciton generation occurs in different layers containing different emitting species, are susceptible to such color shifts if the electron–hole recombination zone shifts with voltage due to unbalanced charge injection and/or transport. Hence, efforts utilizing such approaches need to focus on maintaining a spatially fixed recombination zone. Another, related, lighting requirement is that the device color should not change as the device ages. Any method relying on multiple emissive species could potentially suffer from such color-shifts due to differential aging of the different species during device operation. Some material sets may be more prone to this potential failure than others. Finally, the chosen approach needs to be consistent with other lighting requirements such as the potential for high efficiency and low-cost fabrication. Thus, the cost implications of approaches with complex structures with multiple layers and precisely controlled dopant levels or with precisely controlled microcavity layer thickness need to be addressed.

In 2002, a few demonstrations of white light that specifically addressed illumination quality were published.[20,31,34] The first study[20] continued the trend of placing different dopants in different device layers with the new feature that phosphorescent rather than fluorescent dyes were utilized. The use of phosphorescent dyes holds the potential for increased efficiency due to the ability to make use of normally wasted triplet excitons.[35] Two device structures were investigated. Both structures utilized iridium-based phosphorescent dyes doped in 4,4′-*N*,*N*′-dicarbazole-diphenyl (CBP) layers placed between a hole-transport layer on the anode side of the device and a hole-blocking layer on the cathode side. One structure utilized red and blue phosphorescent dye-doped layers with 10 and 20 nm thicknesses, respectively, that were separated by 3 nm of a hole/exciton-blocking layer. Here the various layer thicknesses were chosen so that the ratio of exciton formation within the two layers as well as exciton diffusion between the layers resulted in a final exciton emission ratio that generated the proper color white light. The color coordinates were reported from this device and from these we can calculate a correlated color temperature of 5000 K that is within 0.01 x, y units of the blackbody as required for illumination-quality light. However, the reported CRI of this particular device was only 50.

The low CRI of this two-phosphor device was low mainly due to the lack of emission in the yellow spectral region. Accordingly, the authors demonstrated a second device structure consisting of blue (20 nm), red (2 nm), and yellow (2 nm) dye-doped layers — this time without any intervening blocking layer. Here a CRI of 83 was obtained at a correlated color temperature of 4400 K and a distance from the blackbody locus of 0.03 x, y units. If not for the large distance from the blackbody locus, this would correspond to high-illumination-quality white light. It is likely that with more optimization of layer thicknesses, a color closer to the blackbody locus could be obtained with this device structure. The efficacy obtained with these device structures was not given at 1000 cd/m^2, but was in the range of 2.5 to 3.5 LPW in the 400 to 800 cd/m^2 region.

The same group presented another phosphorescent approach to white light emission in a separate publication.[31] The exciting result here was that white light could be obtained using only a single blue phosphorescent dopant in a CBP host. The key finding was that, at relatively high doping levels, both monomer blue emission and longer-wavelength emission from excimer states between neighboring monomers could be combined to generate white light. In particular, a device was fabricated that yielded a correlated color temperature (calculated from the CIE coordinates) of 5200 K with a distance from the blackbody locus of less than 0.01 x, y units and a CRI of ~75. This is respectable illumination quality for a light source. Because only a single dopant is utilized in this structure, differential aging should not be an issue and the color should be stable during the operating lifetime. One potential issue is the reported saturation of excimer emission with increasing device current. This could cause a change in color with luminance if the saturation occurs within the expected operating range of the device. The device efficacy at luminance levels appropriate for lighting was not given but, at a display luminance of 100 cd/m^2, a performance of 2.5 LPW was achieved.

Our group at General Electric published an alternative method for generating white light aimed specifically at meeting the needs for illumination.[34] The unique aspect of this work was that white emission was not achieved solely within the active organic layers of the OLED. Instead, a device architecture consisting of a blue-emitting OLED with downconversion layers optically coupled to the OLED substrate was utilized. The downconversion layers were chosen to absorb the blue OLED emission and then re-emit at longer wavelengths. The specific device architecture utilized is illustrated in Figure 10.6. Here, three downconversion layers were

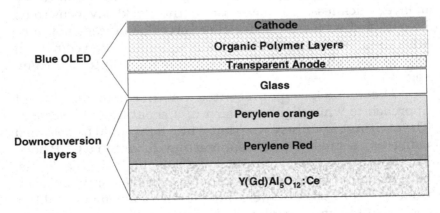

Figure 10.6   Device architecture utilized to demonstrate illumination-quality white light through downconversion of blue light.

attached to a blue light-emitting polymer (LEP)-based OLED. Two of the layers consist of perylene-based organic dyes with high photoluminescence efficiency dispersed within thin films of polymethylmethacrylate, and one layer consists of an inorganic phosphor powder. The electroluminescence spectrum of the blue OLED and the absorption and emission spectra and photoluminescent quantum efficiency of each downconversion layer are depicted in Figure 10.7. The thickness and/or concentration of each layer was chosen to ensure that the combination of non-absorbed blue OLED emission and the downconverted emission results in white light. The resulting spectrum is depicted in the bottom panel of Figure 10.7. It corresponds to a correlated color temperature of 4130 K that is within 0.01 x, y units of the blackbody locus with a CRI of 93. This is an example of high-illumination-quality white light.

The downconversion approach is attractive for lighting applications because only a single color OLED with a single emitting species is required. Thus, to first order, there should be no color change as the luminance or current density is varied. Similarly, color shifts due to differential aging should be avoidable if we ensure that the lifetime of the downconversion layers is substantially greater than that of the underlying blue OLED. This is the case for the materials depicted in Figure 10.6. An added practical benefit of this approach is that a single blue OLED device can be utilized for a wide variety of applications requiring different color temperatures and/or color-rendering capabilities by simply varying the downconversion layers applied. This is illustrated in Figure 10.8, which shows the range of color temperatures and CRIs that can be accomplished using the single blue OLED electroluminescence spectrum depicted in Figure 10.7 combined with varying thicknesses of the same downconversion layers.[36] This ability to vary the final white color of a device while only having to fabricate one underlying blue OLED "engine" has attractive cost implications for commercial production.

There are two losses associated with any downconversion approach. One is the Stokes loss associated with the fact that higher energy photons are converted to lower-energy photons

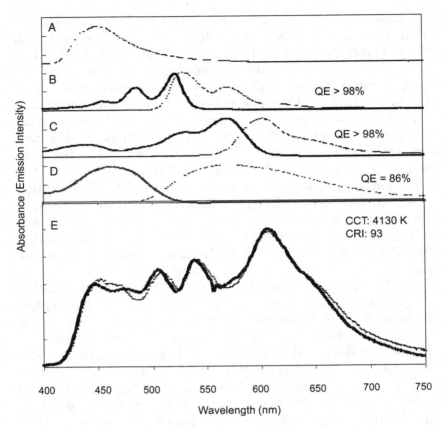

Figure 10.7 Blue OLED device emission spectrum (A) and emission (dotted lines) and absorption spectra (solid lines) and photoluminescent quantum efficiency (QE) of the downconversion materials — perylene orange (B) and perylene red (C) and a Y(Gd)AG:Ce phosphor (D) — used to construct the white device. Panel E shows the emission spectra of the white device (dotted line) along with a fit (solid line) to a downconversion model.

and one is simply the fact that the quantum efficiency for downconversion is typically less than 100%. It is important to realize that the Stokes loss is not unique to the downconversion approach. Although not explicitly defined as a Stokes loss, any white OLED that runs off of a single voltage has the

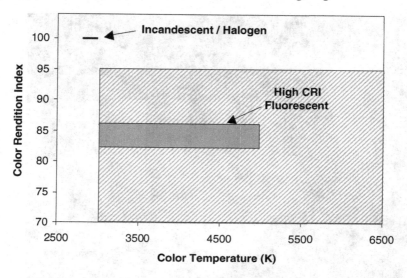

Figure 10.8 Illumination-quality space of interest for lighting. Hatched region depicts area accessible with the device and materials given in Figure 10.7. Also depicted is the performance of incandescent (line) and high-illumination-quality fluorescents (shaded region) available in the market today.

same type of loss because the voltage needs to be high enough to allow the highest energy (blue) emission. This is a loss because, in principle, the lower energy emission colors would require lower voltages if operated alone. The only way around this loss is to develop a device architecture that allows a different voltage to be placed across each color component (e.g., red, green, and blue) of the white emission. The second, quantum efficiency, loss process is unique to downconversion and clearly argues for the use of highly efficient photoluminescent materials. Such materials were used in the published demonstration and these enabled an illumination-quality white efficacy of 3.8 LPW at a lighting luminance level of 1000 cd/m². This corresponds to a wall-plug efficiency of 1.3% W/W. This point is depicted in Figure 10.5. Also depicted in Figure 10.5 is an enhanced performance of 7 LPW (2.4% W/W) obtained by improving the underlying blue device efficiency.

Figure 10.9 A white light OLED research device made at GE Global Research, which provides 70 lm of illumination with a color temperature of 4000 K, a color rendition index of 90, and an efficiency of 7 lm/W.

A picture of an OLED that provides 70 lm of high-illumination-quality light at this enhanced efficiency is depicted in Figure 10.9. Note that all the light utilized for this picture is generated by the OLED.

The various approaches to white light described here clearly show that high-illumination-quality white light can be generated with OLED technology and that there are multiple paths to doing so. In addition, other approaches are likely to be developed. A simple example might be finding a cost-effective way of generating white by mixing the light of separately powered OLEDs with differing output spectra. Such a source has the potential to be color tunable during operation — an attribute greatly desired by lighting designers. To date, as described above, the most efficient illumination-quality white demonstrations have wall-plug efficiencies of less than 3% W/W. This is factor of ~5× worse than the best demonstrated efficiencies for green OLEDs. This is primarily a reflection of the fact that, today, the efficiency of blue emission with OLED technology lags behind green. The relative performance of red OLED devices is less of an issue for white since energy transfer or direct downconversion methods can generate red spectral components from an efficient green device. Given that so far no efficient upconversion method has been demonstrated, this suggests that, to improve illumination-quality white performance, more effort needs to be directed to increasing the efficiency of blue-emitting materials and device structures.

10.4 OUTLOOK

In this chapter, a solid state lighting vision for OLED technology has been proposed. One key aspect of this vision is that an OLED provides a naturally diffuse light source and so effort needs to focus on displacing fluorescent technology, which is currently the premier diffuse lighting source. Another important aspect of this vision is that a key potential advantage of OLED technology vs. other potential solid state lighting technologies is the potential to meet the extremely low manufacturing costs required for lighting. This cost aspect needs to be kept in mind by the technical community as new technical approaches are developed to increase the efficiency, luminance, and life of OLEDs. In particular, solutions, such as self-assembly, that are consistent with low-cost fabrication need to be investigated rather than solutions, such as molecular beam epitaxy, that are not.

The main technical challenges that need to be met for OLED technology to displace fluorescent lighting for general illumination have been laid out in detail. The challenges are indeed formidable and will require a long-term investment in technology development. Because OLEDs possess potential features such as conformability to surfaces that are not possible with current lighting technology, it is likely that products will make it into the lighting market before all of the long-term challenges are met. Such shorter-term applications will help to fuel the necessary long-term development for general illumination.

There are reasons to be optimistic that an OLED-based solid state light source will become a reality. One reason is simply that while the field has demonstrated incredible progress in the last decade, it has been largely constrained into pursuing certain types of device structures due to the needs of display applications. Once this constraint is lifted, new types of device structures and materials that have so far been ignored can be investigated. These extra parallel approaches can only enhance progress. Another, related, reason for optimism has to do with the fact that OLED technology as a whole is still in a very early stage of development. OLEDs utilize organic molecules that are literally blended together into relatively simple device structures that then yield impressive performance. The number of possible organic molecules, each with tunable functions, that can be utilized is virtually unlimited due to the capabilities of modern organic chemistry. In fact, the field is really still in its infancy with regard to understanding what types of molecules should be made. Although the device physics of an OLED is largely understood,[37] the detailed physics of charge transport, exciton spin formation, and energy transfer is not. Similarly, the detailed material science required to understand how molecules interact and produce a characteristic morphology in the solid state is not understood. These details are necessary to guide the development of new organic molecules/polymers and device structures that optimize performance. Thus, there is a good chance that as basic research in OLED technology continues, and as focused research on solid-state lighting accelerates, the exponential rate of progress seen in the last decade will continue into the next. If so, then by the end of the next decade OLEDs will have a good shot at surpassing fluorescents as the premier lighting technology.

REFERENCES

1. A. Bergh, G. Craford, A. Duggal, and R. Haitz, *Phys. Today,* December, 42, 2001.

2. U.S. Lighting Market Characterization, Vol. 1: National Lighting Inventory and Energy Consumption Estimate, U.S. Department of Energy, September 2002.

3. M.R. Krames, M. Ochiai-Holcomb, G.E. Hofler, C. Carter-Coman, E.I. Chen, I.H. Tan, P. Grillot, N.F. Gardner, H.C. Chui, J.W. Huang, S.A. Stockman, F.A. Kish, M.G. Craford, T.S. Tan, C. P. Kocot, and M. Hueschen, *Appl. Phys. Lett.,* 75, 2365, 1999.

4. S. Nakamura and G. Fasol, *The Blue Laser Diode: GaN Based Light Emitters and Lasers,* Springer-Verlag, Berlin, 1997.

5. T. Whitaker, *Compound Semiconduct.,* June 2003.

6. N. Ohta and C.J. Bartleson, "Colorimetry," in *Handbook of Photographic Science and Engineering,* 2nd ed., C. Noel Proudfoot, Ed., IS&T, Springfield, VA, 1997.

7. G. Wyszelki and W.S. Stiles, *Color Science,* 2nd ed., Wiley, New York, 1982.

8. H.F. Ivey, *J. Opt. Soc. Am.,* 53, 1185, 1963.

9. W.A. Thornton, *J. Opt. Soc. Am.,* 61, 1155, 1971.

10. A. Zukauskas, R. Vaicekauskas, F. Ivanauskas, R. Gaska, and M.S. Shur, *Appl. Phys. Lett.,* 80, 234, 2002.

11. M. Ikai, S. Tokito, Y. Sakamoto, T. Suzuki, and Y. Taga, *Appl. Phys. Lett.,* 79, 156, 2001.

12. N. Baynes et. al., unpublished results.

13. M.S. Rea, *The IESNA Lighting Handbook,* Illuminating Engineering Society of North America, New York, 1993.

14. M. Stolka, *Organic Light Emitting Diodes for General Illumination Update 2002,* Optoelectronics Industry Development Association, Washington, D.C., 2002.

15. J. Kido, K. Hongawa, K. Okuyama, and K. Nagai, *Appl. Phys. Lett.,* 64, 815, 1994.

16. J. Kido, H. Shionoya, and K. Nagai, *Appl. Phys. Lett.,* 67, 2281, 1995.

17. J. Kido, M. Kimura, and K. Nagai, *Science,* 267, 1332, 1995.

18. R.H. Jordan, A. Dodabalapur, M. Strukelj, and T.M. Miller, *Appl. Phys. Lett.*, 68, 1192, 1996.

19. R.S. Deshpande, V. Bulovic, and S.R. Forrest, *Appl. Phys. Lett.*, 75, 888, 1999.

20. B. W. D'Andrade, M.E. Thompson, and S.R. Forrest, *Adv. Mater.*, 14, 147, 2002.

21. T. Tsuji, S. Naka, H. Okada, and H. Onnagawa, *Appl. Phys. Lett.*, 81, 3329, 2002.

22. J. Kido, W. Ikeda, M. Kimura, and K. Nagai, *Jpn. J. Appl. Phys.*, 35, L394, 1996.

23. C.W. Ko and Y.T. Tao, *Appl. Phys. Lett.*, 79, 4234, 2001.

24. C.H. Kim and J. Shinar, *Appl. Phys. Lett.*, 80, 2201, 2002.

25. C.H. Chuen and Y.T. Tao, *Appl. Phys. Lett.*, 81, 4499, 2002.

26. B. Hu and F.E. Karasz, *J. Appl., Phys.*, 93, 1995, 2003.

27. M. Granstrom and O. Inganas, *Appl. Phys. Lett.*, 68, 147, 1996.

28. S. Tasch, E.J.W. List, O. Ekstrom, W. Graupner, and G. Leising, *Appl. Phys. Lett.*, 71, 2883, 1997.

29. M. Berggren, G. Gustafsson, O. Inganas, M.R. Andersson, T. Hjertberg, and O. Wennerstrom, *J. Appl. Phys.*, 76, 7530, 1994.

30. J. Thompson, R.I.R. Blyth, M. Mazzeo, M. Anni, G. Gigli, and R. Cingolani, *Appl. Phys. Lett.*, 79, 560, 2001.

31. B.W. D'Andrade, J. Brooks, V. Adamovich, M.E. Thompson, and S.R. Forrest, *Adv. Mater.*, 15, 1032, 2002.

32. A. Dodabalapur, L.J. Rothberg, and T.M. Miller, *Appl. Phys. Lett.*, 65, 2308, 1994.

33. T. Shiga, H. Fujikawa, and Y. Taga, *J. Appl. Phys.*, 93, 19, 2003.

34. A.R. Duggal, J.J. Shiang, C.M. Heller, and D.F. Foust, *Appl. Phys. Lett.*, 80, 3470, 2002.

35. C. Adachi, M.A. Baldo, M.E Thompson, and S.R. Forrest, *J. Appl. Phys.*, 90, 5048, 2001.

36. A.R. Duggal, J.J. Shiang, C.M. Heller, and D.F. Foust, *Proc. SPIE*, 4800, 62, 2003.

37. I.H. Campbell and D.L. Smith, *Solid State Phys.*, 55, 1, 2001.

11

Photoexcited Organic Lasers

ANANTH DODABALAPUR

CONTENTS

11.1 INTRODUCTION

Photoexcited solid-state organic lasers have been made with a wide variety gain media including dye-doped polymers,[1–4] dye-doped molecular single crystals,[5] diluted conjugated polymers,[6] neat films of conjugated polymers,[7–10] and molecular solids.[11–13] These gain media have been included in a fascinating array of resonators to obtain laser action. Molecular solid gain media have included neat films, single crystals as well as doped films deposited from solution[11] or sublimation.[11–13] Laser work on doped molecular solid films has employed charge-transporting hosts such as 8-hydroxyquiolinato aluminum (Alq) or 2-naphthyl-4,5-bis(4-methoxyphenyl)-1,3-oxazole (NAPOXA). The dopant is usually a laser dye such as DCM or an oligomer such as α-sexithiophene (α-6T) and the concentration of the dopant in the host is of order 1% by weight. If the dopant concentration is too high, the luminescence yield is often lowered due to concentration quenching and aggregate formation.

In this chapter, the characteristics of a variety of photopumped lasers with guest-host gain media with single- and multiple-step (cascade) energy transfer are described. Organic and polymeric laser gain media with Förster transfer can result in very low threshold power densities for both amplified spontaneous emission (ASE) and lasing.[11,12] This is due to a combination of favorable factors; one of the most important is a very low optical loss at the emission wavelength. The low losses are a consequence of shifting the emission by >150 nm from the fundamental absorption of the host. Such gain media have been successfully included in many types of resonators including resonators based on whispering gallery modes (WGMs), photonic band gaps, and distributed Bragg reflectors (DBR). Novel patterning and fabrication procedures have been developed for organic-based lasers, which have resulted in the realization of very small lasers (3 μm diameter), lasers formed by self-assembly from a melt, and lasers created using low-cost imprinting techniques.

11.2 MATERIALS

The basic mechanism of laser action in organic molecules as well as polymers can be understood by examining Figure 11.1A. The pump laser excites the molecule from the ground

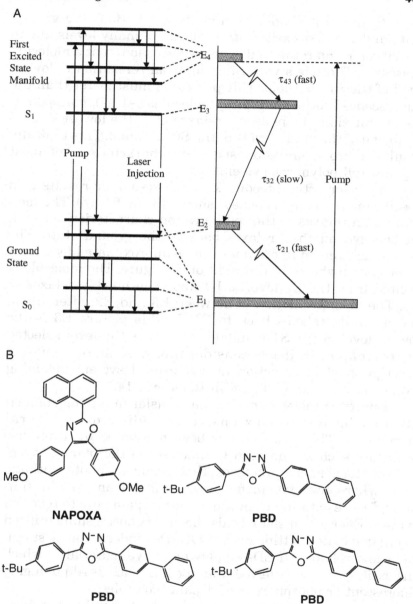

Figure 11.1 Molecular structures of the three host molecules employed. These three materials are electron transporting. Also shown is the molecular structure of the conjugated polymer emitter, which is a soluble derivative of poly-(phenylene vinylene).

state to the first singlet excited state, S1. Both the ground
state and the first excited state consist of many levels due to
the vibronic and rotational degrees of freedom that a molecule
possesses. There is a very rapid internal relaxation to lowest
level of the S1 manifold. This process is must faster than the
fluorescence lifetime of typical molecules (~1 ns). For spontaneous emission, the molecule relaxes from the lowest level of
S1 to one of the many levels of the S0 manifold. This typically
results a broad emission spectrum characteristic of most
organic and polymeric systems.

For laser action, feedback in the resonator results in a
specific energy being selected (shown in the figure). The molecule then relaxes to the configuration characteristic of such
an energy and then relaxes back to the ground state. The
lasing mechanism in organics can be approximated by a four-
level system shown in the right of the figure. The molecule is
excited from the E1 level to E4 and then quickly relaxes to
E3. The laser transition is between E3 and E2, after which
the molecule relaxes back to E1. In this picture E3 is the
lowest level of the S1 manifold, and E2 is the level selected
by resonator feedback considerations. A more complete
description of laser action in a four-level system including
rate equations can be found in Reference 14.

Energy transfer or excitation transfer in organic materials has been quite extensively studied over several
decades.[15,16] This research has been motivated by a number
of factors such as the need to understand the photophysical
and photochemical processes that occur in photosynthetic
units where energy transfer plays an important role in "fun-
neling" the excitation from the antennae pigments to reaction
centers. Energy transfer has also been very successfully utilized
in organic light-emitting diodes (OLEDs); indeed, the most efficient and reliable OLEDs employ an active material in which
the excitation is transferred from the host such as Alq to highly
fluorescent or phosphorescent dopant molecules.[17]

Energy transfer actually refers to a family of related
effects. In energy transfer, the excited state that is generated
in the host molecules by the absorption of light is transferred
through electromagnetic dipole–dipole interactions to guest

molecules. The transfer involves virtual photons, and differs in this important respect from absorption and re-emission which involves real photons. If the distance between the host and the guest is <1 nm, as in some large molecules, then the energy transfer is sometimes referred to as Dexter transfer. If this distance is larger (typically between 1 and 10 nm) the transfer process is often called Förster transfer or resonance energy transfer.[15,16] Besides the critical dimension, another important difference between Förster and Dexter transfer is that in Förster transfer, the excitation is akin to a dipole–dipole interaction whereas in Dexter transfer involves direct electron exchange between closely spaced molecules.

The molecular structures of the host materials used in the experiments described below are shown in Figure 11.1B together with that of a soluble conjugated polymer, which is used as one of the dopants. The weight percentages of the dyes in the host are typically 0.3 to 1%. When either Alq or NAPOXA[18] is used as the host, the host as well as the dopant are co-sublimed. In all other cases, the host and dopant(s) are deposited from solution by spin coating. Conjugated polymer lasers with the lowest threshold pump powers for laser action have also been guest-host systems.[19]

A convenient way of characterizing and comparing various materials for their suitability as laser gain media is the threshold for observing stimulated emission (ASE) in thin films of the material on glass substrates. In this configuration, which is illustrated in Figure 11.2, a planar waveguide is formed with the organic as the core layer and glass/air as the cladding layers, provided that the organic layer is sufficiently thick. Such structures are photoexcited, and the onset of ASE is often observed as a shoulder in the spontaneous emission spectrum, which grows in relative intensity as the pump power is increased.[12] This is illustrated in Figure 11.3. A nitrogen laser ($\lambda = 337$ nm, 2 ns pulse width) is employed as the exciting source in all the measurements.

ASE thresholds as low as 85 W/cm^2 for films with a NAPOXA host and DCMII guest molecules have been measured.[11] This is the lowest value reported for any organic material, and is due to the high luminescence yield of both

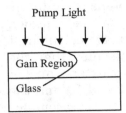

Figure 11.2 Planar waveguide formed by glass ($n = 1.46$), the organic gain layer ($n = 1.65$ to 1.7), and air. Such structures are used by many groups to study some of the stimulated emission characteristics of a material.

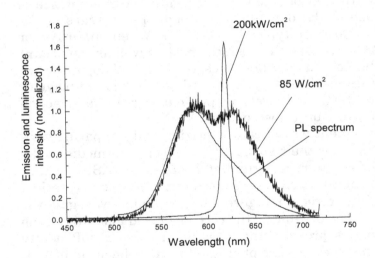

Figure 11.3 Photoluminescence spectrum of a film of NAPOXA on glass (see Figure 11.2) NAPOXA:DCMII at low excitation intensities (PL) when there is only spontaneous emission. Upon photoexcitation at higher power densities with a pulsed light source, a shoulder begins to develop indicating the onset of stimulated emission. Stimulated emission at an incident power density of 85 W/cm² is observed. At higher input powers, stimulated emission dominated the emission spectrum.

NAPOXA and DCMII, and the excellent overlap between the NAPOXA emission spectrum and the DCMII absorption spectrum. The above-mentioned factors result in very efficient Förster transfer of the excited state from NAPOXA molecules to the dye molecules. Conjugated polymer and oligomer emitters also work very well in gain media with Förster transfer.[19] ASE thresholds of 250 W/cm^2 have been reported by us in films containing 10% by weight of a substituted poly(phenylene vinylene) in a 2-(4-biphenylyl)-5-(4-*tert*-butylphenyl)-1,3,4-oxadiazole (PBD) host. These values for ASE thresholds are lower than those reported recently for neat films of conjugated polymers.[7-10]

Low thresholds are also observed in blends of conjugated polymers and the distributed feedback (DFB) lasers fabricated from those blended materials. Optical losses are reduced by spectrally shifting the emission away from the absorption band by employing Förster transfer. Optical losses of about 3 cm^{-1} are observed in the blends, in contrast to losses of 85 cm^{-1} in pure emissive species.[19] The lasing thresholds are reduced from 5000 W/cm^2 for the host to 200 W/cm^2 in the blends.[19] Blends of poly-(*p*-phenylene vinylene) (PPV) and their derivatives help reduce the ASE thresholds.[19] The host–guest blends utilize multiple energy transfer, resulting in the red shift of the emission. The energy transfer in blends, however, has a strong dependence on the concentration of the guest species in the blend.[19]

As a result of multiple energy transfer, it is possible to shift the emission wavelength by a large amount with respect to the pump wavelength (337 nm in our case). Thus, lasers with emission wavelengths ranging from 390 nm (ultraviolet, UV) to 810 nm (infrared, IR) have been fabricated. This wide distribution in emission wavelengths has necessitated innovation in resonator design. With carefully designed DBRs, it is possible to realize resonators that support lasing in the UV, blue, and red wavelengths. For IR wavelengths a laser geometry with feedback provided by a photonic band gap structure is employed. Lasers employing multiple or cascade energy transfer are discussed in Section 11.5.

11.3 WHISPERING GALLERY MODE LASERS

These lasers are based on dielectric resonators with circular symmetry — planar disks, cylinders (rings), and spheres/spheroids. Organic/polymer lasers with all three major categories of WGM resonator designs have been successfully fabricated. Some of the earliest work (which is not described here) was with microdisk lasers based on dye-doped polymer gain layers.[4] Lasers of diameter 7 and 20 µm and possessing well-defined mode structure were demonstrated. These were the first organic lasers to be patterned and also the first that had an all-organic waveguide (the core and both the cladding layers were organic/polymeric). More recently, single-mode microdisk lasers with vacuum-sublimed active layers have been fabricated using a new fabrication procedure, which is outlined below.

The microdisk lasers employed Alq or NAPOXA hosts, which were doped with DCMII (to about 1% by weight). To obtain a smooth periphery, which is a necessity for laser action, a proximity shadow mask (PSM) technique was employed to fabricate the lasers. Glass slides were first coated with about 0.25 µm Au, which was deposited by evaporation, followed by 0.3 µm SiO_2 which was deposited by plasma-enhanced chemical vapor deposition. Holes of circular shape (and diameter 2 to 20 µm) were etched in the SiO_2 with a CF_4 plasma. This etch does not attack the gold, which acts as an etch stop layer. The gold is wet-etched with an etchant that does not attack glass or SiO_2 so as to form a significant undercut of several microns extent. The circular holes in the SiO_2 can be used as apertures or PSMs to deposit microdisk-shaped organic device structures on the glass, as shown in Figure 11.4. The organics also deposit on top of the SiO_2, and to facilitate measurement, the samples are photoexcited through the substrate and the Au acts as a self-aligned shutter, which blocks the pump light over most of the unwanted areas.

The emission spectrum of a single 3-µm-diameter NAPOXA:DCMII microdisk laser is shown in Figure 11.5. At these small dimensions, only a single mode of the cavity overlaps the gain spectrum of the material, resulting in a truly

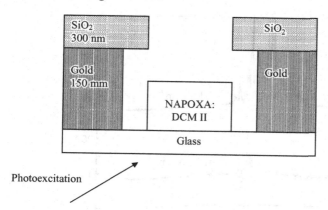

Figure 11.4 Schematic structure of microdisk lasers formed by sublimation through a PSM. The disk diameters were in the range of 2 to 20 µm.

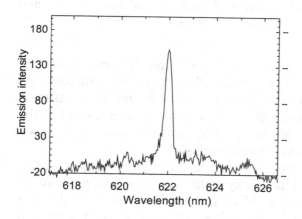

Figure 11.5 Emission spectrum of a single 3-µm microdisk laser with NAPOXA:DCMII. Such lasers are the smallest organic waveguide lasers made so far. There is only a single lasing mode at these small geometries.

single mode laser. The spatial extent of the lasing mode is a ring of approximate width $\lambda/2n$ lying just within the circumference of the disk. The excitons created outside this region can couple into the lasing mode only by diffusion. In most

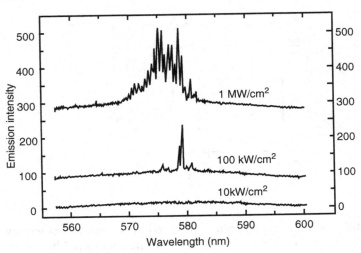

Figure 11.6 Emission spectrum from a ring laser formed by dipping an optical fiber of diameter 125 μm in a mixture of PBD and PPV7 at different excitation power densities.

amorphous organic materials, the exciton diffusion length is very small; in fact, the exciton diffusion behavior in undoped NAPOXA has been previously reported and the diffusion length was found to be <20 nm.[20] This means that most of the excitons created by photoexcitation outside the lasing mode volume result in spontaneous emission.

WGM lasers have also been made with conjugated polymer gain media.[21] The cavity causes a resonating feedback that leads to formation of lasing modes. These modes are observed in microdisks deposited on quartz substrates and show a direct dependence with the cavity. Lasers are also made by coating microrings around thin optical fibers.[21]

The second type of WGM laser that has been fabricated is a "ring" laser. In this laser an optical fiber of diameter 125 μm is coated with the gain medium by dipping it in a solvent containing PBD and PPV7, a soluble conjugated polymer with the structure shown in Figure 11.6. In this structure the circular symmetry about the axis of the rod results in the existence of WGMs around the circumference. As a consequence of the large diameter, many modes lase simultaneously, as seen in Figure 11.6. If the diameter of the fiber is reduced, the individual modes can be distinguished quite clearly.[4]

The spheroid lasers were made by first creating a blend of PBD (90%) and PPV7 (10%), which is melt-processable.[22] Small amounts of this blend were melted on a flexible Teflon substrate in a forming gas atmosphere. The surface tension of the melt is such that a negative contact angle is formed at the interface with the Teflon substrate. Although gravity determines the shape of the droplet in the melted state, it is the surface tension and surface energies that determine the interface contact angle, which must be >90° for the resonator to support WGMs. Cooling the droplet back to room temperature does not result in any significant change in shape. In particular, the very smooth surface formed after the melting appears to be retained. A smooth surface morphology is critical for realizing high-quality factors with such resonators. Melting and cooling will undoubtedly lead to some recrystallization, which would lower the Q due to scattering.

With the procedure outlined above, several sets of solid state droplets with diameters ranging from 30 to 110 μm were made. The lasing spectrum of two devices with diameter = 30 and 60 μm are shown in Figure 11.7. The spacings between the modes scale fairly well with the radius, R_D, according to the expression $M\lambda = 2\pi R_D n_{eff}$, where M is the mode index, and n_{eff} is the effective refractive index. The measured mode spacing is 0.5 nm for $R_D = 55$ μm, 0.9 nm for $R_D = 30$ μm, and 2 nm for $R_D = 15$ μm. These lasers were excited with pump intensities in the range 30 kW/cm² to 1 MW/cm². At 30 kW/cm², laser emission was clearly evident and the threshold is lower; however, the threshold for these lasers could not be accurately determined because the noise level of the detector for incident powers of ~10 kW/cm² was reached.

11.4 PHOTOEXCITED DBR AND DFB LASERS

The DBR-based lasers utilize the reflection properties of surface relief gratings to provide wavelength selective feedback.

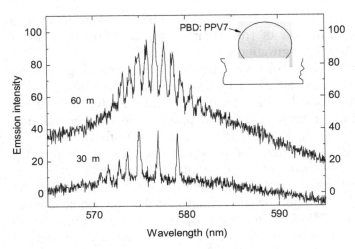

Figure 11.7 Lasing from WGMs around the circumference of two spheroid lasers formed by melting and resolidifying a mixture of PBD (90%) and PPV7 (10%). The diameters are 30 and 60 μm. The inset shows a schematic of the laser shape.

It is also possible to realize DFB and DBR lasers with phase gratings. The gratings are formed by patterning photoresist on thermally grown SiO_2 (on Si substrates). The patterned substrates are etched in a plasma to remove 5 to 50 nm of the oxide. An unpatterned region of 2 to 20 mm length is formed between two DBR mirrors, as shown in the inset of Figure 11.8. The photoresist is removed and the active layer deposited either by spin coating or co-evaporation. The unpatterned region between the two mirrors constitutes the laser cavity and the two mirrors at each end provide wavelength-selective feedback. The grating period is 0.6 μm, which results in reflectivity peaks near 610 nm (third order), 485 nm (fourth order), and 390 nm (fifth order). DBR lasers which emit in the UV, blue, and red have been made with such gratings. The UV laser ($\lambda \sim 392$ nm) uses PBD as the gain medium, and is an example of a small molecule neat film possessing optical gain. The blue laser possesses a gain medium consisting of PBD doped with coumarin 490. Red DBR lasers were

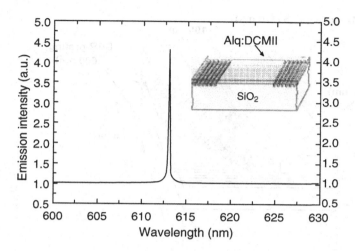

Figure 11.8 Red DBR laser with Alq:DCMII active layer. The length of the cavity is 2 mm. The inset shows the schematic structure of the laser.

made with gain media consisting of hosts such as PBD or Alq doped with either one (simple Förster transfer) or two dyes (cascade Förster transfer). The emission spectrum from a 2-mm-long DBR laser with an Alq:DCMII active layer is shown in Figure 11.8. The single-mode spectrum is a consequence of the high degree of wavelength selectivity, which is an important characteristic of DBR lasers and has led to their widespread use in communication systems. The line width is resolution limited and is <1 nm. The threshold power density required to achieve lasing is 2 to 5 kW/cm² for this material system. With other material combinations, which will be described below, DBR lasers with threshold power densities <1 kW/cm² have been fabricated.

DBR lasers with ridge-waveguide gain regions have been made using the PSM technique outlined earlier. The fabrication of such DBR lasers starts with a SiO₂/Si substrate in which two surface relief DBR mirrors are defined by reactive ion etching. The spacing between the two mirrors is a few millimeters. Such a structure is coated with Au and plasma-deposited SiO₂. A

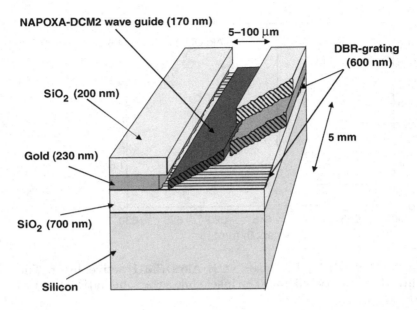

Figure 11.9 Ridge waveguide DBR laser formed by sublimation through a PSM. The width of the ridge is 5 μm and the length of the cavity is several millimeters. The gain material is NAPOXA doped with DCMII.

waveguide mask is then used to pattern photoresist on top of the SiO_2 and create rectangular exposed areas (width 5 to 100 μm). The SiO_2 is etched by RIE to create smooth side walls, and the Au is etched with a selective wet etch, which forms a sufficiently deep undercut. This completes the formation of a PSM for a ridge waveguide layer. The gain medium (NAPOXA doped with DCMII) is sublimed through the PSM to complete the laser, which is schematically illustrated in Figure 11.9. The narrowest ridges (5 μm) that were employed resulted in nearly single-mode lasing as illustrated in Figure 11.10. Wider ridges also resulted in laser action but were multimode.

In DFB lasers the cavity between the two mirrors is eliminated, resulting in a corrugated SiO_2 surface with the active material deposited on it. Laser action results from

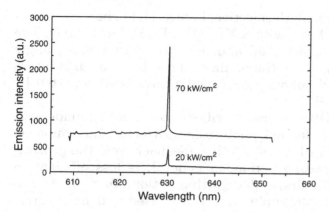

Figure 11.10 Emission spectrum from the ridge waveguide laser shown in Figure 11.10 at two levels of incident power.

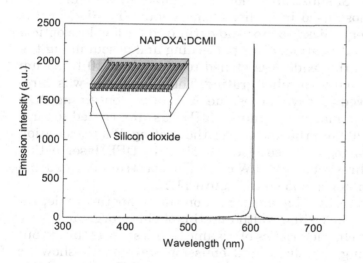

Figure 11.11 Red DFB laser with NAPOXA:DCMII active layer. Lasing occurs as a result of coupling between the forward and backward propagating waves induced by the grating.

grating-induced coupling between the forward and backward propagating waves over length scales of the order of the inverse of the grating coupling constant. This type of laser is also highly wavelength selective. Figure 11.11 shows the spectrum of a DFB laser with a NAPOXA:DCMII gain layer. The laser is highly directional and light is emitted at an angle (~20°) with respect to the plane of the substrate. DFB lasers with conjugated polymer gain media have been reported by McGehee et al.[23]

The DFB/DBR lasers described above were all made with lithographically defined photoresist patterns and employed higher-order (third- to fifth-order) feedback from the grating. Higher-order grating action, in general, is less efficient in comparison with first-order grating action. To realize first-order gratings, holographic techniques used in defining gratings for commercial III–V semiconductor lasers were employed. The grating pitch that was employed is 205 nm, which is designed to result in laser action near 610 nm. The SiO_2-coated Si substrates that were employed were different from all those used in earlier experiments. The SiO_2 in this case was very thick (>3 μm) and optimized for low-loss optical waveguide fabrication. After patterning and descumming the photoresist, the oxide was etched to a depth of 10 to 25 nm to form the surface relief grating. The photoresist was carefully removed to leave no residue. A solution containing PBD and dyes (coumarin 490 and DCMII) was spin-coated to form a film of ~200 nm thickness over the grating. Photoexcitation with the nitrogen pulse laser resulted in DFB laser action with low thresholds (<200 W/cm^2). The spectrum from such a laser resonator is shown in Figure 11.12.

Kozlov et al.[24] have reported on the temperature dependence of the laser line width and threshold in organic lasers. The output characteristics of organic lasers such as the output power, lasing threshold, and emission wavelength show no variation with the changes in temperature.[24] The spectral line width of organic lasers can be smaller than that of their inorganic counterpart. These characteristics of organic lasers are potentially advantageous in a number of applications.[24] The temperature dependence, which is similar to that seen in inorganic quantum dot lasers, is yet another important feature of organic systems.

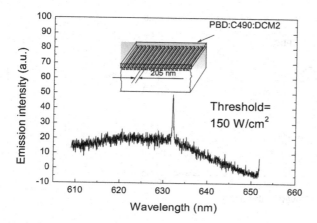

Figure 11.12 Emission spectrum just above threshold of a DFB laser with grating pitch 205 nm formed by holographic techniques. This is an example of a DFB laser with a first-order grating. The gain medium is PBD doped with ~1% each of coumarin 490 and DCMII. The excited state reaches the DCMII molecules through cascade Förster transfer.

11.5 DFB AND DBR LASER BASED ON CASCADE FÖRSTER TRANSFER[11]

Cascade Förster transfer or multiple Förster transfer is akin to optical processes in a photosynthetic unit. In the materials that were examined, the light is absorbed by the host material such as PBD, which is doped with a small percentage of one or more fluorescent dyes with different absorption and emission spectra. The host absorbs the pump light and funnels the excitation to the dye molecules through a Förster-type energy transfer. The average spacing between the dye molecules at the doping concentrations used (3 nm) is less than the Förster distance. The excitation transfer can be extremely efficient, especially if the overlap between the donor (host) emission spectrum and the acceptor (dye) spectrum is good. Figure 11.13 shows the emission and the absorption spectra of the materials used. The spectra of PBD are from a pure film whereas the other spectra have been obtained from solid solutions of the dye

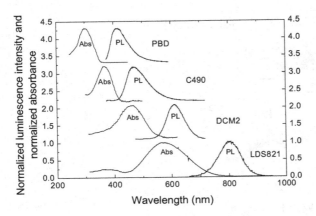

Figure 11.13 Absorbance and photoluminescence spectra of the host material and the fluorescent dyes. The host material is PBD; the dyes are coumarin (C490), DCMII, and LDS821. The spectra for PBD are from a pure film, whereas the other spectra are from solid solutions of the material in polystyrene.

in polystyrene. Efficient transfer is observed between PBD and coumarin 490 (C490) and between PBD and DCM II (with C490 included) and between PBD and LDS821 (with both C490 and DCM II included in the film). The transfer is observed to be poor between PBD and LDS821, when no intermediate dyes such as DCM II or C490 are included. The results concur with the theory of resonance energy transfer. The films used were very thin and were deposited on aluminum to minimize the effects of reabsorption and to eliminate waveguiding. The relative quantum yield of films of the various materials is estimated relative to the quantum yield of a neat film of PBD taken as 1. The quantum yields measured are 1.67 (PCD + C490), 1.49 (PCD + C490 + DCM II), and 0.6 (PCD + C490 + DCM II + LDS821).[11]

The films used to study stimulated emission were deposited on glass substrates. The host PBD with the dopants was deposited by spin coating to form ~200-nm-thick films. The thickness is such that only the lowest-order mode is guided through the structure with PBD as the core and air and glass as the cladding. Unfocused light of wavelength 337 nm from a 2-ns pulsed nitrogen laser was used to photoexcite the films. The pulses had a spot size of 3 mm × 2.3 cm with an energy density of 500 µJ/cm², which corresponds to 250 kW/cm². The pump power level was adjusted using neutral density filters. At low pump powers, the emission spectra are identical to the spontaneous emission spectra shown in Figure 11.13, with the spectrum becoming narrower at high pump powers. The narrowing is characteristic of ASE. The threshold pump power necessary to detect significant stimulated emission is ~10 kW/cm². This is a high value because the ASE is close to the absorption "edge" of the PBD, where the absorption coefficient is high. The gain coefficient must be more than the absorption coefficient to achieve amplification, thus resulting in the requirement of a high pump power.[11]

DBR lasers prepatterned by a thermally oxidized Si wafer were fabricated to from two gratings separated by distances in the range 3 nm to 1 cm. The gratings act as wavelength-selective mirrors, with a grating period of 0.6 µm, which results in reflectivity peaks near 610, 485, and 392 nm, corresponding to the third, fourth, and fifth orders, respectively. The laser is completed by spin coating the active material between the two gratings. The peaks in reflectivity of the gratings are matched with the fluorescence spectra of PBD, PBD doped with C490, and PBD doped with both DCM II and C490. The active gain media together with the DBR resonator was used to fabricate UV, blue, and red lasers, as shown in Figure 11.14.

All measurements were made with spun films. The quality of vacuum-sublimed films is sometimes better than that of spun films. NAPOXA doped with 1% DCMII as an active material in optical amplifiers was also studied. The films were deposited by co-sublimation resulting in good film-forming ability and almost perfect overlap between the host emission and guest absorption spectra as shown in Figure 11.15. Thresholds as low as 85 W/cm² (total power) were observed. This is because the combined absorption and other losses of the NAPOXA host and dye dopants at amplification wavelengths

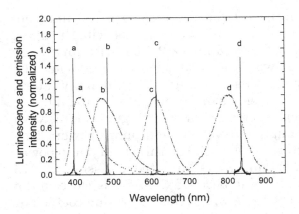

Figure 11.14 Laser emission spectra from devices made from four different active materials, and the spontaneous emission spectra of these materials. Solid lines, laser emission spectra; curve a, PBD; curve b, PBD + C490; curve c, PBD + C490 + DCMII; curve d, PBD + C490 + DCMII + LDS821. Dotted lines, spontaneous emission spectra for these active materials. The lasers producing emission spectra a, b, and c used the same DBR resonator consisting of two 0.6-μm-pitch gratings etched in SiO_2 separated by a gain region 3 to 10 mm long. The third, fourth, and fifth harmonics of the reflectivity of such DBR mirrors match the spontaneous emission spectra of c, b, and a, respectively. The lasing threshold for pure films of PBD is quite high (10 kW/cm²). The measured threshold power densities for the blue (b) and red (c) lasers are 1 to 1.5 kW/cm² (total incident power density). The thresholds for the blue and red lasers are affected by losses in the SiO_2 cladding layer. As a result of these losses, the thresholds for observing ASE on these oxide-coated substrates (without a DBR) is ~10 kW/cm², much higher than the threshold for films on glass substrates. The laser thresholds can be further lowered by reducing optical losses in the SiO_2. For the laser d, a two-dimensional photonic band gap–based distributed feedback geometry is used. This is realized by using a SiO_2-coated Si substrate in which a square lattice of holes (with a circular cross section and radius of 0.15 μm) were etched. The depth of the holes is 10 to 30 nm and the lattice constant is 0.6 μm. The active material was spin-coated on such a patterned substrate to complete the device. The line widths of the lasers a to d are resolution limited and are 0.2 to 0.3 nm.

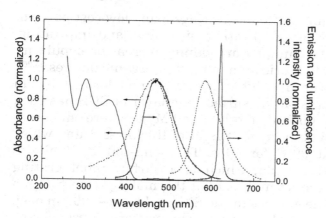

Figure 11.15 Absorbance and spontaneous emission spectra of the compounds used to make vacuum-sublimed films. Solid lines, NAPOXA; dashed lines, DCMII. Also shown is the stimulated emission spectrum from a film of NAPOXA doped with ~1% DCMII when excited at 250 kW/cm^2.

are small (<1 cm^{-1}). The threshold power densities compare very well with the densities for unpatterned films of conjugated polymers. The low threshold powers are as result of low loss at the emission wavelengths, thus increasing gain.

Resonance energy transfer is fundamentally an electromagnetic interaction and the transfer rates can be influenced by strong microcavity effects. The beneficial effects of energy transfer are not restricted to small organic molecules, but also apply in case of longer molecules such as oligomers and polymers. Thus, when excitation is transferred down in energy in organic molecules, the threshold for stimulated emission is lowered along with an enhancement in the waveguiding properties.[11]

11.6 NOVEL FABRICATION METHODS

Inexpensive methods to fabricate organic-based lasers have also been investigated. As noted above, gratings are a very convenient way to provide feedback in waveguide-based lasers. It is possible to define gratings using low-cost patterning techniques such as printing, stamping, and imprinting. Such processing techniques are gaining increasing popularity and acceptance to fabricate a variety of organic devices.[25,26]

A DFB laser in which the grating is formed by an imprinting process on the surface of a layer of BCB has been fabricated. The substrate (both Si and Mylar® were employed) is first coated with a layer of BCB of thickness 4 μm, which is soft-baked at 90°C for 60 s. The semi-soft BCB film is pressed against a mold, which is a surface-relief grating formed by patterning and etching thermally grown SiO_2 on Si substrates. The grooves in the SiO_2 mold are ~100 nm deep and the grating period is 0.6 μm. This pattern is transferred to the BCB, which is then photopolymerized by exposure to UV light. The photopolymerization hardens the BCB and the grating is retained. The gain medium, which is Alq:DCMII in this case, is sublimed over the grating to complete the fabrication. The emission characteristics of such a laser on a Mylar substrate (upon photoexcitation with an N_2 laser) are shown in Figure 11.16. The emission is centered near 657 nm and consists of multiple lines. The threshold pump power for observing laser action was rather high — 12 kW/cm². It is

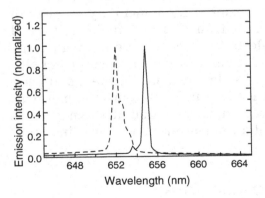

Figure 11.16 Emission spectra from DFB lasers in which the grating is formed by imprinting a thick film of BCB (on Mylar) with a mold. The gain medium is Alq:DCMII. The dashed spectrum is taken when the substrate is bent with a radius of curvature 1 in.

believed that the threshold power can be lowered by more than two orders of magnitude by optimization of the processing steps and choice of the imprint material. This work represents one of the first uses of imprinting techniques in the fabrication of organic/polymer light-emitting devices.

11.7 PHOTONIC CRYSTAL LASERS WITH ORGANIC GAIN MEDIA

The structures employed, although two dimensional, do not possess a complete two-dimensional band gap. The characteristics of lasers formed with such resonators are important prerequisites for the study of lasers formed with a complete two-dimensional band-gap. The lasing properties of resonators formed by a triangular lattice of shallow holes have been examined.[27]

The structure of lasers examined is shown in Figure 11.17. A coating of photoresist on thermally oxidized silicon is patterned to form a triangular lattice of holes with a typical radius of 100 nm. The photolithography was performed with a 248-nm light source giving the pattern a periodicity of 400 nm. Shallow holes of 20- to 40-nm depth are etched in the SiO_2 by reactive ion etching through the photoresist mask. The photoresist is removed and a film of PBD doped with about 1% by weight of coumarin 490 and DCM is deposited over the entire structure by spin coating. The structure resembles a planar waveguide with the organic layer as the core with thickness 150 nm and the SiO_2 as the cladding. Lowest-order transverse electric (TE) and transverse magnetic (TM) modes are supported.[27]

A pulsed nitrogen laser with pulse width 2 ns and 337 nm light is used to photoexcite the structure. A charge-coupled detector (CCD)/spectrometer is used to measure the emission spectra. The PBD molecules absorb the pump and funnel the excitation to the DCM dye molecules through cascade Förster transfer. The gain medium can be approximated by a four-level system.[27]

The typical spectra are shown in Figure 11.18. The emission takes place above a threshold pump power of ~50 kW/cm².

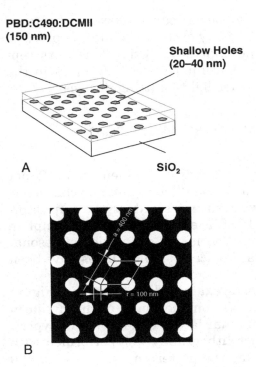

A

B

Figure 11.17 (A) Schematic layer structure of the lasers employed in the study and (B) details of the two-dimensional triangular lattice.

This is significantly higher than the ~1 kW/cm² required for third-order DBR lasers with the same gain medium. This is attributed to the lack of a complete gap that results in a coupling between the lasing mode and other modes localized in the organic.

The dispersion relations for the quasi-two-dimensional structure are calculated for a better understanding. For the first approximation, only a layer of gain medium above the unetched SiO_2 is considered. The thickness is taken to be d with ε_0, ε_1, and ε_2 as the dielectric constants of air, gain medium, and the substrate, respectively. The modes that are localized in the middle layer possessing an in-plane vector of

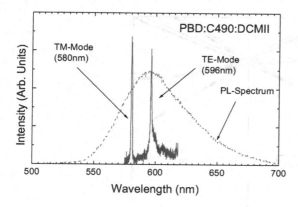

Figure 11.18 Lasers emission spectra from the device shown in Figure 11.17. The two peaks have different polarizations as shown. The spontaneous emission spectrum of the gain medium is also shown with a dashed line.

magnitude k are considered. The transcendental equations are obtained by matching the boundary conditions and obtaining the dispersion relations (k). They are given below:

For TE modes:

$$\tan\left(d\sqrt{\varepsilon_1\omega^2 - k^2}\right) = \frac{\sqrt{\varepsilon_1\omega^2 - k^2}\left(\sqrt{k^2 - \varepsilon_0\omega^2} + \sqrt{k^2 - \varepsilon_2\omega^2}\right)}{\varepsilon_1\omega^2 - k^2\sqrt{k^2 - \varepsilon_0\omega^2}\sqrt{k^2 - \varepsilon_2\omega^2}}$$

For TM modes:[27]

$$\tan\left(d\sqrt{\varepsilon_1\omega^2 - k^2}\right) = \frac{\varepsilon_1\sqrt{\varepsilon_1\omega^2 - k^2}\left(\varepsilon_2\sqrt{k^2 - \varepsilon_0\omega^2} + \varepsilon_0\sqrt{k^2 - \varepsilon_2\omega^2}\right)}{\varepsilon_0\varepsilon_2\left(\varepsilon_1\omega^2 - k^2\right) - \varepsilon_1^2\sqrt{k^2 - \varepsilon_0\omega^2}\sqrt{k^2 - \varepsilon_0\omega^2}}$$

The effect of shallow holes on this system is considered as a perturbation.[27] The two-dimensional grating imposes a periodicity on the system. The resultant photonic band structure is shown in Figure 11.19 for a lattice constant of $a = 400$ nm, $d = 140$ nm, $\varepsilon_0 = 1$, $\varepsilon_1 = 1.7$, and $\varepsilon_2 = 1.46$. As a result of

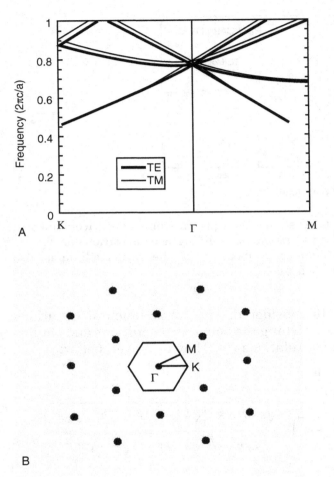

Figure 11.19 (A) Calculated photonic band structure of the device shown in Figure 11.17 for TE and TM polarizations. For TE polarization, the electric field vector is in the plane of the waveguide. (B) The reciprocal lattice and the Brillouin zone showing the high-symmetry points.

the etching of shallow holes, a refractive index variation is created. This opens small band gaps at *k* points where bands of the same symmetry are degenerate. At high symmetry points in the Brillouin zone, the group velocity of the photons may become zero due to the coupling between the bands. At these wave vectors, standing waves are formed, which provide ample feedback necessary for lasing action. This can be verified easily. Degenerate bands at high symmetry points in the range of observed lasing wavelengths need to be determined first. This corresponds to the frequency range $\omega = 0.666$ to 0.689 c/a, which selects the lowest two bands at the M point, where both the TE and TM bands have zero gradient when holes are taken into account. Solving the two equations obtained for the wave vector ($k = 2/a$) results in 584 nm for the TM mode and 594 nm for the TE mode. This is close to values seen in Figure 11.18. There may be a small variation due to discrepancy in measurement of d and ε_1 as well as the frequency shift caused by the introduction of shallow holes. The laser action is caused due to two-dimensional distributed feedback from a triangular photonic crystal.[27]

The two-dimensional nature of a microstructure results in additional TE polarized emission bands that are not observed in corresponding one-dimensional structures.[28] A two-dimensional DFB laser based on poly-[2-methoxy-5-(2'-ethylhexyloxy)-1, 4-phenylene vinylene] exhibited a divergent, cross-shaped far-field emission, which resulted from a distributed feedback occurring over a wide range of wave vectors.[28] The divergent emission is due to high pump energies. The emission is controlled by the photonic dispersion of the structure. The lasing occurs on the low energy band edge, where the emission experiences strong two-dimensional coupling of the waveguide modes.[28] This type of lasing action is novel and the basic device structure can be modified; one of the possible modifications is the creation of a complete two-dimensional photonic band gap.[27]

11.8 SUMMARY

The fabrication and characteristics of a number of optically pumped lasers with organic/polymeric gain media capable of charge transport have been reviewed. The active materials employed were doped films with small molecule hosts and dye, oligomer, and conjugated polymer emitters, in which the excited states created in the host are transferred nonradiatively to the guest molecules, which are the emitters. This energy transfer results in very low absorption losses at the emission wavelength and threshold powers as low as 85 W/cm^2 for the onset of stimulated emission. The laser resonators that have been implemented include resonators based on WGM, photonic band gap, and DBR. The characteristics of microdisk LEDs fabricated using the PSM technique have been described.

ACKNOWLEDGMENT

The author thanks Suvid Nadkarni for assistance in preparing the manuscript.

REFERENCES

1. B.H. Soffer and B.B. McFarland, Continuously tunable narrowband organic dye-lasers, *Appl. Phys. Lett.*, 10, 266–267, 1967.

2. H. Kogelnik and C.V. Shank, Stimulated emission in a periodic structure, *Appl. Phys. Lett.*, 18, 152–154, 1971.

3. I.P. Kaminov, H.P. Weber, and E.A. Chandross, Poly(methyl methacrylate) dye laser with internal diffraction grating, *Appl. Phys. Lett.*, 18, 497–499, 1971.

4. M. Kuwata-Gonokami, R.H. Jordan, A. Dodabalapur, H.E. Katz, M.L. Schilling, R.E. Slusher, and S. Ozawa, Polymer microdisk and microring lasers, *Opt. Lett.*, 20, 2093–2095, 1995.

5. N. Karl, Laser emission from an organic molecular crystal, *Phys. Status Solidi a*, 13, 651–655, 1972.

6. F. Hide, B.J. Schwartz, M.A. Diaz-Garcia, and A.J. Heeger, Laser emission from solutions and films containing semiconducting polymer and titanium dioxide nanocrystals, *Chem. Phys. Lett.*, 256, 424–430, 1996.

7. N. Tessler, G.J. Denton, and R.H. Friend, Lasing from conjugated polymer microcavities, *Nature*, 382, 695–697, 1996.

8. F. Hide, M.A. DiazGarcia, B.J. Schwartz, M.R. Andersson, Q.B. Pei, and A.J. Heeger, Semiconducting polymers: a new class of solid-state laser materials, *Science*, 273, 1833–1836, 1996.

9. S.V. Frolov, W. Gellermann, M. Osaki, and K. Yoshino, Cooperative emission in π-conjugated polymer thin films, *Phys. Rev. Lett.*, 78, 729–733, 1997.

10. X.A. Long, A. Malinowski, D.D.C. Bradley, M. Inbasekharan, and E.P. Woo, Emission processes in conjugated polymer solutions and thin-films, *Chem. Phys. Lett.*, 272, 6–12, 1997; C. Zenz, W. Graupner, S. Tasch, G. Leising, K. Mullen, and U. Scherf, Blue green stimulated emission from a high gain conjugated polymer, *Appl. Phys. Lett.*, 71, 2566–2568, 1997.

11. M. Berggren, A. Dodabalapur, R.E. Slusher, and Z. Bao, Light amplification in organic thin films using cascade energy transfer, *Nature*, 389, 466–469, 1997.

12. M. Berggren, A. Dodabalapur, and R.E. Slusher, Stimulated emission and lasing in dye-doped organic films with Förster transfer, *Appl. Phys. Lett.*, 71, 2230–2232, 1997.

13. V. Kozlov, V. Bulovic, P.E. Burrows, and S.R. Forrest, Laser action in organic semiconductor waveguide and double-heterostructure devices, *Nature*, 389, 362–364, 1997.

14. A.E. Siegman, *Lasers*, University Science Books, Sausalito, CA, 1986.

15. T. Förster, Transfer mechanisms of electronic excitation, *Disc. Faraday Soc.*, 27, 7–17, 1959.

16. B.W. Van Der Meer, G. Coker, and S.Y. Chen, in *Resonance Energy Transfer*, VCH Publishers, New York, 1994.

17. M.A. Baldo, M.E. Thompson, and S.R. Forrest, High-efficiency fluorescent organic light-emitting devices using a phosphorescent sensitizer, *Nature*, 403(6771), 750–753, 2000.

18. A. Dodabalapur, Organic light emitting diodes, *Solid State Commun.*, 102, 259–267, 1997.

19. R. Gupta, M. Stevenson, and A.J. Heeger, Low threshold distributed feedback lasers fabricated from blends of conjugated polymers: reduced losses through Förster transfer, *J. Appl. Phys.*, 92, 4874–4877, 2002.

20. A. Dodabalapur, L.J. Rothberg, R.H. Jordan, T.M. Miller, R.E. Slusher, and J.M. Phillips, Physics and applications of organic microcavity light emitting diodes, *J. Appl. Phys.*, 80, 6954–6964, 1996.

21. Sergey V. Frolov, A. Fujii, D. Chinn, M. Hirohata, R. Hidayat, M. Taraguchi, T. Masuda, K. Yoshino, and Z. Valy Vardeny, Microlasers and micro-LEDs from distributed polyacetylene, *Adv. Mater.*, 10, 869–872, 1998.

22. M. Berggren, A. Dodabalapur, Z. Bao, and R.E. Slusher, Solid-state droplet laser made from an organic blend with a conjugated polymer emitter, *Adv. Mater.*, 9, 968–971, 1997.

23. M. McGehee, F. Hide, M.A. Diaz-Garcia, B.J. Schwartz, D. Moses, and A.J. Heeger, Distributed feedback lasers made with conjugated polymers as the gain material, presented at the Materials Research Society Spring 1997 Meeting, San Francisco, CA, 1997.

24. V.G. Kozlov and S.R. Forrest, Lasing action in organic semi-conductor thin films, *Curr. Opin. Solid State Mater. Sci.*, 4, 203–208, 1999.

25. F. Garnier, R. Hajlaoui, A. Yassar, and P. Srivastava, All polymer field-effect transistor realized by printing techniques, *Science*, 265, 1684–1686, 1994.

26. Z. Bao, Y. Feng, A. Dodabalapur, V.R. Raju, and A.J. Lovinger, High performance plastic transistors fabricated by printing techniques, *Chem. Mater.*, 9, 1299–1301, 1997.

27. M. Meier, A. Mekis, A. Dodabalapur, A. Timko, R.E. Slusher, J.D. Joannopoulos, and O. Nalamasu, Laser action from two-dimensional distributed feedback in photonic crystals, *Appl. Phys. Lett.*, 74, 7–9, 1999.

28. G.A. Turnbull, P. Andrew, W.L. Barnes, and I.D.W. Samuel, Photonic mode dispersion of a two-dimensional distributed feedback polymer laser, *Phys. Rev. B*, 67, 165107-1–165107-8, 2003.

Index

Printed in the United States
by Baker & Taylor Publisher Services